Rock Mechanics and Rock Engineering

Rock Mechanics and Rock Engineering

Volume 1: Fundamentals of Rock Mechanics

Ömer Aydan

Department of Civil Engineering, University of the Ryukyus, Nishihara, Okinawa, Japan

CRC Press
Taylor & Francis Group
Boca Raton London New York

CRC Press is an imprint of the
Taylor & Francis Group, an **informa** business

A BALKEMA BOOK

Published by:
CRC Press/Balkema
Schipholweg 107C, 2316 XC Leiden, The Netherlands

First issued in paperback 2023

© 2020 by Taylor & Francis Group, LLC
CRC Press/Balkema is an imprint of Taylor & Francis Group, an informa business

No claim to original U.S. Government works

ISBN-13: 978-0-367-42162-5 (Volume 1) (hbk)
ISBN-13: 978-0-367-42165-6 (Volume 2) (hbk)
ISBN-13: 978-1-03-265428-7 (pbk)
ISBN-13: 978-0-367-82229-3 (ebk)
ISBN-13: 978-0-367-02935-7 (set)

Typeset by Apex CoVantage, LLC

Library of Congress Cataloging-in-Publication Data
Applied for

Visit the Taylor & Francis Web site at
http://www.taylorandfrancis.com

and the CRC Press Web site at
http://www.crcpress.com

Volume 1
DOI: https://doi.org/10.1201/9780367822293
Volume 2
DOI: https://doi.org/10.1201/9780367822309
Two-volume set
DOI: https://doi.org/10.1201/9780429001239

Contents

Preface

Rock is the main constituent of the crust of the Earth, and its behavior is the most complex one among all materials in the geosphere to be dealt with by humankind. Furthermore, it contains various discontinuities, which make the thermo-hydro-mechanical behavior of rocks more complex. These simply require a higher level of knowledge and intelligence within the Rock Mechanics and Rock Engineering (RMRE) community. Furthermore, the applications of the principles of rock mechanics to mining, civil, and petroleum engineering fields, as well as to earthquake science and engineering, are diverse, and it constitutes rock engineering. Recently, the International Society for Rock Mechanics (ISRM) added "Rock Engineering" in 2017 to its name while its acronym remains "ISRM."

Rock mechanics is concerned with the theoretical and applied science of the mechanical behavior of rock and rock masses, and it is one of branches of mechanics concerned with the response of rock and rock masses to their physical-chemical environment. *Rock engineering* is concerned with the application of the principles of mechanics to physical, chemical, and electromagnetic processes in the uppermost part of the Earth and the design of the rock structures associated with mining, civil, and petroleum engineering. This book is intended to be a fundamental text for younger generations and newcomers, as well as a reference source for experts specialized in rock mechanics and rock engineering.

Due to the wide spectra of rock mechanics and rock engineering, the book is divided into two volumes: *Rock Mechanics and Rock Engineering: Fundamentals of Rock Mechanics* and *Rock Mechanics and Rock Engineering: Applications of Rock Mechanics – Rock Engineering*. In the first volume, the fundamental concepts, theories, analytical and numerical techniques and procedures of rock mechanics and rock engineering, together with some emphasis on new topics, are described as concisely as possible while keeping the mathematics simple.

The second volume is concerned with the applications of rock mechanics and rock engineering in practice. It ranges from classic rock classifications, the response and stability of surface and underground structures, to model testing, monitoring, excavation techniques, and rock dynamics. Particularly, earthquake science and engineering, vibrations and nondestructive techniques are presented as a part of rock dynamics.

Although the overall subject of *Rock Mechanics and Rock Engineering* is presented over two volumes, each volume is complete in its content and should serve the purposes of educators, students, experts, as well as practicing engineers. It is strongly hoped that these two volumes will fulfill the expectations and serve further advances in rock mechanics and rock engineering.

Author biography

Ömer Aydan, born in 1955, studied Mining Engineering at the Technical University of Istanbul, Turkey (BSc, 1979), studied Rock Mechanics and Excavation Engineering at the University of Newcastle upon Tyne, UK (MSc, 1982), and received his PhD in Geotechnical Engineering from Nagoya University, Japan, in 1989.

Professor Aydan has worked at Nagoya University as a research associate (1987–1991) and then at the Department of Marine Civil Engineering at Tokai University, first as Assistant Professor (1991–1993), then as Associate Professor (1993–2001), and finally as Professor (2001–2010). He then became Professor of the Institute of Oceanic Research and Development at Tokai University and is currently Professor at the University of Ryukyus, Department of Civil Engineering & Architecture, Nishihara, Okinawa, Japan. He is also the director of the Disaster Prevention Research Center for Island Region of the University of the Ryukyus.

Ömer Aydan has played an active role on numerous ISRM, JSCE, JGS, SRI, and Rock Mech. National Group of Japan committees and has organized several national and international symposia and conferences.

Chapter 1

Introduction and history of rock mechanics and rock engineering

1.1 Early traces of rock mechanics and rock engineering

The early traces of rock mechanics and rock engineering may be associated with archeological remains left by ancient peoples, such as Hattis, Sumerians, Egyptians, Hittites, Persians, Romans, and Native Americans. The quarries, open-pit mines, castles, underground quarries, semiunderground or underground cities in Anatolia (Anadolu) and underground tombs of Egyptians near Luxor, as well as pyramids, are all well preserved examples of rock engineering structures of the past, even though they did not have the excavation tools of modern times (Aydan, 2008, 2014). In Anadolu (Anatolia), there are traces of open-pit mining dated to 9000 years ago and of underground mines dated to 5000 years ago (Kaptan, 1992; Yener, 1997).

Humankind has constructed underground or semiunderground openings in soft rocks in the past. However, one can also found such structures excavated in limestone in the form of irrigation tunnels. Hard stones (i.e. flints, diorite, obsidian) were used initially, and later, workers started to use metallic tools after gaining the knowledge of extracting metals from ores. Given the 9000-year-old archeological mining and metallurgic traces found in Anatolia, it is likely that the use of metallic tools could be as old as 9000 years (Hatti era). In view of recent findings in Göbekli Tepe in Şanlıurfa, the quarrying of limestone in the region extends at least 11000 BP (Fig. 1.1). It is estimated that Harran City was established at least 5000 years ago, during the period of the Sumerians, who came from Central Asia to Mesopotamia about 7000 years ago and governed the area until

(a) Turkish pyramids near Xianyang (b) Turkish monuments near Xianyang (c) Göbekli monuments

Figure 1.1 Monuments in Central Asia and Anatolia

BCE 2270. Sumerians were the pioneers in all aspects of the modern sciences, engineering, technology, culture and religion of humankind, including cuneiform script (Kramer, 1956).

One can find also some earlier underground quarries in Anatolia and Thebes (Kulaksiz and Aydan, 2010; Kumsar *et al.*, 2003; Aydan and Ulusay, 2003, 2013; Aydan *et al.*, 2008a, 2008b; Aydan and Geniş, 2004; Hamada *et al.*, 2014; Kumsar and Aydan, 2008; Tokashiki *et al.*, 2008). At the Amenophis III Quarry at Qurna of the Thebes region of Egypt, marble mining started probably 3350–3500 years ago. The Bazda Quarry at Harran, Urfa region of Turkey, probably opened 4000 years ago by the Sumerians. The Bazda underground marble mine quarry is the oldest known underground quarry mine in Turkey (Fig. 1.2).

Pyramids made of huge rock blocks to achieve structural stability for thousands of years under both static and dynamic loading conditions, particularly those in Egypt, are well-known worldwide. However, some pyramids have been recently unearthed in Peru, Mexico, Bosnia and present China. Pyramids near Xianyang in present China were constructed by Proto-Turks (Proto-Uygurs) about 3000 BCE, which makes them the oldest pyramids in the world and confirms the hypothesis that pyramids in Egypt were built by people who migrated from Central Asia due to climate change and dried inland seas such as Taklamakan (in Uygur Turkish, is *Döklemegen* means "the point of no return") and Gobi Desert. Besides the good mechanical interlocking of rock blocks, there are caverns within these pyramids. The roofs of these caverns consist of beams of hard rock (mainly granite) with blocks in the sidewalls put together to form inverted V-shaped or trapeze-shaped arches (like Sumerian arches). Of course, the beams were dimensioned in a way that they can resist tensile stresses induced by bending due to surcharge loads for thousands of years.

Friction law, strength of rocks in tension and compressions, was undoubtedly known to ancestral civilizations (i.e. Sumerians, Turanians, Anatolians, Egyptians, Indians, Chinese, Peruvians, Maya, Aztecs, Persians and Roman) and measured by them precisely. The very advanced measurement systems developed by Sumerians have very likely direct connections to modern measurement systems. It is simply our disregard and ignorance of their knowledge and level of their advancement that lead us to quote, for example, Guillaume Amonton and Leonardo da Vinci as the pioneers of modern testing and measurements

(a)   (b)

Figure 1.2 Remnants of open-pit and underground quarries during the early stages of excavations: (a) pathway leading to the quarries, (b) initial underground quarries

techniques. There is a need to initiate a working group from various countries on the history of testing and measurement techniques relevant to rock mechanics and rock engineering and to recognize the actual pioneers with due respect.

1.2   Modern development of rock mechanics and rock engineering

The principles of modern rock mechanics and rock engineering are associated with Talobre (1957), Terzaghi (1946) and Stini (1950), both from Austria (Müller, 1963). Their works were followed by Müller of Karlsruhe University in Germany, Talobre of Électricité de France in France, and Rocha of the Portuguese National Laboratory for Civil Engineering (LNEC), which is still home to the ISRM office. The book published by Jaeger and Cook (1979) was the first theoretical publication in rock mechanics. The books related to rock engineering by E. Hoek and his colleagues (Hoek and Bray, 1977; Hoek and Brown, 1980) at the Imperial College were other milestones in the advancement of rock mechanics and rock engineering.

The International Society for Rock Mechanics (ISRM) has defined rock mechanics as the theoretical and applied science of the mechanical behavior of rock and rock masses; it is that branch of mechanics concerned with the response of rock and rock masses to the force fields of their physical environment. Rock mechanics itself forms part of the broader subject of geomechanics, which is concerned with the mechanical responses of all geological materials, including soils. Rock mechanics is concerned with the application of the principles of engineering mechanics to the design of the rock structures generated by mining, civil and petroleum activity, e.g. tunnels, mining shafts, underground excavations, open pit mines. While the acronym remains the same, the ISRM has changed its name to the International Society for Rock Mechanics and Rock Engineering.

The International Society for Rock Mechanics (ISRM) was founded in Salzburg in 1962 as a result of the enlargement of the so-called Salzburger Kreis. Its foundation is mainly owed to Professor Leopold Müller who acted as president of the Society till September 1966. When one looks at the content of the proceedings of the first Congress, the spectrum of rock mechanics and rock engineering (RMRE) is very wide compared to that these days. In other words, the greater emphasis given to the applications in civil and mining engineering and the relation of rock mechanics with earth science or geoscience is almost nonexistent in the last three decades. The recent decrease of civil engineering constructions and mining activities due to economic reasons and environmental concerns in many countries resulted in the decrease of the interest of academia and the engineering community in RMRE. The overemphasis on nuclear waste disposal problems, which are only relevant to a limited number of countries worldwide, causes further decreases in interest in academia and the engineering community in RMRE.

1.3   Goals and content of this book

Rock is the main constituent of the crust of the Earth, and its behavior is the most complex one among all materials in the geosphere that humankind deals with. Furthermore, it contains various discontinuities, which make the thermo-hydro-mechanical mechanical behavior of rocks more complex. These simply require a higher level of knowledge and intelligence in the RMRE community.

Rock mechanics is concerned with the theoretical and applied science of the mechanical behavior of rock and rock masses, and it is one of the branches of mechanics concerned with the response of rock and rock masses to their physical-chemical environment. Rock mechanics is concerned with the application of the principles of mechanics to physical, chemical and electromagnetic processes in the uppermost part of the Earth and the design of the rock structures associated with mining, civil and petroleum engineering. This book is intended to be a fundamental book for younger generations and newcomers, as well as a reference book for experts specialized in rock mechanics.

The practitioners and experts of rock mechanics should have a profound knowledge of rock-constituting elements, the petrography of rocks, discontinuities and their causes to understand their behavior under various physical and chemical actions in nature. Several chapters are devoted to this issue. First, common rock-forming minerals, rocks, discontinuities and rock mass are explained, and fundamental definitions and their measurement techniques are presented. The governing equations, constitutive laws and experimental techniques are described. The fundamentals of techniques for solving the resulting partial differential equations of rock mechanics are explained, and some specific examples of applications are given. Second, the techniques for the characterization of rock masses, experimental techniques in situ, and the evaluation of the stress state in rock mass using direct and indirect techniques are described, and several specific examples of applications are given. Other chapters are devoted to ice mechanics and extraterrestrial rock mechanics as possible new directions of rock mechanics.

This volume provides the fundamentals as well as many recent and relevant topics for younger generations, newcomers and experts specialized in rock mechanics, with some specific goals such as:

1   Understanding the basic components and features of rocks, discontinuities and rock masses and their physical characterization. This is a quite important aspect as some practitioners of rock mechanics lack this knowledge.
2   The fundamental laws of mechanics for rock and rock masses, constitutive models and associated experimental techniques in laboratory and in situ, numerical techniques. Various physical modeling procedures used in the field of rock mechanics are described to help young generations as well as newcomers understand the fundamentals of rock mechanics.
3   The evaluation of rock masses in nature. Many empirical, experimental and geophysical techniques are developed for this purpose of understanding this very complete subject. These techniques are described, and their applications in the practice are presented. Another important aspect in the design and construction of rock engineering structures is the evaluation of *in-situ* stress state before their construction. This aspect is presented from a broad perspective, and several direct and indirect techniques are explained. Rock excavations techniques are described, and some practical examples are given.
4   The exploration and exploitation of natural resources under extreme climatic conditions on the Earth, Moon, planets, asteroids. Therefore, ice mechanics and extraterrestrial rock mechanics will become important fields of applications of rock mechanics. Current knowledge, findings and techniques are briefly described, and possible future aspects are discussed.

References

Aydan, Ö. (2008) New directions of rock mechanics and rock engineering: Geomechanics and Geoengineering. *5th Asian Rock Mechanics Symposium (ARMS5), Tehran.* pp. 3–21.

Aydan, Ö. (2014) Future advancement of rock mechanics and rock engineering (RMRE). *ROCKMEC'2014-XIth Regional Rock Mechanics Symposium, Afyonkarahisar, Turkey.* pp. 27–50.

Aydan, Ö. & Geniş, M. (2004) Surrounding rock properties and openings stability of rock tomb of Amenhotep III (Egypt). *ISRM Regional Rock Mechanics Symposium, Sivas.* pp. 191–202.

Aydan, Ö. & Ulusay, R. (2003) Geotechnical and geoenvironmental characteristics of man-made underground structures in Cappadocia, Turkey. *Engineering Geology,* 69, 245–272.

Aydan, Ö. & Ulusay, R. (2013) Geomechanical evaluation of Derinkuyu Antique Underground City and its implications in geoengineering. In: *Rock Mechanics and Rock Engineering.* Springer Vienna. pp. 731–754.

Aydan, Ö., Tano, H., Geniş, M., Sakamoto, I. & Hamada, M. (2008a) Environmental and rock mechanics investigations for the restoration of the tomb of Amenophis III. *Japan-Egypt Joint Symposium New Horizons in Geotechnical and Geoenvironmental Engineering, Tanta, Egypt.* pp. 151–162.

Aydan, Ö., Tano, H., Ulusay, R. & Jeong, G.C. (2008b) Deterioration of historical structures in Cappadocia (Turkey) and in Thebes (Egypt) in soft rocks and possible remedial measures. *2008 International Symposium on Conservation Science for Cultural Heritage, Seoul.* pp. 37–41.

Aydan, Ö., Ohta, Y., Daido, M., Kumsar, H., Genis, M., Tokashiki, N., Ito, T. & Amini, M. (2011) Chapter 15: Earthquakes as a rock dynamic problem and their effects on rock engineering structures. In: Zhou, Y. and Zhao, J. (eds.) *Advances in Rock Dynamics and Applications.* CRC Press, London, Taylor and Francis Group, Boca Raton, FL. pp. 341–422.

Geniş, M., Tokashiki, N. & Aydan, Ö. (2009) The stability assessment of karstic caves beneath Gushikawa Castle Remains (Japan). *EUROCK,* 2010, 449–454.

Hamada, M., Aydan, Ö. & Tano, H. (2004) *Rock Mechanical Investigation: Environmental and Rock Mechanical Investigations for the Conservation Project in the Royal Tomb of Amenophis III.* Conservation of the Wall Paintings in the Royal Tomb of Amenophis III, First and Second Phases Report, UNESCO and Institute of Egyptology, Waseda University. pp. 83–138.

Hoek, E. & Bray, J.W. (1977) *Rock Slope Engineering,* 2nd edition. Institution of Mining and Metallurgy, London. 402p.

Hoek, E. & Brown, J.W. (1980) *Underground Excavations in Rock.* Institution of Mining and Metallurgy, London. 527p.

International Society for Rock Mechanics. *History of Society.* www.isrm.net/.

Jaeger, J.C. & Cook, N.G.W. (1979) *Fundamentals of Rock Mechanics,* 3rd edition. Chapman & Hall, London. pp. 79, 311.

Kaptan, E. (1992) Tin and ancient underground tin mining in Anatolia (in Turkish with English abstract). *Geological Engineering,* 40, 15–19, Ankara.

Kramer, S.N. (1956) *History Begins at Sumer: Thirty-Nine Firsts in Man's Recorded History.* University of Pennsylvania Press, Philadelphia.

Kulaksız, S. and Aydan, Ö. (2010) Characteristics of ancient underground quarries of Turkey and Egypt and their comparison. *22nd World Mining Congress, Istanbul.* pp. 607–614.

Kumsar, H. & Aydan, Ö. (2008) Preservation of some ancient cities in Aegean Region of Turkey with an emphasis on Hierapolis, Aphrodisias and Lagina. *2008 International Symposium on Conservation Science for Cultural Heritage, Seoul.* pp. 47–50.

Kumsar, H., Celik, S., Aydan, Ö. & Ulusay, R. (2003) Aphrodisias: Anatolian antique city of building and sculptural stones. *International Symposium on Industrial Minerals and Building Stones, Istanbul.* pp. 301–309.

Müller, L. (1963) *Der Felsbau.* Ferdinand Enke Verlag, Stuttgart.

Stini, I. (1950) *Tunnelbaugeologie.* Springer-Verlag, Vienna. 366p.

Talobre, J. (1957) *La Mechanique des Rocheux*. Dunod, Paris.

Terzaghi, K. (1946) Rock defects and loads on tunnel supports. In: Proctor, R.V. & White, T. (eds.) *Rock Tunneling with Steel Supports*. Commercial Shearing and Stamping Co., Youngstown. pp. 15–99.

Tokashiki, N., Aydan, Ö. & Jeong, G.C. (2008) Stone masonry historical structures in Ryukyu Islands and possible remedial measures. *2008 International Symposium on Conservation Science for Cultural Heritage, Seoul.* pp. 51–55.

Yener, K.A. (1997) *Excavations at Kestel Mine, Turkey.* The Final Season, 1996–1997 Annual Report. Orient Institute, Michigan University.

Chapter 2

Minerals, rocks, discontinuities and rock mass

2.1 Minerals

A mineral is an inorganic natural solid, which is found in nature. Its atoms are arranged in definite patterns (an ordered internal structure), and it has a specific chemical composition that may vary within certain limits. Minerals may be generally subdivided into two major groups (e.g. Goodman, 1989):

1   Silicates
2   Non-silicates

2.1.1 Silicate minerals

Silicate minerals contain silica (SiO_2) either contained or in free form within the mineral lattice structure. The major silicate minerals constituting rocks follow (see Figure 2.1):

(a)   Quartz (SiO_2)

Silica tetrahedra forms a neutral three-dimensional framework structure (trapezohedral) without other cations. This arrangement forms a very stable structure. It is a strong piezo-electric and pyroelectric mineral.

(b)   Olivines (($Mg, Fe)_2SiO_4$) and garnets (($Mg, Fe, Mn)_3(Fe, Al, Cr)_2Si_3O_{12}$)

Olivines and garnets consist of a series of isolated tetrahedra balanced by the cations magnesium (Mg), iron (Fe), and calcium (Ca). The olivines are orthorhombic. However, well formed crystals are rare. Common olivine is usually green or brownish-green in color. Olivine is an unsaturated mineral, and weathering or hydrothermal processes easily alter it. Garnets form crystals of trapezohedra or rhombidodecahedral habit. Metamorphic rocks are the commonest environment for garnets. Garnet is mineral that is resistant weathering.

(c)   Pyroxenes

Pyroxenes have single chains of tetrahedra balanced by similar metal cations and sodium (Na). Most pyroxenes are monoclinic, and they are commonly found in basic or ultrabasic igneous rocks.

Figure 2.1 Views of major silica minerals

(d)   Amphiboles

Amphiboles are characterized by double chains of tetrahedra balanced by similar cations. They are monoclinic or orthorhombic and commonly found in igneous and metamorphic rocks. Asbestos is one of well-known amphibole mineral. Hornblende is a well-known mineral of the amphibole group.

(e)   Micas

Sheets of tetrahedra are building blocks. Aluminum is also involved in these sheet structures, which are charge-balanced by the cations Mg, Na and K. They are divided into the muscovite and biotite groups. Muscovite is transparent and resistant to weathering. Biotite is characterized by shades of brown to black. They are commonly found in igneous and metamorphic rocks.

(f)   Clay minerals

The atomic structure of clay minerals is basically similar to micas, and they generally occur as minute, platy crystals. An important characteristic is their ability to lose or take up water according to temperature and the amount of water present in a system. Some clay minerals contain loosely bonded cations, which can be easily exchanged for others. Clay minerals

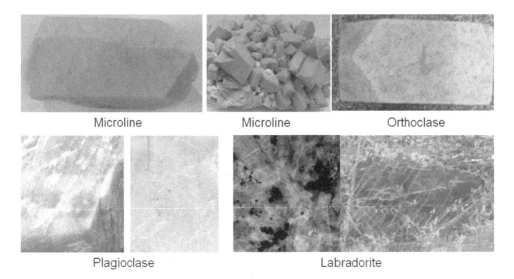

| Microline | Microline | Orthoclase |

| Plagioclase | Labradorite |

Figure 2.2 Major feldspar minerals

are produced by the degradation of silicates or other silicate glasses. Kaolinite, illite, montmorillonite, vermiculite and palygorskite are well-known clay mineral groups. Kaolin is the main constituent of ceramics. Illite is a common clay mineral, and clay in soil mechanics is constituted by illite clay minerals. Montmorillonites (smectite group) are formed by the alteration of basic rocks or other silicates low in K under alkaline conditions. Na-type montmorillonite is especially notable for losing or taking up water and associated volumetric changes.

(g) Feldspars

Feldspars are the most important single group of rock-forming silicate minerals (Fig. 2.2). They consist of a second group of alumino-silicates and the tetrahedra to form three-dimensional frameworks with Ca, Na and K as the balancing cations. They are either monoclinic or triclinic. These very abundant feldspars are subdivided in the K-Na-bearing alkali feldspars and the Ca-Na solid-solution series, called the plagioclase feldspars. Alkali feldspars are found in alkali-rock rocks such as granites, syenites, while plagioclase feldspars are found in intermediate and basic rocks.

(h) Tourmaline

The tourmaline group belongs to cyclosilicates, in which SiO_4 units are linked to form three-, four- or six-membered rings. Tourmaline is the best known six-membered ring silicate, and it is piezoelectric and pyroelectric.

2.1.2 Nonsilicate minerals

Nonsilicate minerals do not contain silica (SiO_2) (Fig. 2.3).

Carbonates	Evaporates	Oxides	Sulfides

Calcite	Rock salt	Hematite	Pyrite

Dolomite	Gypsum	Limonite	Galena

Figure 2.3 Views of various nonsilicate minerals

(a) Carbonates

The important carbonates are the minerals calcite ($CaCO_3$) and dolomite ($CaMg(CO_3)_2$). They are significant rock-forming minerals in limestones and dolomites.

(b) Evaporates

The important groups of evaporate minerals are the halides, including the minerals halite (NaCl, rock salt), sylvite (KCl, potash), and fluorite (CaF_2), as well as the sulfates including the minerals gypsum ($CaSO_4.2H_2O$) and anhydrite ($CaSO_4$). Anhydrite alters to gypsum when it is attacked by watery solutions.

(c) Oxides

Many oxides (hematite and magnetite) and hydroxides (limonite and goethite) of iron are important minor constituents in rocks. The aluminum oxide bauxite can also occur as a rock-forming mineral. Oxides are common in geochemical environments poor in silica. Silicates form easily from magma, so if silica is used up in a magma chamber, then the oxides remain to be formed. Their structure is complex: octahedral and dodecahedral crystals.

(d) Sulfides

The mineral pyrite is the only sulfide that occurs commonly in rocks. Sulfides are most important as economic minerals providing the main sources of elements such as arsenic, copper, lead, nickel, mercury, molybdenum and zinc.

(e) Phosphates

Phosphates are relatively rare. The only important phosphate mineral is apatite.

2.2   Rocks

A rock is an aggregate of one or more minerals and is classified into three major groups: igneous, sedimentary and metamorphic.

2.2.1   *Igneous rocks*

Igneous rocks are created by melting and crystallization of magma. When the magma reaches the surface, the rocks are said to be extrusive (Fig. 2.4). Volcanic lava flows are examples of extrusive igneous rocks. If the magma cools within the Earth, it forms large bodies of crystalline rock called plutons or batholiths. These rocks are called intrusive igneous rocks. Igneous rocks are generally classified on the basis of three factors: (1) grain size and texture, (2) intrusive or extrusive, (3) silica content and mineral composition (Fig. 2.5). However, the classifications based on factors 2 and 3 are commonly used to describe igneous rocks as described next.

2.2.1.1   Intrusive igneous rocks

Igneous rocks slowly cooled inside the crust. (Plutonic rock means formed in the Earth.) They generally consist of large crystals of minerals.

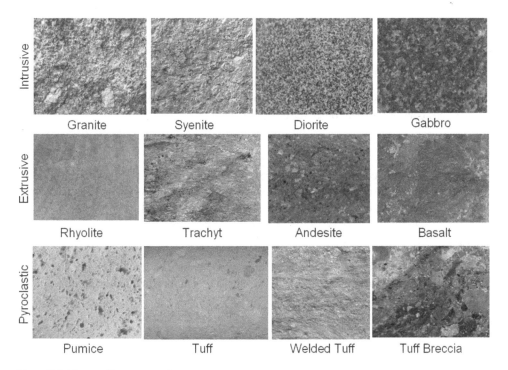

Figure 2.4 Views of igneous rocks

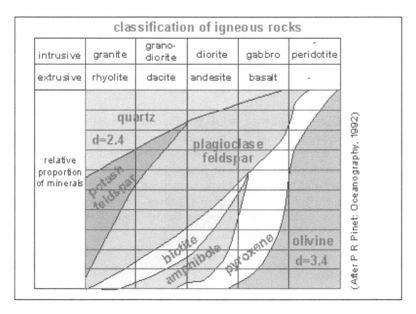

Figure 2.5 Chemical composition of igneous rocks
Source: From Pinet, 1992

(A)   GRANITE

Granite constitutes the continental crust, and its density varies between 2.7 and 2.8 g cm^{-3}.
Granite is an acidic igneous rock and consists of quartz, mica (muscovite and biotite),
potassic feldspar (orthoclase, microcline), plagioclase feldspars and hornblende and tourmaline. Crystals are intermingled, and the amount of quartz is about 30%. Granitic rock
is much less common on the other terrestrial planets.

(B)   SYENITE

Syenite is coarse-grained intrusive igneous rock of the same general composition as granite.
However, the quartz is either absent or present in relatively small amounts (less than 5%).
The name of this rock is related to syenites in the Syene region of Egypt.

(C)   DIORITE

Diorite is a gray to dark gray intermediate intrusive igneous rock composed principally of
plagioclase feldspar, hornblende and/or pyroxene. The amount of quartz is greater than
10%.

(D)   GRANODIORITE

Granodiorite is an intermediate form between granite and diorite. It usually contains abundant biotite and hornblende and orthoclase (potassium feldspar).

(E) GABBRO

Gabbro is a dark, coarse-grained, intrusive igneous rock and contains varied percentages of pyroxene, plagioclase, amphibole and olivine. The amount of quartz is less than 5%. Gabbro is generally coarse grained, with crystal sizes in the range of 1 mm or greater. Gabbro is found in the ocean crust, underneath the basalt layer (0.5–2.5 km), from 2.5 to 6.3 km deep. The lunar highlands have many gabbros.

(F) PERIDOTITE

Peridotite is a dense, coarse-grained rock, consisting mostly of the minerals olivine and pyroxene. Peridotite is ultramafic and ultrabasic, as the rock contains less than 45% silica. Peridotite is the dominant rock of the upper mantle of the Earth.

2.2.1.2   Extrusive Igneous Rocks

These rocks are formed by rapid cooling of magma at the surface. As a result, crystals are either small or glassy.

(A) RHYOLITE

Rhyolite has medium silica content (intermediate), and it is a fine-grained volcanic rock of granitic composition.

(B) TRACHYTE

Trachyte is an igneous, extrusive volcanic rock with an aphanitic to porphyritic texture. The mineral assemblage is predominately potassium feldspar with relatively minor plagioclase. It is a fine-grained volcanic rock of syenitic composition.

(C) ANDESITE

Andesite is considered the extrusive equivalent to diorite. The name "andesite" is derived from the Andes mountain range.

(D) BASALT

Basalt is the extrusive equivalent to gabbro and is made up of feldspars and other minerals common in planetary crusts. It constitutes the ocean crust of the Earth. It is usually fine-grained due to the rapid cooling of lava on the Earth's surface. Unweathered basalt is black or gray. It has been identified as a major surface rock on the dark lunar planes and much of Mars, Venus and the asteroid Vesta.

2.2.1.3   Pyroclastic rocks

Pyroclastic rocks are formed by sedimentation and the welding of debris ejected by volcanoes.

(A) TUFF

Tuff is made of compacted debris from old volcanic ash showers. A tuff of recent origin is generally loose and incoherent. However, the older tuffs are cemented by pressure and the action of infiltrating water, resulting in strong enough yet not very hard material that can be extensively used for building purposes and creating cavities, as seen in Cappadocia region of Turkey.

(B) VOLCANIC BRECCIA

Volcanic breccia (agglomerate) is composed of angular mineral fragments embedded in a matrix, the product of explosive eruptions. Agglomerates are accumulations of large blocks of volcanic material often found around vents. Agglomerates are coarser and less frequently well bedded. The blocks in agglomerates vary greatly in size.

(C) IGNIMBRITES

Welded tuff or ignimbrite is a product of pyroclastic flows hot enough to fuse, or "weld," still hot ash into a single uniform layer called a cooling unit. Ignimbrite is primarily composed of a matrix of volcanic ash, pumice fragments and crystals.

2.2.2 Sedimentary rocks

Sedimentary rocks are formed in layers deposited by wind, water or ice. They are the direct products of the weathering process. As sedimentary layers are buried, they are cemented and lithified. Sediments are subdivided into three types (Fig. 2.6):

1 Clastic sedimentary rocks
2 Chemical sedimentary rocks
3 Organic sedimentary rocks

2.2.2.1 Clastic sedimentary rocks

Clastic sedimentary rocks consist of rock and mineral grains derived from the chemical and mechanical breakdown (weathering) of preexisting rock. They contain rock fragments and, more commonly, particles of quartz and feldspar. The most common cementing materials are silica and calcium carbonate, Clastic rocks are further classified on the basis of grain size. Underneath each rock type, the Wentworth Scale of particle sizes is shown.

(A) CONGLOMERATES AND BRECCIAS

Conglomerates are sedimentary rocks consisting of rounded fragments, whereas breccias consist of angular clasts. Conglomerates and breccias are characterized by clasts larger than 2 mm.

(B) SANDSTONES

Sandstone is a sedimentary rock composed mainly of mineral or rock grains of sand size (0.062–2 mm). Most sandstone is composed of quartz and/or feldspar. Some sandstones are resistant to weathering, and they are porous. When sandstone contains roughly 60%

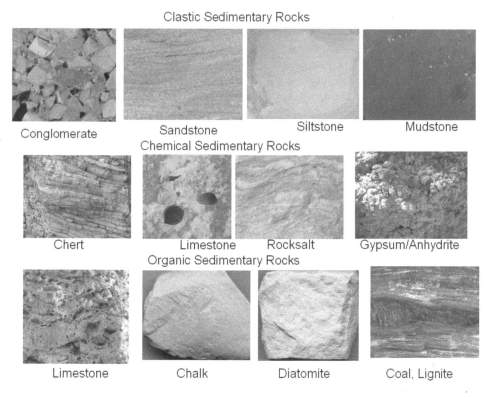

Figure 2.6 Views of various sedimentary rocks

quartz sand and 25% feldspar, it is called Arkose. If the percentage of quartz is greater than 95%, it is called either quartzite or quartzarenite.

(C)   MUDSTONE

Mudstone consists of tiny particles less than 0.0062 mm in diameter. Individual grains are too small to be distinguished without a microscope. With increased pressure and time, the platy clay minerals may become aligned, with the appearance of fissility or parallel layering consolidated mud, rich in organic matter.

(D)   ARGILLITE

Argillite is a sedimentary rock composed of clay particles, which have been hardened and cemented.

2.2.2.2   Chemical and organic sedimentary rocks

These rocks are formed either from minerals that precipitate directly from aqueous (water) solutions or from the accumulation of fossilized remains of organisms that become limestone.

(A)   GYPSUM (CASO$_4$.2H$_2$O)

Gypsum is a very soft chemical rock, and it is composed of calcium sulfate dihydrate.

(B)   ANHYDRITE (CASO$_4$)

Anhydrite, which is an evaporatic rock, is calcium sulfate. Interaction with water produces an increase in volume of the rock layer. This process causes a 30% volumetric expansion, termed as swelling.

(C)   ROCK SALT – HALITE (NACL)

Halite occurs in beds of sedimentary evaporite minerals that result from the drying up of enclosed lakes, playas and seas. It commonly occurs with other evaporite deposit minerals such as several of the sulfates, halides and borates.

(D)   LIMESTONE

Limestone is a sedimentary rock composed largely of the calcite mineral (CaCO$_3$). Limestone may be crystalline, clastic, granular or massive, depending on the method of formation. The primary source of the calcite in limestone may be of organic origin or chemical solution. Limestone makes up about 10% of the total volume of all sedimentary rocks. Travertine is a variety of limestone formed along streams where there are waterfalls around hot or cold springs. Limestone is partially soluble, especially in acid, and therefore forms many erosional landforms, called karsts. Tufa is a thick, rock-like calcium carbonate deposit that forms by precipitation from bodies of water with a high dissolved calcium content.

(E)   CHERT

Chert is a fine-grained, silica-rich cryptocrystalline sedimentary rock. It varies greatly in color from white to black but most often manifests as gray, brown, grayish brown and light green to rusty red.

(F)   CHALK

Chalk is a soft, white, porous form of limestone composed of organic origin. Chalk is formed in shallow waters by the gradual accumulation of the calcite mineral remains of microorganisms such as planktonic green algae, associated with varying proportions of larger microscopic fragments of bivalves, foraminifera and ostracods.

2.2.3   Metamorphic rocks

Metamorphic rocks are rocks formed by the action of pressure (P), temperature (T), and fluids within the Earth. As sediments are deeply buried, they are deformed and new minerals recrystallize at the elevated temperatures and pressures to form metamorphic rocks. Metamorphic rocks are generated by recrystallization of either igneous or sedimentary rocks by the action of any or all of pressure, temperature and pore fluids.

The lower limit of metamorphic temperatures is 150°Celsius. The upper limit is the melting temperature when magma forms. The type of metamorphic rock is determined by the parent rock and the P/T conditions. In general, metamorphism the growth of new

Table 2.1 Relation between parent rock and metamorphic rocks

Rock Name	Type	Parent Rock	Characteristics
Phyllite	Foliated	Silty sandstone	Splitting schistosity surfaces
Slate	Foliated	Shales and mudstone	Prominent splitting surfaces
Schist	Foliated	Fine-grained rocks	Mica minerals, often crinkled or wavy
Gneiss	Foliated	Coarse-grained rocks	Dark and light bands or layers of aligned minerals
Quartzite	Nonfoliated	Sandstone	Interlocking almost fused quartz grains, little or no porosity
Marble	Nonfoliated	Limestone	Interlocking almost fused calcite grains, little or no porosity
Serpentine	Nonfoliated	Peridotite ultramafic rocks	

Gneiss   Micaschist   Slate   Shale

Phyllite   Marble   Quartzite   Hornfels   Serpantine

Figure 2.7 Views of common metamorphic rocks

minerals, the deformation and rotation of mineral grains, the recrystallization of minerals, and the production of anisotropic rock (Table 2.1).

Contact metamorphic rocks are recrystallized and rarely show foliation (Fig. 2.7). Shales baked by igneous contact form very hard fine-grained rocks called hornfels. Calcareous rocks, when subjected to contact metamorphism (an alteration by hot fluids), alter into rocks called skarns.

Metamorphic rocks have been chemically altered by heat, pressure and deformation, while buried deep in the Earth's crust. These rocks show changes in mineral composition or texture or both. This area of rock classification is highly specialized and complex.

(a)   Phyllite

Phyllite was originally a fine-grained sedimentary rock such as sandy mudstone or shale composed mainly of clay minerals in a semirandom orientation. Phyllite is a common metamorphic rock, formed under a low-heat and high-pressure environment.

(b)   Shale

Shale is a fine-grained rock subjected to low-pressure metamorphism. Its original constituent is mudstone. It is characterized by thin laminae breaking with an irregular curving fracture.

(c)   Slate

Slates are foliated rocks representing low-grade metamorphic alteration of shales (laminated clay). Slate is mainly composed of quartz and muscovite or illite, often along with other minerals.

(d)   Schists

Schists are foliated medium-grade metamorphic rock with parallel layers, vertical to the direction of compaction. The schists contain lamellar minerals such as micas, chlorite, talc, hornblende, graphite and others. Quartz often occurs in drawn-out grains to such an extent that a particular form, called quartz schist, is produced. Schist contains more than 50% platy and elongated minerals, often finely interleaved with quartz and feldspar. Schist is characteristically foliated so that individual mineral grains split off easily into flakes or slabs.

(e)   Gneiss

These are banded rocks consisting of alternating layers of quartz and feldspar of high metamorphic grade. The original rock formations may be igneous or sedimentary rocks.

(f)   Quartzites

They represent metamorphosed quartzitic sandstone. Pure quartzite is usually white to gray. Quartzites often occur in various shades of pink and red due to varying amounts of iron oxide. Quartzite is very resistant to chemical weathering.

(g)   Marble

Marble is metamorphosed limestone composed mostly of calcite. The metamorphism causes a complete recrystallization of the original rock into an interlocking mosaic of calcite, aragonite and/or dolomite crystals.

(h)   Serpentinite

Serpentinite is comprised of serpentine minerals. Minerals in this group are formed by a hydration and metamorphic transformation of ultramafic rocks. Serpentinite is formed from olivine via several reactions. It can be easily weathered, resulting in swelling.

2.3   Discontinuities

Rocks, by nature, are geologically classified into three main groups: igneous rocks, sedimantery rocks and metamorphic rocks. Each of these rock classes may be further subdivided into several subclasses. For example, igneous rocks are subdivided into three classes: extrusive, intrusive and semi-intrusive rocks (such as dykes, sills etc.), although the chemical composition of the three types of rocks may be same. The order of minerals and the internal structure of rocks are a result of chemical composition of rising magma, its velocity, and environmental conditions during the cooling process, which greatly affects the discontinuity formation in such rocks.

The term "discontinuity" encompasses all types of interruptions of structural integrity of rock masses. They can be classified into three categories:

1   *Intrinsic discontinuities*: Bedding plane, schistosity, flow plane
2   *Volumetric-strain induced discontinuities*: Sheeting, desiccation, cooling, erosion, freezing-thawing
3   *Plastic deformation-induced discontinuities*: Faults, fracture zones, tension (T), Riedel (R-R′) shear cracks and Skempton (P) fractures.

Discontinuities in rocks are termed as cracks, fractures, joints, bedding planes, schistosity or foliation planes and faults. Discontinuities are products of some certain phenomena to which rocks were subjected in their geological past, and they are expected to be regularly distributed within rock mass. The discontinuities can be classified into the following four groups according to the mechanical or environmental process they were subjected to (Erguvanlı, 1973; Yüzer and Vardar, 1986; Miki, 1986; Ramsay and Huber, 1984; Aydan, 2018, etc.) (Fig. 2.8):

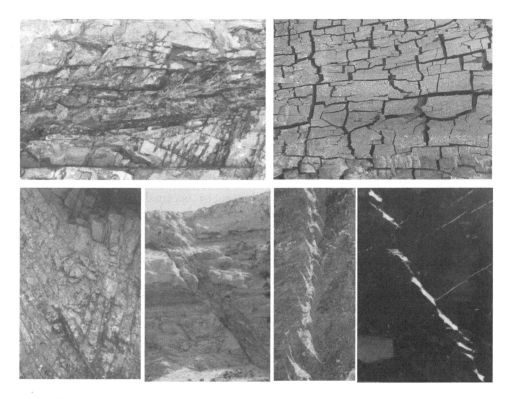

Figure 2.8 Views of discontinuities in situ

1   Tension discontinuities due to:

- Cooling
- Drying
- Freezing
- Bending
- Flexural slip
- Uplifting
- Faulting and stress relaxation due to erosion or glacier retreat or human-made excavation

2   Shear discontinuities due to:

- Folding
- Faulting

3   Discontinuities due to periodic sedimentation
4   Discontinuities due to metamorphism

(a)   Joints

A joint is a discontinuity that is relatively planar and on which there has been no displacement. A series of joints in the same orientation are referred to as a joint set. Joints may be open, healed or filled.

(b)   Bedding joints

Bedding joints that are parallel to the bedding are referred to as bedding joints.

(c)   Foliation joints

Foliation joints are parallel to metamorphic foliation.

(d)   Shear

Shears are structural breaks, where relative movement has occurred. The shear surfaces are characterized by the presence of slickensides, gouge, breccia, mylonite or a combination of these. Shears are in effect small faults and typically have displacements of less than 5 cm.

(e)   Fault

Fault is a shear with significant continuity and evidence of large displacement. A fault can range from centimeters in width to a zone that is tens of meters thick. The fault may contain breccia, gouge and crushed rock. Fault zones are typically conduits for high groundwater flow.

(f)   Contact

Contact is a geologic contact between two distinct lithologic units.

(g) Vein

Vein is an infilling of a discontinuity caused by circulation of mineralized fluid and deposition of minerals. Veins can cause healing of the original discontinuity.

Discontinuities, although they may be viewed as planes in large scale, have undulated surfaces varying in irregularities. As a result, they may be regarded as bands with a certain thickness in association with the amplitude of undulations. The discontinuities may be filled by some infilling materials such as calcite, quartzite or weathering products of host rock or transported materials, or they may exist from the beginning as thin films of clay deposits in sedimentary rocks along bedding planes.

2.4 Rock mass

Rock mass is considered to be the sum total of the rock as it exists in place, taking into account the intact rock material as well as joints, faults and other natural planes of weakness that can divide the rock into interlocking blocks of varying sizes and shapes. Rock mass is classified as intact rock mass, layered rock mass, blocky rock mass and sheared rock mass, which may be jointed.

(a) Intact rock mass

Intact rock mass contains neither joints nor hair cracks. Hence, if it breaks, it breaks across sound rock.

(b) Layered or foliated rock mass

Layered or foliated rock mass consists of individual layers or foliation with little or no resistance against separation along the boundaries between the layers.

(c) Blocky rock mass

Blocky rock mass consists of chemically intact or almost intact rock fragments, which are entirely separated from one another and are perfectly or imperfectly interlocked.

(d) Sheared rock mass (fracture zone)

Sheared (fracture zone) rock mass is rock mass that underwent shearing sufficient to create distinct fault gouge and various degree of fractures such as T, R-R', S fractures. A sheared rock mass generally consists of a crushed rock zone (fault gouge) and shears with slickensides and tension fractures, and the fragment size within the zone will vary with distance from the gouge zone.

Because of the existence of discontinuities as a result of one or combined actions of these processes, the structure of rock mass in nature may look like an assemblage of blocks of some typical shapes (Figs. 2.9 and 2.10). Most common shapes of the blocks are rectangular, rhombohedral, hexagonal or pentagonal prisms. While hexagonal and/or pentagonal prismatic blocks are commonly observed in extrusive basic igneous rocks such as andesite or basalt, and some fine grained sedimentary rocks have undergone

Figure 2.9 views of rock mass in nature

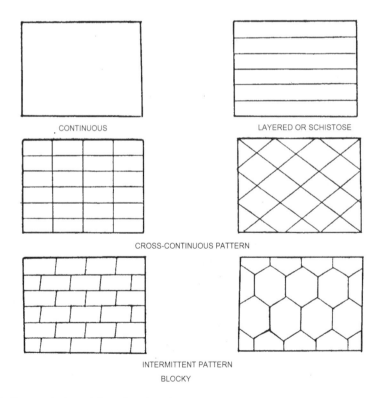

CONTINUOUS

LAYERED OR SCHISTOSE

CROSS-CONTINUOUS PATTERN

INTERMITTENT PATTERN

BLOCKY

Figure 2.10 Geometrical modeling of rock mass

cooling or drying processes, the most common block shapes are between rectangular prism and rhombohedric prism. The lower and upper bases of the blocks are usually limited by planes called flow planes, bedding planes and schistosity or foliation planes in igneous, sedimentary and metamorphic rocks, respectively. These discontinuities can be regarded as very continuous for most of the concerned rock structures. Other discontinuities are usually found in at least two or three sets, crossing these planes orthogonally or obliquely. These secondary sets, if present, may be very continuous or intermittent. As a result, the rock mass may be viewed as shown in Figure 2.10 (Goodman, 1989; Aydan *et al.*, 1989, 2018):

- Continuous medium
- Tabular (layered) medium
- Blocky medium

Blocky medium can be further subdivided into two groups depending upon the continuity of secondary sets as follows (Aydan and Kawamoto, 1987; Aydan *et al.*, 1989):

- Cross-continuously arranged blocky medium
- Intermittently arranged blocky medium

References

Aydan, Ö. (1989) *The Stabilisation of Rock Engineering Structures by Rockbolts*. Doctorate Thesis, Nagoya University, Faculty of Engineering, Nagoya, 204p.

Aydan, Ö. (2018) *Rock Reinforcement and Rock Support*. CRC Press, London. 486p.

Aydan, Ö. & Kawamoto, T. (1987) Toppling failure of discontinuous rock slopes and their stabilisation (in Japanese). *Journal of Japan Mining Society*, 103(1197), 763–770.

Aydan, Ö., Shimizu, Y. & Ichikawa, Y. (1989) The effective failure modes and stability of slopes in rock mass with two discontinuity sets. *Rock Mechanics and Rock Engineering*, 22(3), 163–188.

Erguvanlı, K. (1973) *Engineering Geology (in Turkish)*. ITU Press, Istanbul. No. 966. 552p.

Goodman, R.E. (1989) *Introduction to Rock Mechanics*. Wiley. New York. 576p.

Miki, K. (1986) *Introduction to Rock Mechanics (in Japanese)*. Kajima Press. Tokyo. 317p.

Pinet, P.R. (1992) *Oceanography: An Introduction to the Planet*. West Publishing Company, Saint Paul.

Ramsay, J.G. & Huber, M.I. (1984) *The Techniques of Modern Structural Geology, Vol. 1: Strain Analysis*. Academic Press, London. 306p.

Yüzer, E. & Vardar, M. (1986) *Rock Mechanics (in Turkish)*, ITU Foundation, Istanbul. No. 4. 154p.

Chapter 3

Fundamental definitions and measurement techniques

3.1 Physical parameters of rocks

(a) Bulk density

Bulk density is a property of materials and is defined as the ratio of the mass (m) of rock to the volume (V) it occupies. It is mathematically expressed as follows:

$$\rho = \frac{m}{V} \tag{3.1}$$

(b) Unit weight

Unit weight is defined as the ratio of the weight (W) of rock to the volume (V) it occupies. It is mathematically expressed as follows:

$$\gamma = \frac{W}{V} \tag{3.2}$$

(c) Water content

Water content is a ratio to indicate the amount of water a rock contains. Water content can be either the volumetric (by volume) or the gravimetric (by weight) fraction of the total rock that is filled with liquid water. Volumetric water content is defined as volume of water per unit volume of soil:

$$\theta = \frac{V_w}{V} \tag{3.3}$$

Gravimetric water content is defined as mass of water per unit mass of dry rock:

$$w = \frac{m_w}{m_s} \tag{3.4}$$

(d) Porosity

The porosity of a porous rock describes the fraction of void space in the material to its total volume:

$$n = \frac{V_p}{V} \tag{3.5}$$

where V_p is the nonsolid volume, and V is the total volume of material, including the solid and nonsolid parts.

(e) Longitudinal and traverse seismic wave velocity of rock (V_s, V_p)

Longitudinal and traverse seismic wave velocity of a rock sample with length are defined as follows:

$$V_p = \frac{L}{\Delta t_p} \quad \text{and} \quad V_p = \frac{L}{\Delta t_p} \tag{3.6}$$

where Δt_p and Δt_s are the travel times of longitudinal and traverse waves.

3.2 Physical parameters of discontinuities

Rock discontinuities are characterized by their orientations, spacing and persistency, and surface topography. The methods used to evaluate the parameters are explained in this section.

3.2.1 Discontinuity orientation and its representation

Discontinuity orientation data consists of dip direction (or strike, which is perpendicular to the dip direction) and dip. It is measured either by directly clinometer or indirectly by photogrammetric techniques. If the photogrammetric technique is used, the images of discontinuities on three planes having different normal vectors are necessary. If clinometers are used, it is desirable to use a Clar-type clinometer (compass) as dip direction and dip of a discontinuity can be measured simultaneously. Also, new electronic clinometers store the measured data in digital form.

3.2.2 Discontinuity orientation representation

The stereographic projection method is used for the graphical presentation of orientation data of discontinuities. The method utilizes a sphere of a unit radius. The projections are done by using either equal area or equal angle approach. Equal angle projection is preferred for kinematic assessments of the stability of rock structures. The discontinuities are represented by great circles and/or poles, and they are projected onto equatorial plane using either upper or lower hemisphere projections. When the number of data is too large, the use of pole density projections is desirable.

3.2.3 Discontinuity spacing

Rock discontinuity spacing is also measured though direct measurement of scan lines on the outcrops of rock mass or photogrammetric methods. The number of scan lines should be sufficient to eliminate line bias on measurements, and it is desirable to have at least three outcrop surfaces with different unit vectors. Scan lines should be set up at near right angles to major discontinuity sets in order to avoid scan line–discontinuity set orientation bias. Otherwise, the measurements from scan lines would be apparent, and they have to be converted to true values. In recent years, new techniques of measuring spacing and storage and processing of the data will be added as alternative techniques. Particularly, laser-based techniques could be alternative procedures for dealing with the huge amount of data.

1 Manual techniques (Fig. 3.1)
2 Digital techniques (manual or automated evaluation: Figure 3.2)

   a Photogrammetry (Fig. 3.2(a))
   b Laser scanning (Fig. 3.2(b))

3.2.4 Discontinuity persistency

Discontinuity persistence implies the areal extent or size of discontinuity within a plane. Scan lines must be set up on actual outcrops along the traces of each discontinuity set, and the lengths of rock bridges should be measured. The persistence (T) of the discontinuity is defined as follows:

$$T = (1 - \frac{\sum l_b^i}{L}) \times 100 \qquad (3.7)$$

Where L is scan line length, and l_b^i is length of rock bridge i.

Figure 3.1 An example of line survey of chert layers in Ie island (Okinawa, Japan) as manual technique (scale is 1 m long)

(a) Photogrammetry (Aydan et al. 1999)

(b) Laser technology (Aydan, 2013 unpublished)

Figure 3.2 Digital techniques for spacing measurement

(a) Bedded sandstone

(b) Sedimentary rock

Figure 3.3 (a) Thoroughgoing, (b) intermittent discontinuities

Although the original definition of persistence described is correct, it becomes meaningless for thoroughgoing discontinuities such as bedding planes, schistosity or sheeting joints (Fig. 3.3).

3.2.5 *Discontinuity surface morphology and measurement techniques*

Roughness is a geometrical parameter, and it has directional characteristics (Fig. 3.4). Many methods have been used to characterize the surface topography of rock discontinuities such

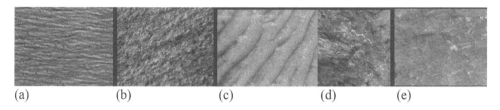

Figure 3.4 Some examples of surface roughness: (a) schistosity surface, (b) sheeting joint, (c) bedding plane in limestone, (d) and (e) striated shear discontinuity

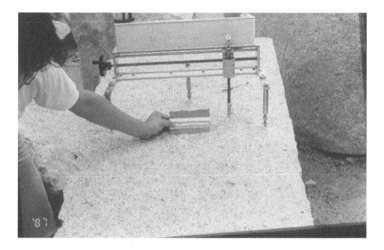

Figure 3.5 Needle-type and roller-stylus for roughness measurements
Source: Aydan *et al.*, 1995

as asperity height, asperity inclination, profile length ratio, autocorrelation function, fractal dimension and so on.

There are different techniques to measure surface morphology of discontinuities. While contact-type profilers are used for this purpose, some new noncontact-type techniques based on laser profiling and photogrammetric techniques will be added as alternative and more accurate systems. Furthermore, it should be noted that Joint Roughness Coefficient (JRC) proposed by Barton and Choubey (1977) is related to an additional friction component associated with the inclination of the asperity wall, and it is one of the parameters.

3.2.5.1   Profiler (needle or roller-pen-recorder)

The earliest profiler used for roughness measurement is the needle profiler. Another profiler utilizes a roller-type stylus. The needle- or roller-type profiler can be manual or automatic (Fig. 3.5). Nevertheless, automatic profilers need a power source, which makes them difficult to use on site.

3.2.5.2 Photogrammetry

The photos of the surface are used to quantify the surface morphology of rock discontinuities. This technique requires the identification of surface profiles from digital images, as shown in Figure 3.6.

A special application of this technique is the shadow profilometry technique, and it is one of the techniques and scale-independent method available for 3-D characterization (Maerz, 1990; Maerz *et al.*, 1990). Shadow profilometry technique is a technique in which an edge of light/shadow is used to trace a surface profile (Fig. 3.7). Multiple profiles produced by moving laterally across a fragment can be used to create a 3-D surface.

Figure 3.6 Profile image of bedding plane in limestone

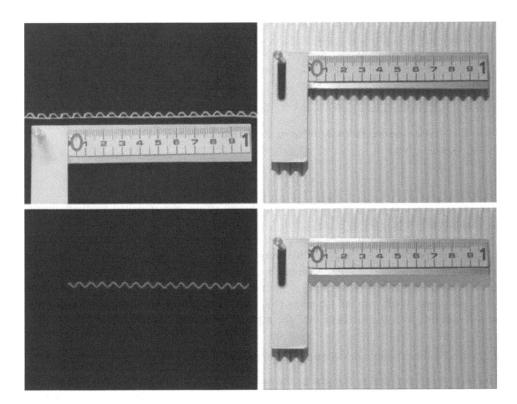

Figure 3.7 Shadow profilometry technique
Source: From Maerz and Hilgers, 2010

3.2.5.3 Laser technology

Laser technology for surface morphology can be used as an automatic modern procedure (Figs. 3.8 and 3.9). The simple technique utilizes 1-D scanning (Fig. 3.8); more sophisticated laser devices can scan for a given bandwidth (Fig. 3.9). Its limitation to *in-situ* applications will be also pointed out.

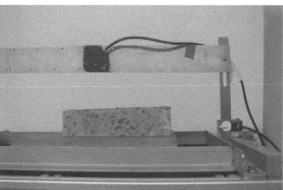

Figure 3.8 1-D scanning devices in laboratory

Source: Aydan's Laboratory

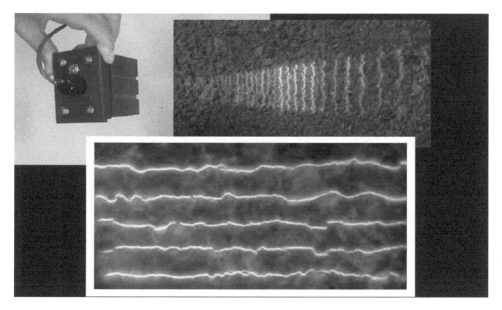

Figure 3.9 Laser scanning device for a given bandwidth in laboratory

Source: From Maerz and Hilgers, 2010

3.2.5.4 Surface morphology characterization

Surface morphology is the geometry of the surface of discontinuities. Appropriate parameters are described here. The parameters associated with linear profiles are height of asperities, inclination of asperity walls, length of asperity wall relative to base length and periodicity of asperities (Aydan and Shimizu, 1995; Aydan *et al.*, 1999; Myers, 1962; Sayles and Thomas, 1977; Tse and Cruden, 1979; Thomas, 1982; Türk *et al.*, 1987).

(A) HEIGHT PARAMETERS

Center-line average height (CLAH) is defined as:

$$CLAH = \frac{1}{L} \int_{x=0}^{x=L} |\varphi| dx \tag{3.8}$$

where L is measurement length; x is distance from origin; φ is height of the profile from the reference base line.

Mean standard variation of height (MSVH) is defined as:

$$MSVH = \frac{1}{L} \int_{x=0}^{x=L} \varphi^2 dx \tag{3.9}$$

Root mean-square of height (RMSH) is defined as:

$$RMSH = \left[\frac{1}{L} \int_{x=0}^{x=L} \varphi^2 dx \right]^{1/2} \tag{3.10}$$

(B) PROFILE LENGTH PARAMETERS

Ratio of profile length (RPL) is defined as:

$$RPL = \frac{1}{L} \int_{x=0}^{x=L} ds = \frac{1}{L} \int_{x=0}^{x=L} \left(1 + \frac{d\varphi}{dx} \right)^{1/2} dx \tag{3.11}$$

(C) ASPERITY INCLINATION PARAMETERS

Weighted asperity inclination (WAI^*) is defined as:

$$WAI^* = \frac{1}{L} \int_{x=0}^{x=L} \left| \frac{d\varphi}{dx} \right| dx \quad 5mm\, WAI = tan^{-1}(WAI^*) \tag{3.12}$$

Weighted asperity inclination difference ($WAID^*$) is defined as:

$$WAID^* = \frac{1}{L_p} \int_{x=0}^{x=L_p} \left| \frac{d\varphi}{dx} \right|_p dx - \frac{1}{L_n} \int_{x=0}^{x=L_n} \left| \frac{d\varphi}{dx} \right|_n dx; WAID = tan^{-1}(WAID^*) \tag{3.13}$$

where p and n stand for positive and negative, respectively. Furthermore, $L = L_p + L_n$.

Mean standard variation of inclination (*MSVI*) is defined as:

$$MSVI = \frac{1}{L} \int_{x=0}^{x=L} \left(\frac{d\varphi}{dx}\right)^2 dx \tag{3.14}$$

Root mean-square of inclination (RMSI) is defined as:

$$RMSI = \left[\frac{1}{L} \int_{x=0}^{x=L} \left\{\frac{d\varphi}{dx}\right\}^2 dx\right]^{1/2} \tag{3.15}$$

(D) PERIODICITY PARAMETERS

Autocorrelation function (*ACF*) is defined as:

$$ACF = \frac{1}{L} \int_{x=0}^{x=L} \varphi(x)\varphi(x+\tau)dx \tag{3.16}$$

Structure function (*SF*) is defined as:

$$SF = \frac{1}{L} \int_{x=0}^{x=L} (\varphi(x) - \varphi(x+\tau))^2 dx \tag{3.17}$$

where τ is a measure of the periodicity of asperities.

(E) FRACTAL DIMENSION

Fractal dimension is defined as:

$$N = Cl^{-D} \tag{3.18}$$

where N is the number of steps, C is a constant, l is step length, and D is fractal dimension. Since the following relations holds between the total length of the profile and the step length:

$$L = Nl \tag{3.19}$$

then the above equation is rewritten as:

$$L = Cl^{1-D} \tag{3.20}$$

(F) ANISOTROPY OF SURFACE MORPHOLOGY PARAMETERS

To characterize the surface morphology of discontinuities, one may introduce an elliptical coordinate system so that the principal axes of the coordinate system coincide with those eigen directions (Fig. 3.10a). Such a coordinate system would be appropriate for many discontinuity types found in rock masses. However, Aydan *et al.* (1996) utilized a Cartesian coordinate system. Let us assume that axis X coincides with the axis of ridges and troughs and axis Y is perpendicular to that of the ridges and troughs. Let us further

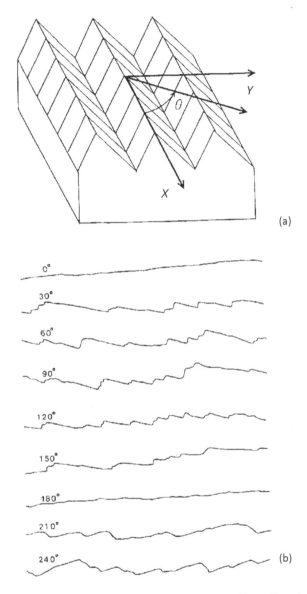

(a)

0°
30°
60°
90°
120°
150°
180°
210°
240°

(b)

Figure 3.10 (a) Notations for measuring profiles of discontinuities, (b) profiles of several discontinuity
types measured by varying measuring direction

Source: Aydan *et al.*, 1996

assume that direction θ is measured from axis X anticlockwise (Fig. 3.10a). A surface morphology parameter F as a function of measuring direction θ may be assumed to be of the following form:

$$F(\theta) = \sum_{i=1}^{n} a_i \cos^i\theta + \sum_{i=1}^{n} b_i \sin^i\theta +$$

$$\sum_{i=1}^{n} c_i \cos^i 2\theta + \sum_{i=1}^{n} d_i \sin^i 2\theta + \cdots + \tag{3.21}$$

$$\sum_{i=1}^{n} y_i \cos^i N\theta + \sum_{i=1}^{n} z_i \sin^i N\theta$$

As a particular form of the preceding equation, we select the following:

$$F(\theta) = a_1\cos\theta + a_2\cos^2\theta + b_1\sin\theta + b_2\sin^2\theta \tag{3.22}$$

Furthermore, we assume that the spectra of the parameter are obtained experimentally along eigen directions as:

$$F(\theta = 0°) = F_0, \quad F(\theta = 90°) = F_{90}, \quad F(\theta = 180°) = F_{180}, \quad F(\theta = 270°) = F_{270} \tag{3.23}$$

With the preceding conditions, constants a_1, a_2, b_1, b_2 are obtained as follows:

$$a_1 = \frac{F_0 - F_{180}}{2}, \quad a_2 = \frac{F_0 + F_{180}}{2}, \quad b_1 = \frac{F_{90} - F_{270}}{2}, \quad b_2 = \frac{F_{90} + F_{270}}{2} \tag{3.24}$$

Nevertheless, it should be noted that constants a_1 and b_1 are expected to be zero for the parameters if the surface morphology is isotropic except for *WAID** and *WAID* unless some errors are caused by the measuring system and digitization.

The length of profiles also influences computed results. For this purpose, the computations were carried out by varying the length of the digitized profile for a sampling interval of 2 mm. As seen from Figure 3.10(b), if the profile length is greater than the wave length of the main asperity of a given discontinuity, the influence of the profile length becomes less pronounced.

The procedure to evaluate the anisotropy of some of fundamental surface morphology parameters; namely, *CLAH*, *WAI* and *RPL* are applied to profiles shown in Figure 3.10b. Figure 3.11 shows the computed parameters as a function of measuring direction θ together with measured results for the sheeting joint in Nakatsukawa granite, whose original profiles were shown in Figure 3.10. Measured spectra of the discontinuity were used in plotting computed results, using Equation (3.22). As seen from the figure, the computed results closely fit the measured ones, although some slight differences between them exist. The differences may be attributable to errors caused while digitizing the measured profiles. Nevertheless, the proposed procedure to evaluate the anisotropy of surface morphology parameters has been concluded to be appropriate. If necessary, better fits to measured parameters can be obtained by using higher-order functions. It is also interesting to note

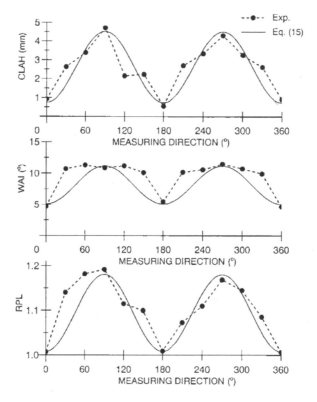

Figure 3.11 Variation of surface morphology parameters *CLAH*, *WAI* and *RPL* as a function of measuring direction (sheeting joint in Nakatsukawa granite)

that asperity inclinations have a minimum along the ridge axis and a maximum perpendicular to the ridge axis.

3.2.5.6 Aperture

Aperture is the separation of between discontinuity surfaces. Aperture can be also measured by borehole cameras (Figs. 3.12 and 3.13).

3.2.5.7 Filling

When discontinuity walls are separated, they may be filled (Fig. 3.14). The filling of discontinuities is directly related to sedimentation of solutions in groundwater. The clayey materials as filling in discontinuities are related either to shearing of discontinuities in the geological past or to weathering or hydrothermal alteration of rock material adjacent to the discontinuity walls. Therefore, when the effect of filling has to be taken into account, the character of the filling and its cause must also be taken into account.

Figure 3.12 Views of apertures of discontinuities in various rocks

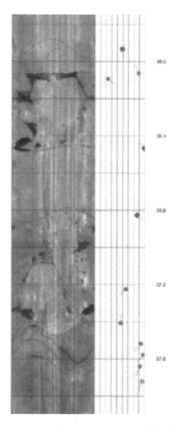

Figure 3.13 Measurements of apertures and orientation using borehole camera
Source: Courtesy of RAAX

Figure 3.14 Views of different infillings of rock discontinuities

3.3   Rock mass

3.3.1   Physical properties of rock mass

The physical properties of rock mass are defined in the same manner as those for intact rocks. However, it is very rare to determine the physical properties of rock masses in practice except the seismic wave properties. As the wave velocity of rock mass is used to infer the properties of rock mass (Ikeda, 1970; Aydan *et al.*, 1993, 1997; Aydan and Kawamoto, 2000), it is often measured in practice during the site exploration as well as during excavation.

3.3.2   Number of discontinuity sets

The number of discontinuity sets plays a major role on the overall behavior of rock mass as well as its stability and thermo-hydro-mechanical properties. In practice, the stereo technique used to plot contouring is carried out to evaluate the number of discontinuity sets, as shown in Figure 3.15(a). However, more descriptive procedures, such as geological interpretation, photogrammetric techniques, geological knowledge–based approach (unit block method), automatic identification of discontinuity sets from stereo projections for evaluating discontinuity sets, are carried out in practice (Aydan *et al.*, 1991) (Fig. 3.15 and 3.16).

Recently, borehole cameras have been improved to determine many fundamental parameters such as the orientation, aperture, infilling, spacing and roughness associated with discontinuities (Fig. 3.17).

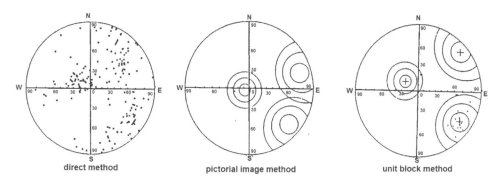

Figure 3.15 Comparison of different techniques for determining number of sets

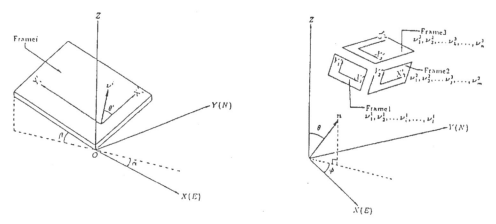

Figure 3.16 Photogrammetric technique

3.3.3 Block size index (I_b) and block volume (V_b)

Block size index and block volume are defined as follows (ISRM, 1978; Peaker, 1990):

$$I_b = \sum_{i=1}^{3} S_i; V_b = S_1 \times S_2 \times S_3 \tag{3.25}$$

where S_i is the average discontinuity spacing.

3.3.4 Volumetric joint count (J_v)

Volumetric joint count is based on discontinuity set spacing (ISRM, 1978), and it is mathematically expressed as:

$$J_v = \sum_{i=1}^{n} \frac{1}{S_i} \tag{3.26}$$

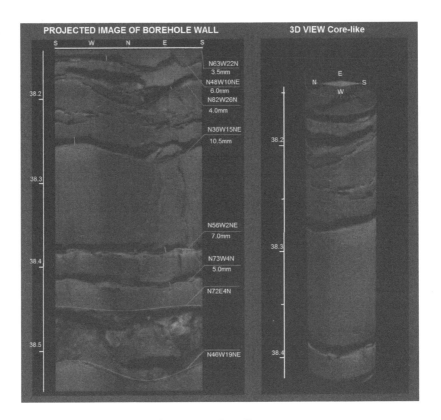

Figure 3.17 Reconstructed core logs from borehole-wall images

Source: Courtesy of RAAX

References

Aydan, Ö. & Kawamoto, T. (2000) The assessment of mechanical properties of rock masses through RMR rock classification system. *GeoEng2000, Melbourne.* p. UW0926.

Aydan, Ö. & Shimizu, Y. (1995) Surface morphology characteristics of rock discontinuities with particular reference to their genesis. *Proc., Int. Meeting on Fractography, Special Publication on Fractography, Geol. Soc. of UK, Geol.* pp. 11–26.

Aydan, Ö., Ichikawa, Y., Shimizu, Y. & Murata, K. (1991) An integrated system for the stability of rock slopes. *The 5th Int. Conf. on Computer Methods and Advances in Geomechanics, Cairns, 1.* pp. 469–465.

Aydan, T., Akagi, T. & Kawamoto, T. (1993) Squeezing potential of rocks around tunnels: Theory and prediction. *Rock Mechanics and Rock Engineering*, 26(2), 137–163.

Aydan, Ö., Shimizu, Y. & Kawamoto, T. (1995) A portable system for in-situ characterization of surface morphology and frictional properties of rock discontinuities. *The 4th Int. Symp. on Field Measurements in Geomechanics, Bergamo.* pp. 463–470.

Aydan, Ö., Shimizu, Y. & Kawamoto, T. (1996) The anisotropy of surface morphology characteristics of rock discontinuities. *Rock Mechanics and Rock Engineering*, 29(1), 47–59.

Aydan, Ö., Ulusay, R. & Kawamoto, T. (1997) Assessment of rock mass strength for underground excavations. *The 36th US Rock Mechanics Symposium.* New York. pp. 777–786.

Aydan, Ö., Tokashiki, N., Shimizu, Y. & Kawamoto, T. (1999) A simple system for measuring the surface morphology characteristics of rock discontinuities (in Japanese). *The 29th Rock Mechanics Symposium of Japan*. Tokyo. 136–140.

Barton, N. & Choubey, V. (1977) The shear strength of rock joints in theory and practice. *Rock Mechanics. Vienna*, 10, 1–54.

Ikeda, K. (1970) A classification of rock conditions for tunnelling. *1st Int. Congr. Eng. Geology, IAEG, Paris*. pp. 1258–1265.

ISRM (1978) Suggested methods for the quantitative description of discontinuities in rock masses. *International Journal of Rock Mechanics and Mining Science Geomechanics Abstracts*, 15, 319–368.

Maerz, N.H. (1990) *Photoanalysis of Rock Fabric*. Ph.D. dissertation, Department of Earth Sciences, University of Waterloo, Waterloo, Ontario, Canada.

Maerz, N.H. & Hilgers, M.C. (2010) A method for matching fractured surfaces using shadow profilometry. *Third International Conference on Tribology and Design 2010, May 11–13 2010, Algarve, Portugal*. pp. 237–248.

Maerz, N.H., Franklin, J.A. & Bennett, C.A. (1990) Joint roughness measurement using shadow profilometry. *International Journal of Rock Mechanics and Mining Science Geomechanics Abstracts*, 27, 329–343.

Myers, M.O. (1962) Characterization of surface roughness. *Wear*, 5, 182–189.

Peaker, S.M. (1990) *Development of a Simple Block Size Distribution Model for the Classification of Rock Masses*. M.Sc. Thesis, Department of Civil Engineering, University of Toronto, Toronto, Ontario, Canada.

Sayles, R.S. & Thomas, T.R. (1977) The spatial representation of surface roughness by means of the structure function: A practical alternative to correlation. *Wear*, 42, 263–276.

Thomas, T.R. (1982) *Rough Surfaces*. Longman. London.

Tse, R. & Cruden, D.M. (1979) Estimating joint roughness coefficients. *International Journal of Rock Mechanics and Mining Science*, 16, 303–307.

Türk, N., Gerig, M.J., Dearman, W.R. & Amin, F.F. (1987) Characterization of rock joint surfaces by fractal dimension. *Proc., 28th US Symposium on Rock Mechanics*. Tucson, Arizona. pp. 1223–1236.

Chapter 4

Fundamental governing equations

Various actual applications of rock mechanics involve mass transportation phenomena such as seepage, diffusion, static and dynamic stability assessment of structures, and heat flow. Principles of fundamental laws presented herein follow basically the laws of continuum mechanics (e.g. Eringen, 1980; Mase, 1970). In this section, the fundamental governing equation of each phenomenon is presented. The governing equations are developed for the one-dimensional case, and they are extended to multidimensional situations.

4.1 Fundamental governing equations for one-dimensional case

4.1.1 Mass conservation law

Mass conservation law is stated as:

gained mass = input flux − output flux + mass generated

Let us consider an infinitely small cubic element as shown in Figure 4.1. The preceding statement can be written in the following form for x-direction as:

$$\Delta m = q_x \Delta y \Delta z \Delta t - q_{x+\Delta x} \Delta y \Delta z \Delta t + g \Delta x \Delta y \Delta z \Delta t \tag{4.1}$$

where m is mass, q is flux, and g is mass generated per unit volume per unit time.

Terms Δm and q are explicitly written as:

$$\Delta m = \Delta \rho \Delta x \Delta y \Delta z \tag{4.2}$$

$$q = \rho v \tag{4.3}$$

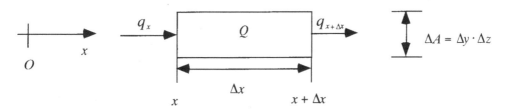

Figure 4.1 Illustration of mass-conservation law

where ρ is density and v is velocity. The quantity $q_{x+\Delta x}$ can be written in the following form using the Taylor expansion:

$$q_{x+\Delta x} = q_x + \frac{\partial q}{\partial x}\Delta x + 0^2 \tag{4.4}$$

Inserting this relation in Equation (4.1) and dividing both sides by $\Delta x \Delta y \Delta z \Delta t$ yields the following:

$$\frac{\Delta \rho}{\Delta t} = -\frac{\Delta(\rho v)}{\Delta x} + g \tag{4.5}$$

Taking the limit results in mass conservation law for 1-D case as:

$$\frac{\partial \rho}{\partial t} = -\frac{\partial(\rho v)}{\partial x} + g \tag{4.6}$$

4.1.2 Momentum conservation law

Momentum balance for a 1-D case can be written for a typical infinitely small control volume (simple momentum concept: $p = m \cdot v$) (Fig. 4.2):

$$-\sigma_x \Delta y \Delta z \Delta t + \sigma_{x+\Delta x}\Delta y \Delta z \Delta t + b\Delta x \Delta y \Delta z \Delta t - \Delta(\rho v)\Delta x \Delta y \Delta z - \Delta(\rho v^2)\Delta y \Delta z \Delta t = 0 \tag{4.7}$$

where σ is stress, b is body force, ρ is density, and v is velocity. Stress $\sigma_{x+\Delta x}$ can be expressed using the Taylor's expansion as:

$$\sigma_{x+\Delta x} = \sigma_x + \frac{\partial \sigma}{\partial x}\Delta x + 0^2 \tag{4.8}$$

Dividing by $\Delta x \Delta y \Delta z \Delta t$ and taking the limits yield the following:

$$\frac{\partial \sigma}{\partial x} + b = \frac{\partial(\rho v)}{\partial t} + \frac{\partial(\rho v^2)}{\partial x} \tag{4.9}$$

The preceding equation may be rewritten as:

$$\frac{\partial \sigma}{\partial x} + b = v\left(\frac{\partial(\rho)}{\partial t} + \frac{\partial(\rho v)}{\partial x}\right) + \rho\left(\frac{\partial v}{\partial t} + v\frac{\partial v}{\partial x}\right) \tag{4.10}$$

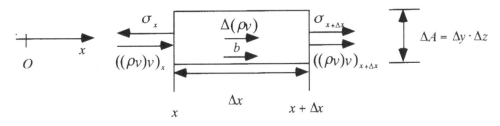

Figure 4.2 Illustration of momentum-conservation law

Introducing the following operator, which is called material derivatization operator:

$$\frac{d}{dt} = \frac{\partial}{\partial t} + v \frac{\partial}{\partial x} \tag{4.11}$$

and using the mass conservation law for no mass generation, Equation (4.10) becomes:

$$\frac{\partial \sigma}{\partial x} + b = \rho \frac{dv}{dt} \tag{4.12}$$

If we define acceleration a as:

$$a = \frac{dv}{dt} \tag{4.13}$$

Equation (4.12) can be rewritten as:

$$\frac{\partial \sigma}{\partial x} + b = \rho a \tag{4.14}$$

Acceleration a may also be expressed in terms of displacement u as:

$$a = \frac{d^2 u}{dt^2} = \ddot{u} \tag{4.15}$$

Accordingly, Equation (4.14) can be reexpressed in terms of u as:

$$\frac{\partial \sigma}{\partial x} + b = \rho \ddot{u} \tag{4.16}$$

4.1.3 Energy conservation laws

Energy balance law for a 1-D case can be written for a typical infinitely small control volume:

$$\Delta(U + K)\Delta x \Delta y \Delta z = ((U + K)v)_x \Delta y \Delta z \Delta t - ((U + K)v)_{x+\Delta x} \Delta y \Delta z \Delta t +$$

$$q_x \Delta y \Delta z \Delta t - q_{x+\Delta x} \Delta y \Delta z \Delta t - \tag{4.17}$$

$$(\sigma v)_x \Delta y \Delta z \Delta t + (\sigma v)_{x+\Delta x} \Delta y \Delta z \Delta t +$$

$$(bv)\Delta x \Delta y \Delta z \Delta t + Q \Delta x \Delta y \Delta z \Delta t$$

where U is internal energy, K is kinetic energy, v is velocity, q is flux, σ is stress, b is body force, and Q is energy generated per unit volume per unit time. Energy $(U + K)_{x+\Delta x}$ and momentum $(\sigma v)_{x+\Delta x}$ can be expressed using the Taylor's expansion as:

$$((U + K)v)_{x+\Delta x} = ((U + K)v)_x + \frac{\partial((U + K)v)}{\partial x}\Delta x + 0^2 \tag{4.18}$$

$$(\sigma v)_{x+\Delta x} = (\sigma v)_x + \frac{\partial(\sigma v)}{\partial x}\Delta x + 0^2 \tag{4.19}$$

Dividing by $\Delta x \Delta y \Delta z \Delta t$ and taking the limits yields the following:

$$\frac{\partial(U+K)}{\partial t} = -\frac{\partial((U+K)v)}{\partial x} + -\frac{\partial q}{\partial x} + \frac{\partial(\sigma v)}{\partial x} + (bv) + Q \tag{4.20}$$

Expressing internal energy U and kinetic energy K as:

$$U = \rho e; \quad K = \frac{1}{2}\rho v^2 \tag{4.21}$$

and rearranging Equation (4.20) yields:

$$\left[\frac{\partial \rho}{\partial t} + \frac{\partial(\rho v)}{\partial x}\right]\left[e + \frac{1}{2}v^2\right] + \rho\frac{\partial(e + \frac{1}{2}v^2)}{\partial t} + \rho v\frac{\partial(e + \frac{1}{2}v^2)}{\partial x} = -\frac{\partial q}{\partial x} + \frac{\partial(\sigma v)}{\partial x} + (bv) + Q \tag{4.22}$$

where e is specific internal energy per unit mass. Using the mass conservation law and momentum balance law, Equation (4.22) becomes:

$$\rho\left(\frac{\partial e}{\partial t} + v\frac{\partial e}{\partial x}\right) = -\frac{\partial q}{\partial x} + \sigma\frac{\partial v}{\partial x} + Q \tag{4.23}$$

Introducing the following operator:

$$\frac{d}{dt} = \frac{\partial}{\partial t} + v\frac{\partial}{\partial x} \tag{4.24}$$

Equation (4.23) can be rewritten as

$$\rho\frac{de}{dt} = -\frac{\partial q}{\partial x} + \sigma\frac{\partial v}{\partial x} + Q \tag{4.25}$$

Denoting the gradient of velocity by strain rate as:

$$\dot{\varepsilon} = \frac{\partial v}{\partial x} = \frac{\partial \dot{u}}{\partial x}; \quad \dot{u} = \frac{dv}{dt} \tag{4.26}$$

Equation (4.25) is rewritten in the following form:

$$\rho\frac{de}{dt} = -\frac{\partial q}{\partial x} + \sigma\dot{\varepsilon} + Q \tag{4.27}$$

If we express free energy e by cT where c is specific heat capacity and T is temperature, we have the following form for energy balance law:

$$\rho\frac{d(cT)}{dt} = -\frac{\partial q}{\partial x} + \sigma\frac{\partial v}{\partial x} + Q \tag{4.28}$$

If we further introduce Fourier's law as a constitutive law between heat flux q and temperature T as:

$$q = -k\frac{\partial T}{\partial x} \tag{4.29}$$

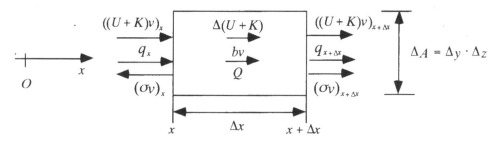

Figure 4.3 Illustration of energy conservation law

Equation (4.28) takes the following well-known 1-D energy balance law in thermodynamics by assuming that heat conductivity coefficient and specific heat capacity are constant:

$$\rho c \frac{dT}{dt} = k \frac{\partial^2 T}{\partial x^2} + \sigma \dot{\varepsilon} + Q \tag{4.30}$$

See Figure 4.3.

4.1.5 Fundamental governing equations for coupled hydromechanical behavior

4.1.5.1 1-D mass conservation law for a mixture of solid and fluid

The mass conservation law for a mixture without mass generation is stated as it was previously:

gained mass = input flux − output flux

Let us consider an infinitely small cubic element as shown in Figure 4.4. The final equation for mass conservation law of mixture for 1-D case is fundamentally same:

$$\frac{\partial p}{\partial t} = \frac{\partial q}{\partial x} \tag{4.31}$$

However, the average density of mixture is defined as:

$$\rho = (1 - n)\rho_s + n\rho_f \tag{4.32}$$

where n is the volume fraction of fluid, ρ_f is density of fluid, and ρ_s is density of solid.

$$q = \rho v = (1 - n)\rho_s v_s + n\rho_f v_f \tag{4.33}$$

Let us also define a relative velocity between the solid phase and fluid phase as:

$$v_r = n(v_f - v_s) \quad or \quad n_v = v_r - nv_s \tag{4.34}$$

The mass conservation laws for each constituent can be written as:
Solid phase

$$\frac{\partial}{\partial t}((1 - n)\rho_s) = -\frac{\partial}{\partial x}((1 - n)\rho_s v_s) \tag{4.35}$$

Defining the following operator:

$$\frac{d_s}{dt} = \frac{\partial}{\partial t} + v_s \frac{\partial}{\partial x} \tag{4.36}$$

The preceding equation becomes:

$$\frac{d_s n}{dt} = \frac{(1-n)}{\rho_s} \frac{d_s \rho_s}{dt} + (1-n) \frac{\partial v_s}{\partial x} \tag{4.38}$$

Fluid phase

$$\frac{\partial}{\partial t}(n\rho_f) == -\frac{\partial}{\partial x}(\rho_f v_r + n\rho_f v_s) \tag{4.39}$$

The preceding equation becomes:

$$\frac{d_s n}{dt} = -\frac{n}{\rho_f} \frac{d_s \rho_f}{dt} - \frac{1}{\rho_f} \frac{\partial(\rho_f v_r)}{\partial x} - n\frac{\partial}{\partial x}(\rho_f v_f) \tag{4.40}$$

Equating Equations (4.38) and (4.40), we have:

$$\frac{(1-n)}{\rho_s} \frac{d_s \rho_s}{dt} + \frac{\partial v_s}{\partial x} = -\frac{n}{\rho_f} \frac{d_s \rho_f}{dt} - \frac{1}{\rho_f} \frac{\partial(\rho_f v_r)}{\partial x} \tag{4.41}$$

Let us write constitutive laws for the volumetric response of each constituent in the following forms:

$$\frac{1}{\rho_s} \frac{d_s \rho_s}{dt} = \frac{1}{K_s} \frac{d_s p}{dt} \tag{4.42}$$

$$\frac{1}{\rho_f} \frac{d_s \rho_f}{dt} = \frac{1}{K_f} \frac{d_s p}{dt} \tag{4.43}$$

Inserting these constitutive laws in Equation (4.41), we obtain:

$$\frac{\partial \varepsilon_v}{\partial t} = -\left(\frac{(1-n)}{K_s} + \frac{n}{K_f}\right) \frac{d_s p}{dt} - \frac{1}{\rho_f} \frac{\partial}{\partial x}(\rho_f v_r) \tag{4.44}$$

where

$$\frac{\partial v_s}{\partial x} = \frac{\partial}{\partial x}\left(\frac{\partial u_s}{\partial t}\right) = \frac{\partial}{\partial t}\left(\frac{\partial u_s}{\partial x}\right) = \frac{\partial \varepsilon_v}{\partial t} \tag{4.45}$$

Let us introduce Darcy's law given by:

$$v_r = -\frac{k}{\mu} \frac{\partial p}{\partial x} = -K \frac{\partial p}{\partial x} \tag{4.46}$$

Thus Equation (4.44) becomes:

$$\frac{\partial \varepsilon_v}{\partial t} = -\left(\frac{(1-n)}{k_s} + \frac{n}{K_f}\right)\frac{d_s p}{dt} + \frac{1}{\rho_f}\frac{\partial}{\partial x}\left(\rho_f K \frac{\partial p}{\partial x}\right) \tag{4.47}$$

Some particular cases of the preceding equation are:
Undrained condition and slow deformation: $v_r = 0$ and $v_s = 0$

$$\frac{\partial \varepsilon_v}{\partial t} = -\left(\frac{(1-n)}{K_s} + \frac{n}{K_f}\right)\frac{\partial p}{\partial t} = -C_u \frac{\partial p}{\partial t} \tag{4.48}$$

Drained condition and slow deformation: $v_s = 0$

$$\frac{\partial \varepsilon_v}{\partial t} = -\left(\frac{(1-n)}{K_s}\right)\frac{\partial p}{\partial t} = -C_d \frac{\partial p}{\partial t} \tag{4.49}$$

If the solid and fluid are incompressible materials, Equation (4.42) or (4.43) becomes:

$$\frac{\partial \varepsilon_v}{\partial t} = -\frac{\partial v_r}{\partial x} = \frac{\partial}{\partial x}\left(K \frac{\partial p}{\partial x}\right) = K \frac{\partial^2 p}{\partial x^2} \tag{4.50}$$

Equation 4.50 becomes Terzaghi's or Biot's equation used in coupled problems in which densities are assumed to be constant.

4.1.5.2   1-D force conservation law for a mixture of solid and fluid

Total force equilibrium in terms of total stress without inertia form yields the following relation:

$$\frac{\partial \sigma}{\partial x} + b = 0 \tag{4.51}$$

Let us introduce the concept of effective stress as:

$$\sigma = \sigma' - \alpha p \tag{4.52}$$

where α is a physical nondimensional quantity. When α is equal to 1, it corresponds to Terzaghi's effective stress law. On the other hand, if it is not equal to 1, it corresponds to Biot's effective stress law. Accordingly, the equilibrium equation becomes:

$$\frac{\partial \sigma'}{\partial x} - \alpha \frac{\partial p}{\partial x} + b = 0 \tag{4.53}$$

Taking the time derivative of the preceding expression, we have:

$$\frac{\partial \dot{\sigma}'}{\partial x} - \alpha \frac{\partial \dot{p}}{\partial x} + \dot{b} = 0 \tag{4.54}$$

Equation 4.54 is used together with Equation 4.47 for the hydromechanical response of rock masses.

Under dynamic conditions, which require the consideration of inertia components, Equation (4.51) is replaced by the following equation:

$$\frac{\partial \sigma}{\partial x} + \rho g = (1 - n)\rho_s \frac{d_s v_s}{dt} + n\rho_f \frac{d_f v_f}{dt} \tag{4.55}$$

$$\frac{\partial \sigma_f}{\partial x} + \rho_f g = \rho_f \ddot{u}_f + \frac{1}{n}\tau_{sf} \tag{5.56}$$

where

$$\sigma = (1 - n)\sigma_s + n\sigma_f \tag{4.57}$$

$$\tau_{sf} = -\frac{n}{K} v_r \tag{4.58}$$

The coefficient K is called the hydraulic conductivity coefficient (wrongly called permeability in many publications), which appears in Darcy's law. Equations (4.55) and (4.56) can be rewritten as:

Total system

$$\frac{\partial \sigma}{\partial x} + \rho g = \rho \frac{\partial^2 u_s}{\partial t^2} + \rho_f \frac{\partial^2 w}{\partial t^2} \tag{4.59}$$

Fluid phase

$$\frac{\partial \sigma_f}{\partial x} + \rho_f g = \rho_f \frac{\partial^2 u_s}{\partial t^2} + \frac{\rho_f}{n}\frac{\partial^2 w}{\partial t^2} + \frac{1}{K}\frac{\partial w}{\partial t} \tag{4.60}$$

Where $w = n(v_f - v_s)$.

4.1.5.3 1-D seepage

If fluid density does not change with position and no mass is gained or lost, the preceding equations can be rewritten as:

$$\frac{\partial \rho}{\partial t} = -\rho \frac{\partial}{\partial x}(\bar{v}) \tag{4.61}$$

Although it is known that the relationship between average velocity ($\bar{v}$) and pressure gradient ($-\partial p/\partial x$) becomes nonlinear with increasing values of Reynolds numbers, Darcy's law is widely used for analyzing test results on soil and rock. Darcy's law can be given as:

$$\bar{v} = -\frac{k}{\eta}\frac{dp}{dx} \tag{4.62}$$

where k is permeability, η is viscosity, and p is pressure. Inserting Equation (4.62) into Equation (4.61) yields the governing equations for fluid flow as:

$$\frac{\partial \rho}{\partial t} = \frac{\rho k}{\eta}\frac{\partial}{\partial x}\left(\frac{\partial p}{\partial x}\right) \tag{4.63}$$

If the following relation exists between fluid density and pressure as:

$$\rho = cp \tag{4.64}$$

Equation (4.63) takes the following form, which is the governing equation used in seepage:

$$\frac{\partial p}{\partial t} = \frac{k}{c\eta} \frac{\partial^2 p}{\partial x^2} \quad \text{or} \quad \frac{\partial p}{\partial t} = \beta \frac{\partial^2 p}{\partial x^2} \tag{4.65}$$

where $\beta = \frac{k}{c\eta}$.

4.2 Multidimensional governing equations

4.2.1 Mass conservation laws for seepage and diffusion phenomena

Water is always present in rock mass, and it strongly affects the stability of rock engineering structures. Furthermore, rock excavations, which are generally of large scale, may disturb groundwater regime and may have some environmental impacts. In any case, the governing equation for seepage flow in rock mass is derived from the mass conservation law if the rock mass is considered a porous medium. The final form of the governing equation for seepage flow may be written as:

$$S \frac{\partial h}{\partial t} = -\nabla \cdot \mathbf{q} \pm Q \tag{4.66}$$

where $S, h, \mathbf{q}$ and Q are the storativity coefficient, water head, flux vector and source or sink, respectively. ∇ is the directional derivative operator.

As for the diffusion phenomenon of a certain substance, one may obtain the governing equation as:

$$\frac{d\phi}{dt} = -\nabla \cdot \mathbf{f} \pm P \tag{4.67}$$

where $\phi, \mathbf{f}$ and P are concentration of substance, diffusion flux vector and source or sink, respectively.

4.2.2 Momentum conservation law

Momentum conservation law for rock mass can be derived as done in continuum mechanics, and the final form can be written as:

$$\rho \frac{\partial \mathbf{v}}{\partial t} = -\nabla \cdot \boldsymbol{\sigma} + \mathbf{b} \tag{4.68}$$

where $\rho, \mathbf{v}, \boldsymbol{\sigma}$ and $\mathbf{b}$ are density, velocity, stress tensor and body force, respectively.

4.2.3 Angular momentum conservation law

The final form of the angular momentum equation indicates that the stress tensor is a symmetric tensor, that is:

$$\sigma_{ij} = \sigma_{ji} \tag{4.69}$$

4.2.4 Energy conservation law

The energy conservation law of rock mass may be written in the following form:

$$\rho c \frac{\partial T}{\partial t} = -\nabla \cdot \mathbf{q}_h + \boldsymbol{\sigma} \cdot \dot{\boldsymbol{\varepsilon}} + Q_h \tag{4.70}$$

where $c, T, \mathbf{q}_h, \dot{\boldsymbol{\varepsilon}}$ and Q_h are specific heat, temperature, strain rate tensor and heat source or sink, respectively.

4.2.5 Fundamental equation of fluid flow in porous media

The mass conservation law for fluid flowing through the pores within rock may be given in the following form with the use of the mixture theory and assuming that a coordinate system is fixed to the solid phase (i.e. Aydan, 1998, 2001a, 2001b):

$$\frac{\partial (\phi \rho_f)}{\partial t} = -\nabla \cdot (\phi \mathbf{q}_f) \tag{4.71}$$

where $\nabla = \frac{\partial}{\partial x_i} \mathbf{e}_i$, $i = 1, 3$; ρ_f is fluid density, ϕ is porosity, and $\mathbf{q}_f$ is fluid flux.

One may write the following relation for fluid flux in terms of relative velocity $\mathbf{v}_r$ of the fluid and the velocity $\mathbf{v}_s$ of solid phase as:

$$\phi \mathbf{q}_f = \rho_f (\mathbf{v}_r + \phi \mathbf{v}_s) \tag{4.72}$$

Let us assume that the flow of fluid obeys Darcy's law. Thus we have:

$$\mathbf{v}_r = -\frac{k}{\eta} \nabla p \tag{4.73}$$

where k is permeability, and η is viscosity of fluid. Inserting Equations (4.73) and (4.72) into Equation (4.71) yields the following:

$$\frac{\partial (\phi \rho_f)}{\partial t} = \nabla \cdot (\rho_f (\frac{k}{\eta} \nabla p - \phi \mathbf{v}_s)) \tag{4.74}$$

The material derivative operator according to Eulerian description may be written as (Eringen, 1980):

$$\frac{d_s}{dt} = \frac{\partial}{\partial t} + \mathbf{v}_s \cdot \nabla () \tag{4.75}$$

Introducing this operator into Equation (4.74), we have the following relation:

$$\frac{d_s \phi}{dt} + \frac{\phi}{\rho_f} \frac{d_s \rho_f}{dt} = \nabla \cdot (\frac{k}{\eta} \nabla p) - \frac{1}{\rho_f} \nabla \cdot (\phi \mathbf{v}_s) \tag{4.76}$$

The following constitutive relations are assumed to hold among porosity, fluid and solid densities and pressure (i.e. Zimmerman et al., 1986):

$$\frac{d_s \phi}{dt} = (C_b - (1 + \phi) C_s) \frac{d_s p}{dt}; \frac{1}{\rho_f} \frac{d_s \rho_f}{dt} = C_f \frac{d_s p}{dt} \tag{4.77}$$

If the velocity of solid phase is assumed to be small so that it can be neglected, Equation (4.76) takes the following form with the use of Equation (4.77):

$$\beta \frac{\partial p}{\partial t} = \nabla \cdot (\nabla p) \tag{4.78}$$

where

$$\beta = [(C_b - C_s) + \phi(C_f - C_s)] \frac{\eta}{k} \tag{4.79}$$

(a) Special forms of Equation 4.78 for longitudinal and axisymmetric cases

Equation (4.78) can be rewritten for one-dimensional longitudinal flow as:

$$\beta \frac{\partial p}{\partial t} = \frac{\partial^2 p}{\partial x^2} \tag{4.80}$$

Similarly Equation (5.8) can be also written for axisymmetric radial flow as:

$$\beta \frac{\partial p}{\partial t} = \frac{1}{r} \frac{\partial}{\partial r} \left(r \frac{\partial p}{\partial r} \right) \tag{4.81}$$

(b) Governing equations of fluid in reservoirs attached to sample

Using the mass conservation law and the constitutive relation between pressure and fluid density, the velocities v_1, v_2 of fluid contained in reservoirs numbered (1) and (2) and attached to the ends of a sample can be written as:

$$v_1 = -C_f V_1 \frac{\partial p_1}{\partial t}, v_2 = -C_f V_2 \frac{\partial p_2}{\partial t} \tag{4.82}$$

where V_1 and V_2 are volumes of reservoirs, and p_1 and p_2 are pressures acting on the fluid reservoirs.

4.2.6 Modeling of water absorption/desorption processes and associated volumetric changes in rocks

Some rocks such fine-grain sandstone, mudstone and siltstone start to fracture when losing its water content, as observed in many laboratory tests and in-situ. The situation is similar to the reverse problem of swelling. It is considered that rock shrinks as it loses its water content. This consequently results in shrinkage strain, leading to the fracturing of rock in tension. Therefore, a coupled formulation of the problems is necessary.

The water content variation in rock can be modeled as a diffusion problem. Thus the governing equation is written as:

$$\frac{d\theta}{dt} = -\nabla \cdot \mathbf{q} + Q \tag{4.83}$$

where θ, $\mathbf{q}$, Q and t are water content, water content flux, water content source and time, respectively. If the water content migration obeys Fick's law, the relation between flux $\mathbf{q}$

and water content is written in the following form:

$$\mathbf{q} = -k\nabla\theta \tag{4.84}$$

where k is the water diffusion coefficient. If some of water content is transported by the groundwater seepage or airflow in open space, this may be taken into account through the material derivative operator in Equation (4.83). However, it would be necessary to describe or evaluate the seepage velocity or airflow.

If the stress variations occur at slow rates, the equation of motion without inertial term may be used in incremental form as:

$$\nabla \cdot \dot{\boldsymbol{\sigma}} = \mathbf{0} \tag{4.85}$$

The simplest constitutive law for rock between stress and strain fields would be a linear law, in which the properties of rocks may be related to the water content in the following form (i.e. Aydan et al., 2006):

$$\dot{\boldsymbol{\sigma}} = \mathbf{D}(\theta)\dot{\boldsymbol{\varepsilon}}_e \tag{4.86}$$

The volumetric strain variations associated with shrinkage (inversely swelling) may be related to the strain field in the following form:

$$\dot{\boldsymbol{\varepsilon}}_e = \dot{\boldsymbol{\varepsilon}} - \dot{\boldsymbol{\varepsilon}}_s \tag{4.87}$$

4.2.7 Thermo-mechanical modeling heat transport in rocks

The well-known governing equation of energy conservation (4.70), rewritten for porous media, takes the following form:

$$\rho c \frac{\partial T}{\partial t} = -\nabla \cdot (k\nabla T) + \boldsymbol{\sigma} \cdot \dot{\boldsymbol{\varepsilon}} + Q_h \tag{4.88}$$

The incremental form of the equation of motion without inertia from Equation (4.68) is given by:

$$\nabla \cdot \dot{\boldsymbol{\sigma}} = \mathbf{0} \tag{4.89}$$

The well-known equation of strain component induced by temperature variation is given by the following equation:

$$\dot{\boldsymbol{\varepsilon}}_T = \lambda \Delta T \boldsymbol{I} \tag{4.90}$$

where $\boldsymbol{I}$ is the Kronecker delta tensor. The constitutive law in terms of net-strain is generally written in the following incremental form:

$$\dot{\boldsymbol{\sigma}} = \mathbf{D}(\dot{\boldsymbol{\varepsilon}} - \dot{\boldsymbol{\varepsilon}}_T) \tag{4.91}$$

4.3 Derivation of governing equations in integral form

In this section, fundamental conservation laws are derived using the approach of integral form.

4.3.1 *Mass conservation law*

Mass is defined as:

$$m = \int_\Omega \rho(\mathbf{x}, t) d\Omega \tag{4.92}$$

Mass conservation law requires:

$$\frac{dm}{dt} = \frac{d}{dt} \int_\Omega \rho d\Omega + \int_\Omega \rho \frac{d(d\Omega)}{dt} \tag{4.93}$$

as

$$d\Omega = J d\Omega_o \quad \text{and} \quad \frac{d(d\Omega)}{dt} == (\nabla \cdot \mathbf{x}) d\Omega \tag{4.94}$$

With the use of the Reynolds transport theorem (Appendices 4 and 6) together with Equation (4.94), Equation (4.93) becomes:

$$\frac{dm}{dt} = \int_\Omega \left(\frac{d\rho}{dt} + \rho \nabla \cdot \mathbf{v} \right) d\Omega \tag{4.95}$$

To satisfy this condition, the integrand should be zero, so that we have the following:

$$\frac{d\rho}{dt} + \rho \nabla \cdot \mathbf{v} = 0 \tag{4.96}$$

The time derivative in Lagrangian and Eulerian descriptions are given in the following forms:

Lagrangian description

$$\frac{d()}{dt} = \frac{\partial()}{\partial t} \tag{4.97a}$$

Eulerian description

$$\frac{d()}{dt} = \frac{\partial()}{\partial t} + v \cdot \nabla() \tag{4.97b}$$

With the use of Equation (4.97a) in Equation (4.96), we obtain the following relation:

$$\frac{\partial \rho}{\partial t} + \mathbf{v} \cdot \nabla \rho + \rho \nabla \cdot \mathbf{v} = 0 \quad \text{or} \quad \frac{\partial \rho}{\partial t} + \nabla \cdot (\rho \mathbf{v}) = 0 \tag{4.98}$$

Using index notation, we have:

$$\mathbf{v} = v_k \mathbf{e}_k \quad \text{and} \quad \nabla = \frac{\partial}{\partial x_k} \mathbf{e}_k \tag{4.99}$$

Equation 4.98 can be rewritten as:

$$\frac{\partial \rho}{\partial t} + \frac{\partial(\rho v_k)}{\partial x_k} = \frac{\partial \rho}{\partial t} + (\rho v_k)_{,k} = 0 \tag{4.100}$$

if $\frac{d\rho}{dt} = 0$ media is incompressible so that $\rho(\nabla \cdot \mathbf{v}) = 0$ or $\nabla \cdot \mathbf{v} = 0$.

4.3.2   Momentum conservation law

The definition of momentum is:

$$\mathbf{p} = \int_{\Omega} \rho \mathbf{v} d\Omega \tag{4.101}$$

Preliminary relations

$$\int_{\Omega} \nabla \cdot \boldsymbol{\sigma} d\Omega = \int_{\Gamma} \boldsymbol{\sigma} \cdot \mathbf{n} d\Gamma, \frac{d(d\Omega)}{dt} == (\nabla \cdot \mathbf{x}) d\Omega, \mathbf{t} = \boldsymbol{\sigma} \cdot \mathbf{n} \tag{4.102}$$

Conservation of momentum is written in the following form in view of Equation (4.102), which is also known Reynolds transport theorem:

$$\frac{d}{dt} \int_{\Omega} \rho \mathbf{v} d\Omega = \int_{\Gamma} \mathbf{t} d\Gamma + \int_{\Omega} \mathbf{b} d\Omega \tag{4.103}$$

Equation (4.103) may be rewritten as:

$$\int_{\Omega} \left(\frac{d(\rho \mathbf{v})}{dt} + (\rho \mathbf{v}) \nabla \cdot \mathbf{v} \right) d\Omega = \int_{\Omega} \nabla \cdot \boldsymbol{\sigma} d\Omega + + \int_{\Omega} \mathbf{b} d\Omega \tag{4.104}$$

Carrying out the derivation in Equation (4), we have the following:

$$\int_{\Omega} \left(\left(\frac{d\rho}{dt} + \rho(\nabla \cdot \mathbf{v}) \right) \mathbf{v} \right) d\Omega + \int_{\Omega} \rho \frac{d\mathbf{v}}{dt} d\Omega = \int_{\Omega} (\nabla \cdot \boldsymbol{\sigma} + \mathbf{b}) d\Omega \tag{4.105}$$

The first term on left-hand side disappears by virtue of mass conservation law, and the equation takes the following form:

$$\int_{\Omega} \rho \frac{d\mathbf{v}}{dt} d\Omega = \int_{\Omega} (\nabla \cdot \boldsymbol{\sigma} + \mathbf{b}) d\Omega \tag{4.106}$$

Equation (4.106) may be rewritten as:

$$\int_{\Omega} \left[\rho \frac{d\mathbf{v}}{dt} - (\nabla \cdot \boldsymbol{\sigma} + \mathbf{b}) \right] d\Omega = 0 \tag{4.107}$$

To satisfy Equation (4.107), the integrand should be zero, so that we have the following relation:

$$\rho \frac{d\mathbf{v}}{dt} = \nabla \cdot \boldsymbol{\sigma} + \mathbf{b} \tag{4.108}$$

Furthermore, the derivation on the left-hand side may be related to acceleration or displacement vectors as follows:

$$\frac{d\mathbf{v}}{dt} = \mathbf{a} \quad \text{or} \quad \frac{d\mathbf{v}}{dt} = \frac{d^2\mathbf{u}}{dt^2} \tag{4.109}$$

4.3.3 Angular momentum conservation law

Some preliminary relations are:

$$\mathbf{v} \times \mathbf{v} = 0, \int_{\Gamma} \mathbf{r} \times \mathbf{n} d\Gamma = \int_{\Omega} \nabla \times \mathbf{r} d\Omega = \int_{\Omega} \mathbf{I} d\Omega, \mathbf{r} \times \boldsymbol{\sigma} \cdot \mathbf{n} = (\mathbf{r} \times \boldsymbol{\sigma}) \cdot \mathbf{n}, \frac{d\mathbf{r}}{dt} = \mathbf{v} \tag{4.110}$$

$$\int_{\Gamma} \mathbf{r} \times \mathbf{t} d\Gamma = \int_{\Gamma} \mathbf{r} \times (\boldsymbol{\sigma} \cdot \mathbf{n}) d\Gamma = \int_{\Omega} \nabla \cdot (\mathbf{r} \times \boldsymbol{\sigma}) d\Omega$$

$$= \int_{\Omega} (\nabla \times \mathbf{r}) \cdot \boldsymbol{\sigma} d\Omega + \int_{\Omega} \mathbf{r} \times (\nabla \cdot \boldsymbol{\sigma}) d\Omega \tag{4.111}$$

Angular momentum law requires the following condition to be valid:

$$\frac{d}{dt} \int_{\Omega} \mathbf{r} \times \rho \mathbf{v} d\Omega = \int_{\Gamma} \mathbf{r} \times \mathbf{t} d\Gamma + \int_{\Omega} \mathbf{r} \times \mathbf{b} d\Omega \tag{4.112}$$

$$\int_{\Omega} \frac{d\mathbf{r}}{dt} \times \rho \mathbf{v} d\Omega + \int_{\Omega} \mathbf{r} \times \frac{d(\rho\mathbf{v})}{dt} d\Omega + \int_{\Omega} \mathbf{r} \times (\rho\mathbf{v}) \frac{d(d\Omega)}{dt} \tag{4.113}$$

$$= \int_{\Omega} (\nabla \times \mathbf{r}) \cdot \boldsymbol{\sigma} d\Omega + \int_{\Omega} \mathbf{r} \times (\nabla \cdot \boldsymbol{\sigma} + \mathbf{b}) d\Omega$$

or

$$\int_{\Omega} \mathbf{v} \times \rho \mathbf{v} d\Omega + \int_{\Omega} \mathbf{r} \times \left(\frac{d\rho}{dt} \mathbf{v} + \rho \frac{d\mathbf{v}}{dt} \right) d\Omega + \int_{\Omega} (\mathbf{r} \times \rho\mathbf{v}) \nabla \cdot \mathbf{v} d\Omega \tag{4.114}$$

$$= \int_{\Omega} (\nabla \times \mathbf{r}) \cdot \boldsymbol{\sigma} d\Omega + \int_{\Omega} \mathbf{r} \times (\nabla \cdot \boldsymbol{\sigma} + \mathbf{b}) d\Omega$$

Thus

$$\int_{\Omega} \mathbf{v} \times \rho \mathbf{v} d\Omega + \int_{\Omega} \mathbf{r} \times \left(\frac{d\rho}{dt} + \rho \nabla \cdot \mathbf{v} \right) \mathbf{v} d\Omega + \int_{\Omega} \left(\mathbf{r} \times \left(\rho \frac{d\mathbf{v}}{dt} - \nabla \cdot \boldsymbol{\sigma} - \mathbf{b} \right) \right) d\Omega \tag{4.115}$$

$$- \int_{\Omega} (\nabla \times \mathbf{r}) \cdot \boldsymbol{\sigma} d\Omega = 0$$

By virtue of the mass conservation law, momentum conservation law, and preliminary relations, Equation (4.115) reduces to the following form:

$$\int_{\Omega} (\nabla \times \mathbf{r}) \cdot \boldsymbol{\sigma} d\Omega = \int_{\Omega} \boldsymbol{\varepsilon} \cdot \boldsymbol{\sigma} d\Omega = 0 \tag{4.116}$$

where ε is known as the permutation symbol (see Appendix 1), and it is a rank 3 tensor. It is given in index notation as:

$$\varepsilon_{ijk}\sigma_{jk} = 0 \tag{4.117}$$

The rank 3 permutation tensor has the following properties:

$$\varepsilon_{ijk} = \begin{cases} i \neq j \neq k;\,;1,2,3,1,2(\varepsilon_{ijk} = 1) \\ i \neq j \neq k;\,;3,2,1,3,2(\varepsilon_{ijk} = -1) \\ i = j \neq k, i = k \neq j, i \neq j = k; \varepsilon_{ijk} = 0 \end{cases} \tag{4.118}$$

With this property, it is shown that the stress tensor is symmetric:

$$\sigma_{jk} = \sigma_{kj} \tag{4.119}$$

4.4.4 Energy conservation law

The time variation of internal and kinetic energies of a body with a given volume and surface area should be equal to the energy input from surface traction and body force and heat flux and volumetric heat production:

$$\frac{d}{dt}(U + K)) = W + H \tag{4.120}$$

where

$$U = \int_{\Omega} \rho e d\Omega, K = \frac{1}{2} \int_{\Omega} \rho \mathbf{v} \cdot \mathbf{v} d\Omega, W = \int_{\Gamma} \mathbf{t} \cdot \mathbf{v} d\Gamma + \int_{\Omega} \mathbf{b} \cdot \mathbf{v} d\Omega, \tag{4.121}$$

$$H = -\int_{\Gamma} \mathbf{q} \cdot \mathbf{n} d\Gamma \pm \int_{\Omega} Q d\Omega$$

With the use of Equation (4.121), one may write the following relations:

$$\frac{dU}{dt} = \int_{\Omega} \left(\frac{d(\rho e)}{dt} + \rho e \nabla \cdot \mathbf{v} \right) d\Omega \tag{4.122}$$

$$\frac{dK}{dt} = \frac{1}{2} \int_{\Omega} \left(\frac{d(\rho \mathbf{v})}{dt} \cdot \mathbf{v} + \rho \mathbf{v} \cdot \frac{d\mathbf{v}}{dt} + (\rho \mathbf{v} \cdot \mathbf{v}) \nabla \cdot \mathbf{v} \right) d\Omega \tag{4.123}$$

$$W = \int_{\Omega} (\nabla \cdot \boldsymbol{\sigma} + \mathbf{b}) \cdot \mathbf{v} d\Omega + \int_{\Omega} \boldsymbol{\sigma} : \nabla \mathbf{v} d\Omega \tag{4.124}$$

$$H = -\int_{\Omega} (\nabla \cdot \mathbf{q} \pm Q) d\Omega \tag{4.125}$$

Inserting Equation (4.122) to Equation (4.123) into the left-hand side of Equation (4.120) results in:

$$\frac{d}{dt}(U+K)) = \frac{1}{2} \int_{\Omega} (\frac{d\rho}{dt} + \rho \nabla \cdot \mathbf{v}) \mathbf{v} \cdot \mathbf{v} d\Omega + \int_{\Omega} (\frac{d\rho}{dt} + \rho \nabla \cdot \mathbf{v}) e d\Omega + \int_{\Omega} \rho \frac{de}{dt} d\Omega + \int_{\Omega} \rho \mathbf{v} \cdot \frac{d\mathbf{v}}{dt} \tag{4.126}$$

Inserting Equations (4.124) and (4.125) into the right-hand side of Equation (4.120) and requiring that the mass conservation and momentum conservation laws are satisfied, the final for Equation (4.120) takes the following form:

$$\int_{\Omega} \left(\rho \frac{de}{dt} - \boldsymbol{\sigma} : \nabla \mathbf{v} + \nabla \cdot \mathbf{q} \pm Q \right) d\Omega = 0 \tag{4.127}$$

To satisfy the Equation (4.127), the integrand should be zero so that the following relation is obtained:

$$\rho \frac{de}{dt} = -\nabla \cdot \mathbf{q} + \boldsymbol{\sigma} : \nabla \mathbf{v} \pm Q \tag{4.128}$$

As $\dot{\boldsymbol{\varepsilon}} = \nabla \mathbf{v}$, Equation (4.128) is rewritten as:

$$\rho \frac{de}{dt} = -\nabla \cdot \mathbf{q} + \boldsymbol{\sigma} : \dot{\boldsymbol{\varepsilon}} \pm Q \tag{4.129}$$

If internal energy e is related to temperature (T) together with the specific heat coefficient c as cT, Equation (4.129) becomes:

$$\rho c \frac{dT}{dt} = -\nabla \cdot \mathbf{q} + \boldsymbol{\sigma} : \dot{\boldsymbol{\varepsilon}} \pm Q \tag{4.130}$$

Equation (4.130) is known as the first law of thermodynamics.

References

Aydan, Ö. (1998) Finite element analysis of transient pulse method tests for permeability measurements. *The 4th European Conf. on Numerical Methods in Geotechnical Engineering-NUMGE98, Udine.* pp. 719–727.

Aydan, Ö. (2001a) Modelling and analysis of fully coupled hydro-thermo-diffusion phenomena. *Int. Symp. On Clay Science for Engineering, Balkema, IS-SHIZUOKA.* pp. 353–360.

Aydan, Ö. (2001b) A finite element method for fully coupled hydro-thermo-diffusion problems and its applications to geoscience and geoengineering. *10th IACMAG Conference, Austin.*

Aydan, Ö., Daido, M., Tano, H., Nakama, S. & Matsui, H. (2006) The failure mechanism of around horizontal boreholes excavated in sedimentary rock. *50th US Rock mechanics Symposium, Paper No. 06-130 (on CD)*.

Eringen, A.C. (1980) *Mechanics of Continua*. Robert E. Krieger Publishing Co., Huntington, NY, 606p.

Mase, G. (1970) *Theory and Problems of Continuum Mechanics*, Schaum Outline Series. McGraw Hill Co., New York, 230p.

Zimmerman, R.W., Somerton, W.H. & King, M.S. (1986) Compressibility of porous rocks. *Journal of Geophysical Research Atmospheres*, 91, 12765–12777.

Chapter 5

Constitutive laws

The fundamental governing equations cannot be solved in their original forms as the number of equations is lower than the number of variables to be determined. If constitutive laws, which are fundamentally determined from experiments, are introduced, their solution becomes possible. In this chapter, the well-known constitutive laws are introduced

5.1 One-dimensional constitutive laws

5.1.1 1-D Linear constitutive laws

In this subsection, linear constitutive laws heat, seepage, diffusion and mechanical behavior of rocks and rock masses are given.

5.1.1.1 Fourier's law

Fourier's law states that heat flux q is linearly proportional to the gradient of temperature T, that is (Fig. 5.1):

$$q = -k\frac{\partial T}{\partial x} \tag{5.1}$$

where k is called thermal heat conductivity.

5.1.1.2 Fick's law

Fick's law is essentially the same as that of Fourier's, and it is used in diffusion problems. Fick's law states that mass flux q is linearly proportional to the gradient of mass concentration C, that is (Fig. 5.2):

$$q = -k\frac{\partial C}{\partial x} \tag{5.2}$$

where k is called the diffusion coefficient.

5.1.1.3 Darcy's law

Darcy's law is also essentially the same as that of Fourier's, and it is used in ground-water seepage problems. Darcy's law states that seepage flux q is linearly proportional to the

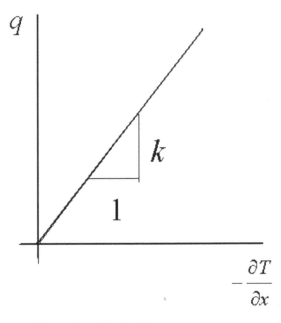

Figure 5.1 One-dimensional illustration of Fourier's law

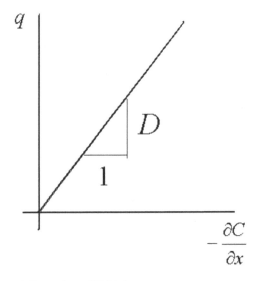

Figure 5.2 One-dimensional illustration of Fick's law

gradient of groundwater pressure p, that is (Fig. 5.3):

$$q = -k\frac{\partial p}{\partial x} \tag{5.3}$$

where k is called the hydraulic conductivity, and it is sometimes wrongly used as the permeability coefficient.

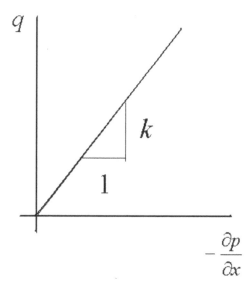

Figure 5.3 One-dimensional illustration of Darcy's law

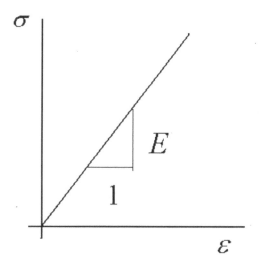

Figure 5.4 One-dimensional illustration of Hooke's law

5.1.1.4 Hooke's law

Hooke's law is used in the theory of elasticity of solids. Hooke's law states that stress σ is linearly proportional to strain ε, that is (Fig. 5.4 and Fig. 5.6(a)):

$$\sigma = E\varepsilon \tag{5.4}$$

where E is the elasticity modulus.

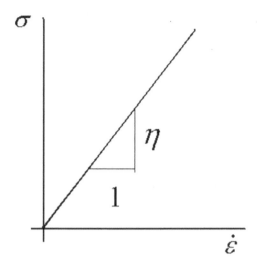

Figure 5.5 One-dimensional illustration of Newton's law

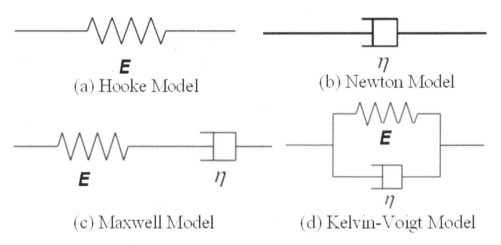

(a) Hooke Model

(b) Newton Model

(c) Maxwell Model

(d) Kelvin-Voigt Model

Figure 5.6 Simple rheological models

5.1.1.5   Newton's law

Newton's law (Fig. 5.5 and Fig. 5.6(b)) is linear and given in the following form:

$$\sigma = \eta\dot\varepsilon \qquad (5.5)$$

If this law is integrated over the time, it takes the following form with a condition, that is, $\varepsilon = 0$ at $t = 0$:

$$\varepsilon = \frac{\sigma}{\eta}t \qquad (5.6)$$

If we assume that the strain rate is given in the following form:

$$\dot{\varepsilon} = \frac{\sigma}{\eta} \tag{5.7}$$

The preceding equation can be written as:

$$\varepsilon = \dot{\varepsilon} t \tag{5.8}$$

This has a similarity to the steady-state creep response.

5.1.1.6 Maxwell's law

Substance in Maxwell's law (Fig. 5.6(c)) is assumed to consist of elastic and viscous components connected in series. Therefore, total strain and its derivative are given as:

$$\varepsilon = \varepsilon_e + \varepsilon_v \text{ and } \dot{\varepsilon} = \dot{\varepsilon}_e + \dot{\varepsilon}_v \tag{5.9}$$

The constitutive relations for elastic and viscous responses are:

$$\varepsilon_e = \frac{\sigma}{E} \text{ and } \dot{\varepsilon}_v = \frac{\sigma}{\eta} \tag{5.10}$$

If $\sigma = \sigma_o$ for $t \geq 0$ and $\varepsilon = \varepsilon_o$ with $\varepsilon_o = \sigma_o/E$, the preceding function becomes:

$$\varepsilon = \frac{\sigma_o}{E} + \frac{\sigma_o}{\eta} t \tag{5.11}$$

This equation also has a similarity to the steady-state creep response.

5.1.1.7 Kelvin-Voigt law

Substance in the Kelvin-Voigt law (Fig. 5.6(d)) is assumed to be elastic, and viscous components are connected in parallel. Therefore, total stress is given as:

$$\sigma = E\varepsilon + \eta\dot{\varepsilon} \tag{5.12}$$

If stress is applied σ_o at $t = 0$ with $\varepsilon = 0$ and is sustained thereafter, the following relation is obtained:

$$\varepsilon = \frac{\sigma_o}{E}(1 - e^{-t/t_r}) \text{ with } t_r = \frac{E}{\eta} \tag{5.13}$$

It is interesting to note that the preceding response is similar to the transient creep stage. Figure 5.7 shows the creep strain responses for different simple rheological models.

5.1.1.8 Generalized Kelvin model

The model (Fig. 5.8(a)) has a Hookean element and a Kelvin element connected in series. The total strain of the model is:

$$\varepsilon = \varepsilon_h + \varepsilon_k \tag{5.14}$$

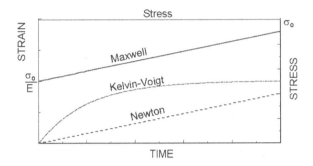

Figure 5.7 Creep strain response of simple rheological models

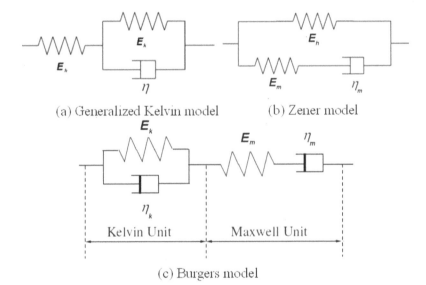

Figure 5.8 More complex rheological models

The stress relations of each element are given as:

$$\varepsilon_h = \frac{\sigma}{E_h} \text{ and } \sigma = E_k \varepsilon_k + \eta \dot{\varepsilon}_k \tag{5.15}$$

Thus, one gets the following equation:

$$\sigma = \eta \left(\dot{\varepsilon} - \frac{\dot{\sigma}}{E_h} \right) + E_k \left(\varepsilon - \frac{\sigma}{E_h} \right) \tag{5.16}$$

If stress σ_o is applied at $t = 0$ with $\varepsilon = 0$ and $\varepsilon_e = \sigma_o/E_h$ and is sustained thereafter, the following relation is obtained:

$$\varepsilon = \frac{\sigma_o}{E_h} + \frac{\sigma_o}{E_k} (1 - e^{-t/t_r}) \text{ with } t_r = \frac{\eta}{E_k} \tag{5.17}$$

As noted from this relation, instantaneous strain due to elastic response and transient creep stage can be modeled.

5.1.1.9 Zener model

The Zener model (Fig. 5.8(b)) is also known as the standard linear solid model, and it consists of a Hooke element and Maxwell element connected to each other in parallel.

Total stress may be given in the following form:

$$\sigma = \sigma_h + \sigma_m \tag{5.18}$$

The constitutive laws of Hooke and Maxwell elements are:

$$\sigma_h = E_h \varepsilon, \ \varepsilon = \varepsilon_s + \varepsilon_d, \ \dot{\varepsilon} = \dot{\varepsilon}_s + \dot{\varepsilon}_d, \ \dot{\varepsilon}_s = \frac{\dot{\sigma}_m}{E_m}, \ \dot{\varepsilon}_d = \frac{\sigma_m}{\eta_m} \tag{5.19}$$

Thus, one can easily get the following differential equation:

$$\frac{d\varepsilon}{dt} + \frac{1}{\eta_m} \cdot \frac{E_h E_m}{E_h + E_m} \varepsilon = \frac{1}{E_h + E_m} \left(\frac{d\sigma}{dt} + \frac{E_m}{\eta_m} \sigma \right) \tag{5.20}$$

If stress is applied σ_o at $t = 0$ with and $\varepsilon_o = \sigma_o / (E_h + E_m)$ and is sustained thereafter, the following relation is obtained:

$$\varepsilon = \frac{\sigma_o}{E_h} \left[1 - \frac{E_m}{E_h + E_m} e^{-t/t_r} \right] \text{ with } t_r = \eta_m \frac{E_m + E_h}{E_m E_h} \tag{5.21}$$

The creep response to be determined from this model involves the instantaneous strain and transient creep.

5.1.1.10 Burgers model

Burgers model (Fig. 5.8(c)) consists of Maxwell and Kelvin elements connected to each other in series. The constitutive relations for each element can be given as:

$$\dot{\varepsilon}_m = \frac{\dot{\sigma}}{E_m} + \frac{\sigma}{\eta_m} \text{ and } \sigma = E_k \varepsilon_k + \eta \dot{\varepsilon}_k \tag{5.22}$$

The total strain is given by:

$$\varepsilon = \varepsilon_m + \varepsilon_k \tag{5.23}$$

If stress is applied σ_o at $t = 0$ with $\varepsilon = 0$ and $\varepsilon_m = \sigma_o / E_m$ and is sustained thereafter, the following relation is obtained:

$$\varepsilon = \frac{\sigma_o}{E_m} + \frac{\sigma_o}{E_k} (1 - e^{-t/t_k}) + \frac{\sigma_o}{\eta_m} t \text{ with } t_k = \frac{\eta_k}{E_k} \tag{5.24}$$

As noted, this model can simulate the instantaneous strain due to elastic response and transient and steady state creep stages. Figure 5.9 shows and compares the creep strain responses for different more complex rheological models.

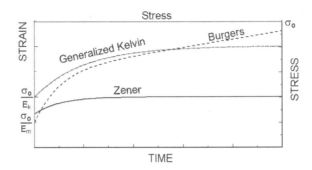

Figure 5.9 Creep responses from more complex rheological models

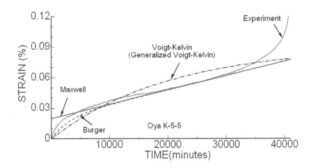

Figure 5.10 Comparison rheological models with experimental responses

Figure 5.10 compares the experimental responses with those from intuitive and rheological models. As noted from these figures, each model has its own merits and drawbacks (Aydan, 2016).

5.1.2 1-D nonlinear constitutive laws for solids

5.1.2.1 Elasto-plastic law

In the following discussion, a constitutive law is derived based on the concepts of classical plasticity theory. The elasto-plastic behavior of materials is illustrated in Figure 5.11. The classical plasticity theory is based upon the following assumptions.

- Yield function is of the following form:

$$F(\sigma, \kappa) = f(\sigma) - K(\kappa) = 0 \tag{5.25}$$

- Flow rule, given here, holds:

$$d\varepsilon^p = \lambda \frac{\partial G}{\partial \sigma} \tag{5.26}$$

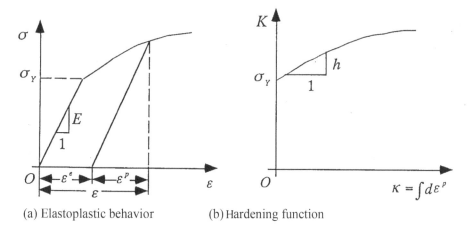

(a) Elastoplastic behavior   (b) Hardening function

Figure 5.11 Illustration of elasto-plastic behavior and some fundamental parameters

- Prager's consistency condition, given here, holds:

$$dF = \frac{\partial F}{\partial \sigma} d\sigma + \frac{\partial F}{\partial \kappa} \frac{\partial \kappa}{\partial \varepsilon^p} d\varepsilon^p = 0 \tag{5.27}$$

- Strain increment is a linear sum of elastic and plastic increments:

$$d\varepsilon = d\varepsilon^e + d\varepsilon^p \tag{5.28}$$

- For elastic component, Hooke's law holds:

$$d\sigma = D^e d\varepsilon^e \tag{5.29}$$

Inserting Equation (5.26) into Equation (5.27) and denoting the following by h:

$$h = -\frac{\partial F}{\partial \kappa} \frac{\partial \kappa}{\partial \varepsilon^p} \frac{\partial G}{\partial \sigma} \tag{5.30}$$

yields the following between plastic strain increment $d\varepsilon^p$ and stress increment $d\sigma$ as

$$d\varepsilon^p = \frac{1}{h} \frac{\partial G}{\partial \sigma} \left(\frac{\partial F}{\partial \sigma} d\sigma \right) = C^p d\sigma \tag{5.31}$$

The preceding relation is called Melan's formula. The following can be written:

$$d\sigma = D^e d\varepsilon - \frac{1}{h} D^e \frac{\partial G}{\partial \sigma} \left(\frac{\partial F}{\partial \sigma} d\sigma \right) \tag{5.32}$$

Multiplying the preceding expression by $\frac{\partial F}{\partial \sigma}$ yields:

$$\frac{\partial F}{\partial \sigma} d\sigma = \frac{\frac{\partial F}{\partial \sigma} D^e d\varepsilon}{1 + \frac{1}{h} \frac{\partial F}{\partial \sigma} \left(D^e \frac{\partial G}{\partial \sigma} \right)} \tag{5.33}$$

Utilizing the preceding relation in Equation (5.32) yields the incremental elasto-plastic law as:

$$d\sigma = \left(D^e - \frac{D^e \frac{\partial G}{\partial \sigma} \frac{\partial F}{\partial \sigma} D^e}{h + \frac{\partial F}{\partial \sigma}\left(D^e \frac{\partial G}{\partial \sigma}\right)} \right) d\varepsilon \tag{5.34}$$

5.1.2.2 Visco-plastic models

(A) BINGHAM MODEL-ELASTIC PERFECTLY VISCO-PLASTIC MODEL

The visco-plastic model of the Bingham type assumes that the material behaves elastically below the yield stress level and visco-plastically above the yield stress level, given as:

$$\varepsilon = \frac{\sigma}{E} \ \text{if} \ \sigma < \sigma_o \tag{5.35}$$

$$\varepsilon = \frac{\sigma - \sigma_o}{\eta} t + \frac{\sigma}{E} \ \text{if} \ \sigma > \sigma_o \tag{5.36}$$

This equation corresponds to the perfectly visco-plastic material if σ_o corresponds to yield threshold value of stress. Furthermore, the fluidity coefficient is defined as:

$$\gamma = \frac{1}{\eta} \tag{5.37}$$

(B) ELASTIC-VISCO-PLASTIC MODEL OF HARDENING TYPE (PERZYNA TYPE)

Elastic-visco-plastic model of hardening type (Perzyna type) (Fig. 5.12) assumes that the material behaves elastically below the yield stress level and visco-plastically above the yield stress level σ_Y. The yield strength of visco-plastic material in relation to the visco-plastic strain of hardening type can be written as:

$$Y = \sigma_Y + H\varepsilon_{vp} \tag{5.38}$$

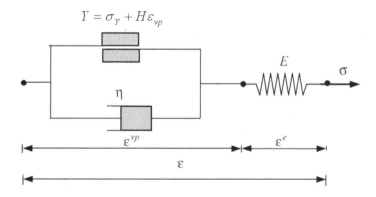

Figure 5.12 Elastic-visco-plastic model

Furthermore, total strain is assumed to be a sum of elastic strain and visco-plastic strain as:

$$\varepsilon = \varepsilon_e + \varepsilon_{vp} \tag{5.39}$$

Thus the stress–strain relations are given in the following form:

$$\sigma_p = \sigma = E\varepsilon \text{ if } \sigma_p < Y \tag{5.40}$$

$$\sigma_p = \sigma_Y + H\varepsilon_{vp} \text{ if } \sigma_p > Y \tag{5.41}$$

Total stress at any time can be written as:

$$\sigma = \sigma_p + \sigma_d \tag{5.42}$$

The viscous component of stress is related to the visco-plastic strain rate as follows:

$$\sigma_d = C_p \frac{d\varepsilon_{vp}}{dt} \tag{5.43}$$

Thus, one can obtain the following differential equation for visco-plastic response:

$$\sigma = \sigma_Y + H\varepsilon_{vp} + C_p \frac{d\varepsilon_{vp}}{dt} \tag{5.44}$$

Replacing the visco-plastic strain with the use of total strain and elastic strain in the preceding equation, one can easily obtain the following relation:

$$H \, E\varepsilon + \frac{1}{C_p} E \frac{d\varepsilon}{dt} = H \, \sigma + E(\sigma - \sigma_Y) + \frac{1}{C_p} \frac{d\sigma}{dt} \tag{5.45}$$

Let us assume that a constant stress $\sigma = \sigma_A$ is applied and kept constant (creep test). The preceding differential equation is reduced to the following form:

$$\frac{d\varepsilon}{dt} + \frac{H}{E}\varepsilon = \frac{H}{C_p E}\sigma_A + \frac{1}{C_p}(\sigma_A - \sigma_Y) \tag{5.46}$$

The solution of the differential equation is obtained as follows:

$$\varepsilon = Ce^{-\frac{H}{C_p}t} + \frac{1}{E}\sigma_A + \frac{1}{H}(\sigma_A - \sigma_Y) \tag{5.47}$$

when $t = 0$, $\varepsilon = \varepsilon_e = \sigma_A/E$.

The final form of the preceding equation becomes:

$$\varepsilon = \frac{\sigma_A}{E} + \frac{(\sigma_A - \sigma_Y)}{H}\left(1 - e^{-\frac{H}{C_p}t}\right) \tag{5.48}$$

Figure 5.13 shows the elastic-visco-plastic strain response for visco-plastic hardening and Bingham-type visco-plastic behaviors.

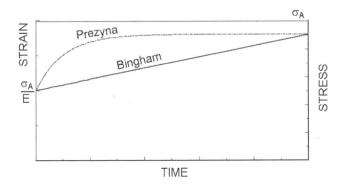

Figure 5.13 Responses obtained from elastic-visco-plastic models

(C)   ELASTO-VISCO-PLASTIC MODEL OF HARDENING TYPE

Instead of using the elasticity model for linear (recoverable) response, some of the rheological models described in the previous section can be adopted. For the nonlinear (permanent) response, the models described can be utilized. For example, if the linear response is modeled using the Kelvin-Voigt–type model, the following relation would hold for the linear part ($\sigma < \sigma_y$):

$$\varepsilon_r = \frac{\sigma}{E} \text{ and } \sigma = E\varepsilon_r + \eta\dot{\varepsilon}_r \text{ with } \varepsilon = \varepsilon_r \tag{5.49}$$

As for the nonlinear (permanent) part $\sigma \geq \sigma_y$, the following can be written:

$$\sigma = \sigma_Y + H\varepsilon_p + C_p \frac{d\varepsilon_p}{dt} \tag{5.50}$$

Total strain is assumed to consist of linear (recoverable) and nonlinear (permanent) components:

$$\varepsilon = \varepsilon_r + \varepsilon_p \tag{5.51}$$

(D)   AYDAN-NAWROCKI–TYPE ELASTO-VISCO-PLASTIC CONSTITUTIVE LAW

In an analogy to the derivation of an incremental elasto-plastic constitutive law, we start with the following:

- Yield function

$$F(\sigma,\kappa_p,\kappa_v) = f(\sigma) - K(\kappa_p,\kappa_v) = 0 \tag{5.52}$$

It should be noted here that the yield function is a function of permanent plastic and visco-hardening parameters (Fig. 5.14).

- Flow rule

$$d\varepsilon^p = \lambda\frac{\partial G}{\partial \sigma}, \quad d\varepsilon^p = \dot{\lambda}\frac{\partial G}{\partial \sigma} + \lambda\frac{\partial \dot{G}}{\partial \sigma} \tag{5.53}$$

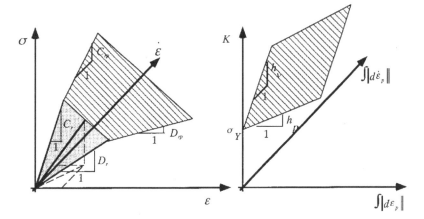

Figure 5.14 The elasto-visco-plastic model for one-dimensional response

- Prager's consistency condition

$$dF = \frac{\partial F}{\partial \sigma} \cdot d\sigma + \frac{\partial F}{\partial \kappa_p} \frac{\partial \kappa_p}{\partial \varepsilon} \cdot d\varepsilon^p + \frac{\partial F}{\partial \kappa_v} \frac{\partial \kappa_v}{\partial \dot{\varepsilon}^p} \cdot d\varepsilon^p = 0 \qquad (5.54)$$

- Linear decomposition of the strain increment $d\varepsilon$ and strain rate increment $d\dot{\varepsilon}$ into their reversible and permanent components $d\varepsilon^r$ and $d\varepsilon^p$

$$d\varepsilon = d\varepsilon^r + d\varepsilon^p, \quad d\dot{\varepsilon} = d\dot{\varepsilon}^r + d\dot{\varepsilon}^p, \qquad (5.55)$$

- Voigt-Kelvin's law

$$d\sigma = D^r d\varepsilon^r + C^r d\dot{\varepsilon}^r \qquad (5.56)$$

where σ is the stress tensor, $K(\kappa_p, \kappa_v)$ is the hardening function, G is the plastic potential, λ is the proportionality coefficient, κ_p is the plastic hardening parameter, κ_v is the viscos hardening parameter, ε is the strain tensor, ε^r is the reversible strain tensor, $\dot{\varepsilon}^r$ is the reversible strain rate tensor, ε^p is the permanent strain tensor, $\dot{\varepsilon}^p$ is the permanent strain rate tensor, D^r is the elasticity tensor, C^r is the viscosity tensor, and $(\cdot)$ denotes the dot product.

In elastic-visco-plastic formulations of the Perzyna type, *flow rule* is always assumed to be of the following form:

$$d\varepsilon^p = \dot{\lambda} \frac{\partial G}{\partial \sigma} \qquad (5.57)$$

The preceding equation implies that any plastic straining is always time dependent.

Here, permanent strain rate increment, given by Equation (5.53), is simplified to the following form by assuming that $\dot{\lambda} = 0$:

$$d\varepsilon^p = \lambda \frac{\partial \dot{G}}{\partial \sigma} \tag{5.58}$$

The preceding equation implies that the plastic potential function shrinks (or expands) in time domain while keeping its original form in stress space.[1]

Substituting Equations (5.53) and (5.58) into Equation (5.54) and rearranging the resulting expression, together with the denotation of its denominator by h_{rp} (hardening modulus):

$$h_{rp} = -\left[\frac{\partial F}{\partial \kappa_p} \frac{\partial \kappa_p}{\partial \varepsilon^p} \cdot \frac{\partial G}{\partial \sigma} + \frac{\partial F}{\partial \kappa_v} \frac{\partial \kappa_v}{\partial \varepsilon^p} \cdot \frac{\partial \dot{G}}{\partial \sigma} \right] \tag{5.59}$$

we have λ as:

$$\lambda = \frac{1}{h_{rp}} \frac{\partial F}{\partial \sigma} \cdot d\sigma \tag{5.60}$$

Now let us insert the preceding relation into Equations (5.53) and (5.58). We have the constitutive relations between the permanent strain increment $d\varepsilon^p$, the permanent strain rate increment $d\dot{\varepsilon}^p$, and the stress increment $d\sigma$ as:

$$d\varepsilon^p = \frac{1}{h_{rp}} \frac{\partial G}{\partial \sigma} \left(\frac{\partial F}{\partial \sigma} \cdot d\sigma \right) = \frac{1}{h_{rp}} \left(\frac{\partial G}{\partial \sigma} \otimes \frac{\partial F}{\partial \sigma} \right) d\sigma \tag{5.61}$$

$$d\dot{\varepsilon}^p = \frac{1}{h_{rp}} \frac{\partial \dot{G}}{\partial \sigma} \left(\frac{\partial F}{\partial \sigma} \cdot d\sigma \right) = \frac{1}{h_{rp}} \left(\frac{\partial \dot{G}}{\partial \sigma} \otimes \frac{\partial F}{\partial \sigma} \right) d\sigma \tag{5.62}$$

where $(\otimes)$ denotes the tensor product. The inverse of the preceding relations could not be obtained, that is, whether the plastic potential G is of the associated or nonassociated type. Therefore, the following technique is used to establish the relation between $d\sigma$ and $d\varepsilon$, $d\dot{\varepsilon}$. Using the relations (5.55), (5.56), (5.61) and (5.62), one can write the following:

$$d\sigma = D^r d\varepsilon - D^r \frac{1}{h_{rp}} \frac{\partial G}{\partial \sigma} \left(\frac{\partial F}{\partial \sigma} \cdot d\sigma \right) + C^r d\varepsilon - C^r \frac{1}{h_{rp}} \frac{\partial \dot{G}}{\partial \sigma} \left(\frac{\partial F}{\partial \sigma} \cdot d\sigma \right) \tag{5.63}$$

Taking the dot products of the both sides of the preceding expression by $\partial F / \partial \sigma$ yields:

$$\frac{\partial F}{\partial \sigma} \cdot d\sigma = \frac{\frac{\partial F}{\partial \sigma} \cdot (D^r d\varepsilon) + \frac{\partial F}{\partial \sigma} \cdot (C^r d\varepsilon)}{1 + \frac{1}{h_{rp}} \frac{\partial F}{\partial \sigma} \cdot \left(D^r \frac{\partial G}{\partial \sigma} \right) + \frac{1}{h_{rp}} \frac{\partial F}{\partial \sigma} \cdot \left(C^r \frac{\partial \dot{G}}{\partial \sigma} \right)} \tag{5.64}$$

Substituting the preceding relation in (5.63) gives the incremental elasto-visco-plastic constitutive law as:

$$d\sigma = D^{rp} d\varepsilon + C^{rp} d\varepsilon \tag{5.65}$$

---

1 It should be noted that permanent strain rate increment consists of time-independent and time-dependent parts.

where

$$D^{rp} = D^r - \cfrac{D^r\dfrac{\partial G}{\partial \sigma} \otimes \dfrac{\partial F}{\partial \sigma}D^r}{h_{rp} + \dfrac{\partial F}{\partial \sigma} \cdot \left(D^r\dfrac{\partial G}{\partial \sigma}\right) + \dfrac{\partial F}{\partial \sigma} \cdot \left(C^r\dfrac{\partial \dot{G}}{\partial \sigma}\right)}$$

$$C^{rp} = C^r - \cfrac{C^e\dfrac{\partial \dot{G}}{\partial \sigma} \otimes \dfrac{\partial F}{\partial \sigma}C^r}{h_{rp} + \dfrac{\partial F}{\partial \sigma} \cdot \left(D^r\dfrac{\partial G}{\partial \sigma}\right) + \dfrac{\partial F}{\partial \sigma} \cdot \left(C^r\dfrac{\partial \dot{G}}{\partial \sigma}\right)}$$

5.2 Multidimensional constitutive laws

5.2.1 Fourier's law

Fourier's law states that heat flux q_i is linearly proportional to the gradient of temperature T, that is:

$$q_i = -K_{ij}\frac{\partial T}{\partial x_i} \tag{5.66}$$

where K_{ij} is called the thermal heat conductivity tensor

5.2.2 Fick's law

Fick's law is essentially the same as that of Fourier's, and it is used in diffusion problems. Fick's law states that mass flux q_i is linearly proportional to the gradient of mass concentration C, that is:

$$q_i = -D_{ij}\frac{\partial C}{\partial x_i} \tag{5.67}$$

where D_{ij} is called the diffusion coefficient tensor

5.2.3 Darcy's law

Darcy's law is essentially the same as that of Fourier's, and it is used in groundwater seepage problems. Darcy's law states that seepage flux q_i is linearly proportional to the gradient of groundwater pressure p, that is:

$$q_i = -K_{ij}\frac{\partial p}{\partial x_i} \tag{5.68}$$

where K_{ij} is called the hydraulic conductivity tensor.

5.2.4 Hooke's law

When rock or rock mass behaves linearly without any rate dependency, the simplest constitutive law is Hooke's law. This law is written in the following form:

$$\sigma_{ij} = D_{ijkl}\varepsilon_{kl} \tag{5.69}$$

where $\sigma_{ij}, \varepsilon_{kl}$ and D_{ijkl} are stress, strain and elasticity tensors, respectively.

If material is homogeneous and isotropic, Equation (5.69) may be written as:

$$\sigma_{ij} = 2\mu\varepsilon_{ij} + \lambda\delta_{ij}\varepsilon_{kk} \tag{5.70}$$

Where δ_{ij} is Kronecker delta tensor, λ and μ are Lamé coefficients, which are given in terms of elasticity (Young's) modulus (E) and Poisson's ratio (v) as:

$$\lambda = \frac{Ev}{(1+v)(1-2v)}, \mu = \frac{E}{2(1+v)} \tag{5.71}$$

5.2.5 Newton's law

Newton's law is used in fluid mechanics. Newton's law states that stress σ_{ij} is linearly proportional to strain rate $\dot{\varepsilon}_{kl}$, that is:

$$\sigma_{ij} = C_{ijkl}\dot{\varepsilon}_{kl} \tag{5.72}$$

where C_{ijkl} is the viscosity tensor.

5.2.6 Kelvin-Voigt's law

Kelvin-Voigt's law is used in the field of visco-elasticity. Voigt's law states that stress σ_{ij} is linearly proportional to strain ε_{ij} and strain rate $\dot{\varepsilon}_{ij}$, that is:

$$\sigma_{ij} = D_{ijkl}\varepsilon_{kl} + C_{ijkl}\dot{\varepsilon}_{kl} \tag{5.73}$$

where $\dot{\varepsilon}_{kl}$ and C_{ijkl} are the strain rate and viscosity tensors, respectively.

If material is homogeneous and isotropic, Equation (5.73) may be written in analogy to Equation (5.70) as:

$$\sigma_{ij} = 2\mu\varepsilon_{ij} + \lambda\delta_{ij}\varepsilon_{kk} + 2\mu * \dot{\varepsilon}_{ij} + \lambda * \delta_{ij}\dot{\varepsilon}_{kk} \tag{5.74}$$

Coefficients $\lambda*$ and $\mu*$ may be called viscous Lamé coefficients.

5.2.7 Navier-Stokes law

Navier-Stokes constitutive law can be visualized as a simple case of Kelvin-Voigt's law. If the material behavior is associated with volumetric strain (ε_v) and pressure (p) under the steady-state condition without any shear resistance like fluids, Equation (5.74) can be rewritten as:

$$\sigma_{ij} = -p\delta_{ij} + 2\mu * \dot{\varepsilon}_{ij} + \lambda * \delta_{ij}\dot{\varepsilon}_{kk} \tag{5.75}$$

If the coefficient $\lambda^* = 0$, the preceding relation reduces to the following form:

$$\sigma_{ij} = -p\delta_{ij} + 2\mu * \dot{\varepsilon}_{ij} \quad \text{or} \quad \sigma_{ij} = -p\delta_{ij} + \mu * \dot{\gamma}_{ij} \tag{5.76}$$

The preceding constitutive law corresponds to the constitutive law commonly used in fluid mechanics known as Navier-Stokes law.

There are different visco-elasticity models in literature. As all the models cannot be covered here, the reader is advised to consult the available literature on the topic of visco-elasticity (e.g. Farmer, 1983; Jaeger and Cook, 1979; Mirza, 1978; Owen and Hinton, 1980; Serata *et al.*, 1968).

5.3 Nonlinear behavior (elasto-plasticity and elasto-visco-plasticity) for solids

Every material in nature starts to yield under a certain stress or strain state and rock or rock mass is a no exception. The terms used to describe the material behavior such as "elasticity" and "visco-elasticity" are replaced by the terms of "elasto-plasticity" or "elasto-visco-plasticity" as soon as material behavior deviates from linearity. The relation between total stress and strain or strain rate can no longer be used and every relation must be written in incremental form. For example, if the conventional plasticity models are used, the elasto-plastic constitutive law between incremental stress and strain tensors takes the following form:

5.3.1 Elasto-plastic law

The derivation of an incremental elasto-plastic constitutive law, based on the conventional elasto-plastic theory, starts with the following:

- Yield function

$$F(\boldsymbol{\sigma},\kappa) = f(\boldsymbol{\sigma}) - K(\kappa) = 0 \tag{5.77}$$

- Flow rule

$$d\boldsymbol{\varepsilon}^p = \lambda \frac{\partial G}{\partial \boldsymbol{\sigma}} \tag{5.78}$$

- Prager's consistency condition (Drucker and Prager, 1952)

$$dF = \frac{\partial F}{\partial \boldsymbol{\sigma}} \cdot d\boldsymbol{\sigma} + \frac{\partial F}{\partial \kappa}\frac{\partial \kappa}{\partial \boldsymbol{\varepsilon}^p} \cdot d\boldsymbol{\varepsilon}^p = 0 \tag{5.79}$$

- Linear decomposition of the strain increment $d\boldsymbol{\varepsilon}$ into its elastic and plastic components $d\boldsymbol{\varepsilon}^e$ and $d\boldsymbol{\varepsilon}^p$

$$d\boldsymbol{\varepsilon} = d\boldsymbol{\varepsilon}^e + d\boldsymbol{\varepsilon}^p \tag{5.80}$$

- Hooke's law

$$d\boldsymbol{\sigma} = \mathbf{D}^e d\boldsymbol{\varepsilon}^e \tag{5.81}$$

where $\boldsymbol{\sigma}$ is the stress tensor, $K(\kappa)$ is hardening function, G is plastic potential, λ is proportionality coefficient, κ is hardening parameter, $d\boldsymbol{\varepsilon}$ is strain tensor, $d\boldsymbol{\varepsilon}^e$ is elastic strain tensor, $d\boldsymbol{\varepsilon}^p$ is plastic strain tensor, $\mathbf{D}^e$ is elasticity tensor, and $(\cdot)$ denotes the dot product.

Substituting Eqn. (5.78) into Eqn. (5.79) and rearranging the resulting expression together with the denotation of its denominator by h (hardening modulus):

$$h = -\frac{\partial F}{\partial \kappa}\frac{\partial \kappa}{\partial \boldsymbol{\varepsilon}^p} \cdot \frac{\partial G}{\partial \boldsymbol{\sigma}} \tag{5.82}$$

we have λ as:

$$\lambda = \frac{1}{h}\frac{\partial F}{\partial \boldsymbol{\sigma}} \cdot d\boldsymbol{\sigma} \tag{5.83}$$

Now, let us insert the preceding relation into Eqn. (24). We have the constitutive relation between the plastic strain increment $d\boldsymbol{\varepsilon}^p$ and the stress increment $d\boldsymbol{\sigma}$, which is also known as Melan's formula, as:

$$d\boldsymbol{\varepsilon}^p = \frac{1}{h}\frac{\partial G}{\partial \boldsymbol{\sigma}}\left(\frac{\partial F}{\partial \boldsymbol{\sigma}} \cdot d\boldsymbol{\sigma}\right) = \frac{1}{h}\left(\frac{\partial G}{\partial \boldsymbol{\sigma}} \otimes \frac{\partial F}{\partial \boldsymbol{\sigma}}\right)d\boldsymbol{\sigma} = \mathbf{C}^p d\boldsymbol{\sigma} \tag{5.84}$$

where $(\otimes)$ denotes the tensor product. The inverse of the preceding relation cannot be obtained as the determinant of the plasticity matrix is $|\mathbf{C}^p| = 0|$, irrespective of whether the plastic potential G is of the associated or nonassociated type. Therefore, the following technique is used to establish the relation between $d\boldsymbol{\varepsilon}$ and $d\boldsymbol{\sigma}$. Using the relations (5.82), (5.83) and (5.84), one can write the following:

$$d\boldsymbol{\sigma} = \mathbf{D}^e d\boldsymbol{\varepsilon} - \mathbf{D}^e \frac{1}{h}\frac{\partial G}{\partial \boldsymbol{\sigma}}\left(\frac{\partial F}{\partial \boldsymbol{\sigma}} \cdot d\boldsymbol{\sigma}\right) \tag{5.85}$$

Taking the dot products of the both sides of the preceding expression by $\partial F/\partial \boldsymbol{\sigma}$ yields:

$$\frac{\partial F}{\partial \boldsymbol{\sigma}} \cdot d\boldsymbol{\sigma} = \frac{\frac{\partial F}{\partial \boldsymbol{\sigma}} \cdot (\mathbf{D}^e d\boldsymbol{\varepsilon})}{1 + \frac{1}{h}\frac{\partial F}{\partial \boldsymbol{\sigma}} \cdot \left(\mathbf{D}^e \frac{\partial G}{\partial \boldsymbol{\sigma}}\right)} \tag{5.86}$$

Substituting the preceding relation in (5.85) gives the incremental elasto-plastic constitutive law as:

$$d\boldsymbol{\sigma} = \left(\mathbf{D}^e - \frac{d\mathbf{D}^e\frac{\partial G}{\partial \boldsymbol{\sigma}} \otimes \frac{\partial F}{\partial \boldsymbol{\sigma}}\mathbf{D}^e}{h + \frac{\partial F}{\partial \boldsymbol{\sigma}} \cdot \left(\mathbf{D}^e\frac{\partial G}{\partial \boldsymbol{\sigma}}\right)}\right)d\boldsymbol{\varepsilon} \tag{5.87}$$

The hardening modulus h is generally determined as a function of a hardening parameter κ by employing either a work-hardening model or a strain-hardening model. The hardening parameter κ is defined for both cases as follows:

$$\kappa = W^p = \int \boldsymbol{\sigma} \cdot d\boldsymbol{\varepsilon}^p \quad \text{work-hardening} \tag{5.88}$$

$$\kappa = \int \|d\boldsymbol{\varepsilon}^p\| \quad \text{strain hardening} \tag{5.89}$$

where W^p is plastic work.

The materials (i.e. steel, glass fibers) exhibit a nondilatant plastic behavior and isotropically harden. Therefore, a work-hardening model is generally used together with the effective stress–strain concept, defined as:

$$\sigma_e = \sqrt{\frac{3}{2}(\mathbf{s} \cdot \mathbf{s})} \quad d\varepsilon_e^p = \sqrt{\frac{2}{3}(d\boldsymbol{\varepsilon}^p \cdot d\boldsymbol{\varepsilon}^p)} \tag{5.90}$$

where $\mathbf{s}$ is the deviatoric stress tensor.

As the volumetric plastic strain increment $d\bar{\varepsilon}_v^p = 0$, together with the coaxiality of the stress and plastic strain, the hardening parameter κ of work-hardening type can be rewritten in the following form:

$$d\kappa = dW^p = \boldsymbol{\sigma} \cdot d\boldsymbol{\varepsilon}^p = \mathbf{s} \cdot de^p = \sigma_e d\varepsilon_e^p \tag{5.91}$$

where de^p is the deviatoric strain increment. The hardening modulus h for this case takes the following form with the use of Euler's theorem[2] by taking a homogeneous plastic potential G of order m:

$$h = -\frac{\partial F}{\partial \kappa}\frac{\partial \kappa}{\partial \boldsymbol{\varepsilon}^p} \cdot \frac{\partial G}{\partial \boldsymbol{\sigma}} = \frac{\partial K}{\partial W^p}\boldsymbol{\sigma} \cdot \frac{\partial G}{\partial \boldsymbol{\sigma}} = m\frac{\partial K}{\partial W^p}G \tag{5.92}$$

If $F = G$ and $f(\boldsymbol{\sigma}) = \sigma_e$, then the hardening modulus h becomes:

$$h = m\frac{\partial K}{\partial W^p}f(\boldsymbol{\sigma}) = m\frac{\partial \sigma_e}{\partial \varepsilon_e} \tag{5.93}$$

The hardening modulus can then be easily obtained from a gradient of the plot of a uniaxial test in σ_1 and ε_1^p space, since the effective stress and strain in the uniaxial state become:

$$\sigma_e = \sigma_1 \quad \varepsilon_e^p = \varepsilon_1^p \tag{5.94}$$

5.3.2 Elastic-visco-plasticity

These approaches assume that the materials are assumed to be elastic before yielding and behave in a visco-plastic manner following yielding. In visco-plastic evaluations, $\mathbf{e}_p$ is replaced by $\mathbf{e}_{vp}$.

5.3.2.1 Power-type models

When Norton-type constitutive law is used for creep response, the visco-plastic strain rate $(\mathbf{e}_{vp} = \boldsymbol{\varepsilon}_{vp})$ is expressed as follows:

$$\frac{d\boldsymbol{\varepsilon}_{vp}}{dt} = \left(\frac{\sigma_{eq}}{\sigma_o}\right)^n \frac{\partial \sigma_{eq}}{\partial \boldsymbol{\sigma}} \tag{5.95}$$

Perzyna-type elastic-visco-plastic laws are used for representing nonlinear rate dependency involving plasticity:

$$\frac{d\boldsymbol{\varepsilon}_{vp}}{dt} = \lambda\mathbf{s} \tag{5.96}$$

2 $x \cdot \partial f/\partial x = mf.$

where λ is proportionality coefficient and is interpreted as the fluidity coefficient. This parameter is obtained from uniaxial creep experiments as:

$$\lambda = \frac{\dot{\varepsilon}_c}{\sigma} \tag{5.97}$$

5.3.2.2   Elasto-visco-plasticity

Another approach was proposed by Aydan and Nawrocki (1998), in which the material behavior is visco-elastic before yielding and becomes visco-plastic after yielding. The derivation of this constitutive law involves the following:
   Yield function

$$F(\boldsymbol{\sigma}, \kappa_p, \kappa_v) = f(\boldsymbol{\sigma}) - K(\kappa_p, \kappa_v) = 0 \tag{5.98}$$

It should be noted that the yield function is a function of permanent plastic and visco-hardening parameters (Fig. 5.14).

   Flow rule

$$d\boldsymbol{\varepsilon}^p = \lambda \frac{\partial G}{\partial \boldsymbol{\sigma}}, d\dot{\boldsymbol{\varepsilon}}^p = \dot{\lambda} \frac{\partial G}{\partial \boldsymbol{\sigma}} + \lambda \frac{\partial \dot{G}}{\partial \boldsymbol{\sigma}} \tag{5.99}$$

   Prager's consistency condition

$$dF = \frac{\partial F}{\partial \boldsymbol{\sigma}} \cdot d\boldsymbol{\sigma} + \frac{\partial F}{\partial \kappa_p} \frac{\partial \kappa_p}{\partial \boldsymbol{\varepsilon}_p} \cdot d\boldsymbol{\varepsilon}_p + \frac{\partial F}{\partial \kappa_v} \frac{\partial \kappa_v}{\partial \dot{\boldsymbol{\varepsilon}}_p} \cdot d\dot{\boldsymbol{\varepsilon}}_p = 0 \tag{5.100}$$

Linear decomposition of the strain increment $(d\boldsymbol{\varepsilon})$ and strain rate increment $(d\dot{\boldsymbol{\varepsilon}})$ into their reversible $(d\boldsymbol{\varepsilon}^r)$ and permanent components $(d\boldsymbol{\varepsilon}^p)$

$$d\boldsymbol{\varepsilon} = d\boldsymbol{\varepsilon}^r + d\boldsymbol{\varepsilon}^p, d\dot{\boldsymbol{\varepsilon}} = d\dot{\boldsymbol{\varepsilon}}^r + d\dot{\boldsymbol{\varepsilon}}^p \tag{5.101}$$

Incremental Kelvin-Voigt law

$$d\boldsymbol{\sigma} = \mathbf{D}^r d\boldsymbol{\varepsilon}^r + \mathbf{C}^r d\dot{\boldsymbol{\varepsilon}}^r \tag{5.102}$$

where $\boldsymbol{\sigma}$ is the stress tensor, $\boldsymbol{\varepsilon}$ is the strain tensor, $K(\kappa_p, \kappa_v)$ is the hardening function, G is the plastic potential, λ is the proportionality coefficient, κ_p is the plastic hardening parameter, κ_v is the viscos hardening parameter, $d\boldsymbol{\varepsilon}^r$ is the reversible incremental strain tensor, $d\dot{\boldsymbol{\varepsilon}}^r$ is the reversible incremental strain rate tensor, $d\boldsymbol{\varepsilon}^p$ is the permanent incremental strain tensor, $d\dot{\boldsymbol{\varepsilon}}^p$ is the permanent incremental strain rate tensor, $\mathbf{D}^r$ is the elasticity tensor, $\mathbf{C}^r$ is the viscosity tensor, and $(\cdot)$ denotes dot product.
   In elastic-visco-plastic formulations of the Perzyna type, the flow rule is assumed to be of the following form:

$$d\dot{\boldsymbol{\varepsilon}}^p = \dot{\lambda} \frac{\partial G}{\partial \boldsymbol{\sigma}} \tag{5.103}$$

The flow rule implies that any plastic straining is time dependent. Aydan and Nawrocki (1998) have suggested the following form:

$$d\dot{\boldsymbol{\varepsilon}}^p = \lambda \frac{\partial \dot{G}}{\partial \boldsymbol{\sigma}} \tag{5.104}$$

This Aydan and Nawrocki (1998) flow rule implies that the plastic potential function shrinks (or expands) in time domain while keeping its original form in stress space and that the permanent strain increment consists of time-dependent and time-independent parts.

Substituting Equation (5.104) in Equation (5.100) and rearranging the resulting equations yields the following:

$$dF = \frac{1}{h_{rp}} \frac{\partial F}{\partial \boldsymbol{\sigma}} \cdot d\boldsymbol{\sigma} \tag{5.105}$$

where h_{rp} is called the hardening modulus and is given specifically as follows:

$$h_{rp} = -\left[\frac{\partial F}{\partial \kappa_p} \frac{\partial \kappa_p}{\partial \boldsymbol{\varepsilon}_p} \cdot \frac{\partial G}{\partial \boldsymbol{\sigma}} + \frac{\partial F}{\partial \kappa_v} \frac{\partial \kappa_v}{\partial \dot{\boldsymbol{\varepsilon}}_p} \cdot \frac{\partial \dot{G}}{\partial \boldsymbol{\sigma}} \right] \tag{5.106}$$

Inserting these relations into Equation (5.105) yields the constitutive relations of permanent strain increment and permanent strain rate increment in relation to stress increment as

$$d\boldsymbol{\varepsilon}_p = \frac{1}{h_{rp}} \frac{\partial G}{\partial \boldsymbol{\sigma}} \left(\frac{\partial F}{\partial \boldsymbol{\sigma}} \cdot d\boldsymbol{\sigma} \right) = \frac{1}{h_{rp}} \left(\frac{\partial G}{\partial \boldsymbol{\sigma}} \otimes \frac{\partial F}{\partial \boldsymbol{\sigma}} \right) \cdot d\boldsymbol{\sigma} \tag{5.107}$$

$$d\dot{\boldsymbol{\varepsilon}}_p = \frac{1}{h_{rp}} \frac{\partial \dot{G}}{\partial \boldsymbol{\sigma}} \left(\frac{\partial F}{\partial \boldsymbol{\sigma}} \cdot d\boldsymbol{\sigma} \right) = \frac{1}{h_{rp}} \left(\frac{\partial \dot{G}}{\partial \boldsymbol{\sigma}} \otimes \frac{\partial F}{\partial \boldsymbol{\sigma}} \right) \cdot d\boldsymbol{\sigma} \tag{5.108}$$

where ($\otimes$) denotes the tensor product. Similar to the argument associated with Equation (5.84), the inverse of these relations cannot be determined. Therefore the following technique is used to establish the relation between stress increment and strain and strain rate increments. Using the preceding relations, one can write the following:

$$d\boldsymbol{\sigma} = \mathbf{D}^r d\boldsymbol{\varepsilon} - \mathbf{D}^r \frac{1}{h_{rp}} \frac{\partial G}{\partial \boldsymbol{\sigma}} \left(\frac{\partial F}{\partial \boldsymbol{\sigma}} \cdot d\boldsymbol{\sigma} \right) + \mathbf{C}^r d\dot{\boldsymbol{\varepsilon}} - \mathbf{C}^r \frac{1}{h_{rp}} \frac{\partial \dot{G}}{\partial \boldsymbol{\sigma}} \left(\frac{\partial F}{\partial \boldsymbol{\sigma}} \cdot d\boldsymbol{\sigma} \right) \tag{5.109}$$

Taking the dot products of the both sides of the preceding expression by $\partial F / \partial \boldsymbol{\sigma}$ yields:

$$\frac{\partial F}{\partial \boldsymbol{\sigma}} \cdot d\boldsymbol{\sigma} = \frac{\frac{\partial F}{\partial \boldsymbol{\sigma}} \cdot (\mathbf{D}^r d\boldsymbol{\varepsilon}) + \frac{\partial F}{\partial \boldsymbol{\sigma}} \cdot (\mathbf{C}^r d\dot{\boldsymbol{\varepsilon}})}{1 + \frac{1}{h_{rp}} \frac{\partial F}{\partial \boldsymbol{\sigma}} \cdot \left(\mathbf{D}^r \frac{\partial G}{\partial \boldsymbol{\sigma}} \right) + \frac{1}{h_{rp}} \frac{\partial F}{\partial \boldsymbol{\sigma}} \cdot \left(\mathbf{C}^r \frac{\partial \dot{G}}{\partial \boldsymbol{\sigma}} \right)} \tag{5.110}$$

Substituting the preceding equation in Equation (5.102) gives the incremental elasto-visco-plastic constitutive law as:

$$d\boldsymbol{\sigma} = \mathbf{D}^{rp} d\boldsymbol{\varepsilon} + \mathbf{C}^{rp} d\dot{\boldsymbol{\varepsilon}} \tag{5.111}$$

where

$$\mathbf{D}^{rp} = \mathbf{D}^{r} - \frac{\mathbf{D}^{r}\frac{\partial G}{\partial \boldsymbol{\sigma}} \otimes \frac{\partial F}{\partial \boldsymbol{\sigma}}\mathbf{D}^{r}}{h_{rp} + \frac{\partial F}{\partial \boldsymbol{\sigma}} \cdot \left(\mathbf{D}^{r}\frac{\partial G}{\partial \boldsymbol{\sigma}}\right) + \frac{\partial F}{\partial \boldsymbol{\sigma}} \cdot \left(\mathbf{C}^{r}\frac{\partial \dot{G}}{\partial \boldsymbol{\sigma}}\right)} \tag{5.112}$$

$$\mathbf{C}^{rp} = \mathbf{C}^{r} - \frac{\mathbf{C}^{r}\frac{\partial \dot{G}}{\partial \boldsymbol{\sigma}} \otimes \frac{\partial F}{\partial \boldsymbol{\sigma}}\mathbf{C}^{r}}{h_{rp} + \frac{\partial F}{\partial \boldsymbol{\sigma}} \cdot \left(\mathbf{D}^{r}\frac{\partial G}{\partial \boldsymbol{\sigma}}\right) + \frac{\partial F}{\partial \boldsymbol{\sigma}} \cdot \left(\mathbf{C}^{r}\frac{\partial \dot{G}}{\partial \boldsymbol{\sigma}}\right)} \tag{5.113}$$

Figure 5.14 illustrates the elasto-visco-plastic model for a one-dimensional response.

5.3.3 Yield/failure criteria

Nonlinear behavior requires the existence of yield functions. These yield functions are also called failure functions at the ultimate state when rocks rupture. For a two-dimensional case, it is common to use the Mohr-Coulomb yield criterion given by:

$$\tau = c + \sigma_n \tan\phi \quad or \quad \sigma_1 = \sigma_c + q\sigma_3 \tag{5.114}$$

where c, φ and σ_c are cohesion, friction angle and uniaxial compressive strength. σ_c and q are related to cohesion and friction angle in the following form:

$$\sigma_c = \frac{2c\cos\phi}{1 - \sin\phi} \text{ and } q = \frac{1 + \sin\phi}{1 - \sin\phi} \tag{5.115}$$

Since the intermediate principal stress is indeterminate in the Mohr-Coulomb criterion and there is a corner-effect problem during the determination of incremental elasto-plasticity tensor, the use of Drucker-Prager criterion (Drucker and Prager, 1952) is quite common in numerical analyses, which is given by:

$$\alpha I_1 + \sqrt{J_2} = k \tag{5.116}$$

where

$$I_1 = \sigma_I + \sigma_{II} + \sigma_{III}, J_2 = \frac{1}{6}\left((\sigma_I - \sigma_{II})^2 + (\sigma_{II} - \sigma_{III})^2 + (\sigma_{III} - \sigma_I)^2\right)$$

Nevertheless, it is possible to relate the Drucker-Prager yield criterion with the Mohr-Coulomb yield criterion. On the π-plane, if the inner corners of the Mohr-Coulomb yield surface are assumed to coincide with the Drucker-Prager yield criterion, the following relations may be derived (Fig. 5.15):

$$\alpha = \frac{2\sin\phi}{\sqrt{3(3 + \sin\phi)}}, k = \frac{6c\cos\phi}{\sqrt{3(3 + \sin\phi)}} \tag{5.117}$$

where c, φ are cohesion and friction angle, respectively.

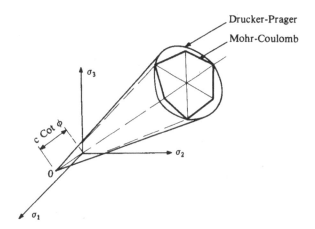

Figure 5.15 Illustration of yield criteria in principal stress space
Source: From Owen and Hinton, 1980

In rock mechanics, a recent yield criterion is Hoek-Brown's criterion (1980), which is written as:

$$\sigma_1 = \sigma_3 + \sqrt{m\sigma_c\sigma_3 + s\sigma_c^2} \qquad (5.118)$$

where m and s are some coefficients. While the value of s is 1 for intact rock, the values of m and s change when they are used for rock mass.

Aydan (1995) proposed a yield function for the thermo-plasticity yielding of rock as given by:

$$\sigma_1 = \sigma_3 + [S_\infty - (S_\infty - \sigma_c)e^{-b_1\sigma_3}]e^{-b_2 T} \qquad (5.119)$$

where S_∞ is the ultimate deviatoric strength while coefficients b_1, b_2 are empirical constants.

A number of examples of applications of yield (failure) criteria to actual experimental results involving igneous, metamorphic and sedimentary rocks are described and compared with one another as well as with experimental results.

(a) Sedimentary rocks

A series of uniaxial and triaxial compression and Brazilian tensile tests were carried out on Oya tuff, which is a well-known volcanic sedimentary rock in Japan (Seiki and Aydan, 2003). Figure 5.16 shows the failure criteria of Mohr-Coulomb, Hoek and Brown, and Aydan applied to experimental results. The best fits with experimental results were obtained for Aydan's criterion and Mohr-Coulomb criterion for $\sigma_3 > 0$. However, if the yield criterion is required to evaluate both uniaxial compressive strength and tensile strength, the criteria of Hoek-Brown and Mohr-Coulomb cannot evaluate the triaxial strength of rocks.

The next application is concerned with very weak sandstone from Tono mine in Central Japan Aydan *et al.* (2006). Figure 15.17 shows the fitted failure criteria of Mohr-Coulomb, Hoek and Brown, and Aydan applied to experimental results of sandstone of Tono. As seen from the Figure 5.4, the best fits to experimental results were obtained for Aydan's criterion

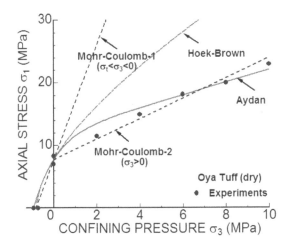

Figure 5.16 Comparisons of yield criteria for Oya tuff (dry)

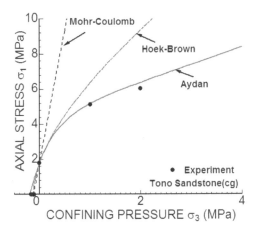

Figure 5.17 Comparisons of yield criteria for experimental results on Tono sandstone

and the Hoek and Brown criterion. Nevertheless, the criterion of Hoek-Brown deviates from experimental results when the confining pressure is greater than 1 MPa.

A series of uniaxial and triaxial compression and Brazilian tests carried out on samples from a limestone formation in which Gökgöl karstic cave in Zonguldak province, Turkey, is located (Aydan *et al.*, 2012). Triaxial compression experiments were carried under confining pressures up to 40 MPa. Figure 5.18 shows the experimental results together with several fitted failure criteria of bilinear Mohr-Coulomb, Hoek-Brown, and Aydan. The best fits to experimental results were obtained from the applications of the bilinear Mohr-Coulomb criterion and Aydan's criterion. If the Hoek-Brown criterion is required to represent both the tensile strength and the uniaxial compression strength, it is well fitted to the lower-bound values. However, the estimated curve by Aydan's criterion can better represent the bilinear Mohr-Coulomb criterion as well as all experimental results.

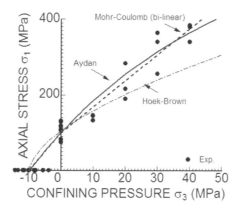

Figure 5.18 Comparisons of yield criteria for experimental results on limestone

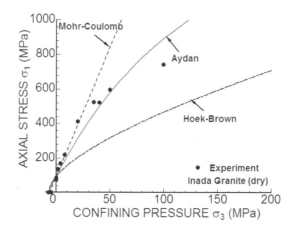

Figure 5.19 Comparisons of yield criteria for experimental results on Inada granite

(b) Igneous rocks

Inada granite is a well-known igneous hard rock, and its uniaxial compressive strength is generally greater than 100 MPa. Figure 5.19 shows the fitted failure criteria of Mohr-Coulomb, Hoek and Brown, and Aydan applied to experimental results carried out under confining pressures up to 100 MPa. For confining pressures up to 50 MPa, the best fits to experimental results are those of Aydan and Mohr-Coulomb. If the parameters of the Hoek-Brown criterion are determined to represent tensile strength and uniaxial compressive strength, the estimated triaxial strengths for high confining pressures are entirely different from experimental results.

Hirth and Tullis (1994) reported the results of triaxial experiments on quartz aggregates under very high confining pressures, which is almost 6 times its uniaxial compressive strength. The best fit to the experimental results is obtained for Aydan's criterion, as seen in Figure 5.20. Up to a confining pressure of 1000 MPa, the estimation by the Mohr-Coulomb criterion is better than that by the Hoek-Brown criterion. Again, very high discrepancy is observed among the estimations by Hoek-Brown criterion and experimental results.

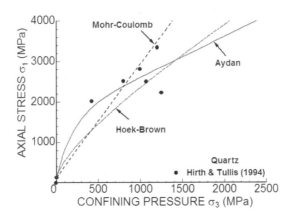

Figure 5.20 Comparisons of yield criteria for experimental results on quartz

(c) Metamorphic rocks

The yield (failure) criterion of metamorphic rocks must consider the anisotropy caused by the orientation of schistosity in relation to the applied principal stresses during experiments. Jaeger and Cook (1979) developed a procedure involving yield conditions for shearing along a schistosity plane and shearing through rock, based on the Mohr-Coulomb yield (failure) criterion. There have been some attempts by several researchers to express the dependency of yield function to the schistosity orientation (McLamore and Gray, 1967; Donath, 1964; Nasseri *et al.*, 2003). When the yield (failure) criterion is evaluated, one should take into account the effects of characteristics of rocks as well as the schistosity orientation with respect to applied stresses. Nevertheless, the overall functional form of yield criterion should be similar for a given orientation except the specific values of their parameters.

Nasseri *et al.* (2003) reported extensive experimental research on the failure characteristics of Himalayan schists subjected to confining pressures up to 100 MPa. They also reported the tensile strength characteristics of all rock types. Figure 5.21 shows the failure criteria of bilinear Mohr-Coulomb, Hoek-Brown, and Aydan applied to experimental results of Himalayan chlorite schist for the orientation angle of 90 degrees. In fitting relations of the criteria of Hoek-Brown and Aydan, the failure criteria are required to represent uniaxial compressive strength and tensile strength. We note that experimental results are well represented by the bilinear Mohr-Coulomb criterion. Once again the Hoek-Brown criterion deviates from triaxial compressive experimental results if it is required to represent the tensile strength and compressive strength. Aydan's criterion provides a best continuous fit to experimental results as noted in Figure 5.21.

Waversik and Fairhurst (1970) presented results of triaxial compressive tests on Tennessee marble, while Haimson and Fairhurst (1970) reported the tensile and uniaxial compressive strength of the same rock. Again we fitted the failure criteria of bilinear Mohr-Coulomb, Hoek-Brown, and Aydan to a combined set of experiments on Tennessee marble with the requirement of representing its tensile and uniaxial compressive strength as shown in Figure 15.22. The overall tendency of fitted criteria to experimental results remains the same. Particularly, the requirement of representing both tensile and compressive strength

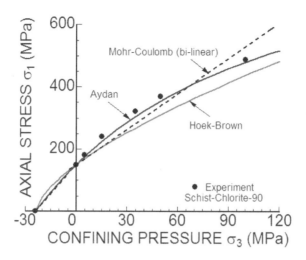

Figure 5.21 Comparisons of yield criteria for experimental results on chlorite schist

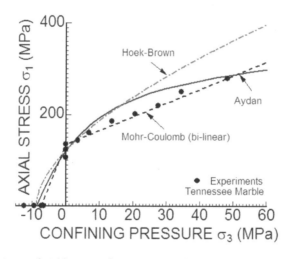

Figure 5.22 Comparisons of yield criteria for experimental results on Tennessee marble

by the criterion of Hoek-Brown result in different values of parameter m, as also noted by Betournay *et al.* (1991). The value of m is roughly equal to the ratio of uniaxial compressive strength to tensile strength, which is known as the brittleness index.

(d)   Application to thermal triaxial compression experiments

In geomechanics, there is almost no yield (failure) criterion–incorporating effect of temperature on yield (failure) properties of rocks, although there are some experimental researches

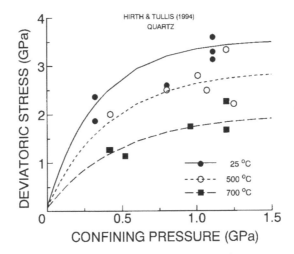

Figure 5.23 Experimental results of Hirth and Tullis (1994) on quartz for three different ambient temperatures

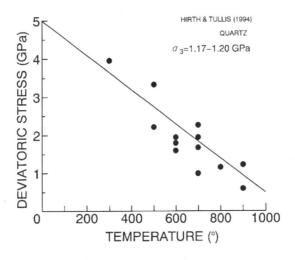

Figure 5.24 The reduction of deviatoric strength of quartz as a function of temperature for a confining pressure of 1.17–1.2 GPa

(Hirth and Tullis, 1994). Aydan's criterion is the only criterion known to incorporate the temperature, and it was used to study the stress state of the Earth (Aydan, 1995). Figure 5.23 shows the experimental results for three different values of ambient temperature reported by Hirth and Tullis (1994), while Figure 5.24 shows the reduction of strength with temperature for a given confining pressure of 1.17–1.2 GPa.

Aydan's yield (failure) criterion is applied to experimental results, as shown in Figures 5.23 and 5.24, and results are shown in Figure 5.25.

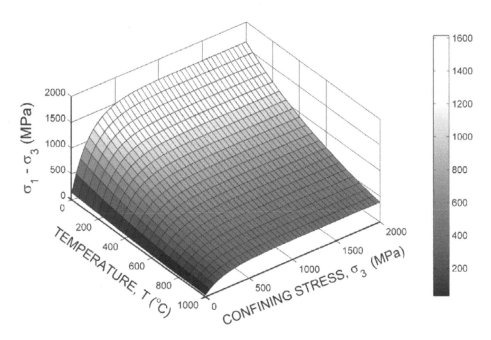

Figure 5.25 Three-dimensional representation of Aydan's failure criterion for experimental results of Hirth and Tullis (1994)

5.4 Equivalent models for discontinuum

Rock masses in nature contain numerous discontinuities in the form of cracks, joints, faults, bedding planes and the like. Therefore, various continuum equivalent models of discontinuum have been proposed and used since the beginning of 1970 to assess the stability of rock tunnels (e.g. Budiansky and O'Connel, 1976; Hill, 1963; Kachanov, 1958). Discontinuum is distinguished from continuum by the existence of contacts or interfaces between the discrete bodies that comprise the system. Relative sliding or separational movements in such localized zones present an extremely difficult problem in mechanical modeling and numerical analysis. The formulation for representing contacts is very important when a system of interacting blocks is considered, and it has been receiving a considerable interest among researchers. The main characteristics of the models are described in this section.

5.4.1 Equivalent elastic compliance model (EECM) (Singh's model)

Singh's model (1973) is based on the theory proposed by Hill (1963) for composite materials. The elastic constitutive law of the rock mass is obtained by making the following assumptions (Fig. 5.26):

1 Discontinuities are distributed in sets in the rock mass.
2 The geometry of discontinuities (area and orientation) are known.

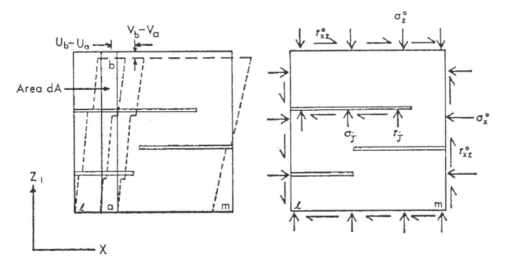

Figure 5.26 Singh's model

3   The constitutive law of discontinuities is expressed in terms of relative normal and shear displacements and applied shear and normal stresses and the shear and normal stiffnesses are used to express the behavior of discontinuities.
4   The stress tensor acting on discontinuities is related to that acting on the representative volume through a tensor called stress-concentration tensor.
5   The strain tensor of the representative element is a linear sum of the strain tensor of the intact rock and the additional strain tensor due to discontinuities.
6   The volume of discontinuities is assumed to be negligible as compared with that of the rock so that the stress tensor acting on the representative volume of rock mass is the same as that on the intact rock.

The constitutive law derived in a local coordinate system is then transformed to that in the global coordinate system. The formulations given by Goodman and Amadei and by Goodman are the simplified form of Singh's model. Application of these models to tunnels in jointed media with a cross-continuous pattern and intermittent pattern is given and compared with a discrete model. This model is the first equivalent model to be applied to rock engineering problems.

5.4.2   Crack tensor model (CTM)

This model proposed by Oda (1982) for rock masses follows basically the same steps of Singh's model in order to obtain the elastic constitutive law of the rock mass. The main differences are as follows:

1   Constitutive law is directly derived in a global coordinate system.
2   The geometry of discontinuities (area and orientation) are represented by a series of even-order tensors (up to fourth-order tensors).

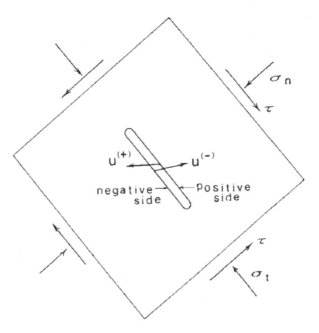

Figure 5.27 Crack tensor model
Source: Oda *et al.*, 1993

3   The additional strain tensor due to discontinuities is determined from a procedure
     utilizing an analogy to the theoretical solutions for penny-shaped or elliptical inclu-
     sions embedded in an elastic medium used in linear elastic fracture mechanics
     (Fig. 5.27).

The application of this model to tunnels, particularly branching tunnels, is described in a
recent paper by Oda *et al.* (1984, 1993).

5.4.3   Damage model (DM)

The damage model is based upon the theory proposed by Kachanov (1958) for creeping
metals. It is elaborated by Murakami (1985) by introducing a second-order tensor called
the damage tensor, and it is applied to a rock mass by Kyoya (1989) and Kawamoto *et
al.* (1988). Assumptions 1 and 2 of Singh are also the same in this model. However, this
model differs from other models, and it is based on the following additional assumptions:

1   The discontinuities are assumed to be not transmitting any stress across, implying the
     discontinuities have no stiffness at all.
2   The stresses are assumed to be acting only on the intact parts, implying the parallel
     connection principle for the stress field. The average stress (Cauchy stress) is related to
     the stress (net stress or intensified stress) on the intact part through the second-order
     damage tensor, which represents a tensorial area reduction in the mass (Fig. 5.28).

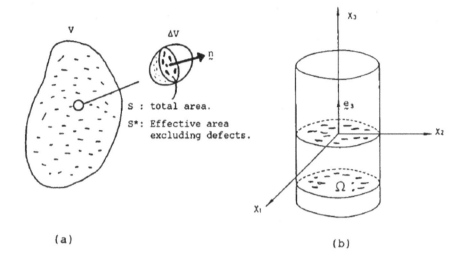

(a)

S : total area.

S*: Effective area excluding defects.

(b)

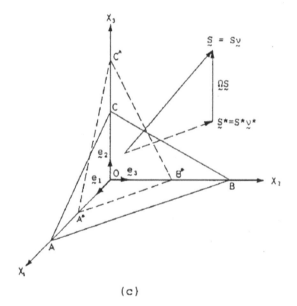

(c)

Figure 5.28 Damage model

Source: Kyoya, 1989

3   The strain tensor of the representative elementary volume is the same as that of the intact rock.
4   The constitutive law is introduced between the net stress tensor and the strain tensor.

It should be noted that this model could not be directly used for thoroughgoing discontinuity sets because of Assumption 1. Nevertheless, Kyoya (1989) introduces some coefficients for normal and shear responses to differentiate the behavior under tension and compression. The applications of this model to tunnels and underground caverns are described by Kyoya (1989) and Kawamoto *et al.* (1988). Swoboda and Ito (1992) extended this model to model crack propagations in jointed media and gave several examples of its application.

5.4.4   Micro-structure models

Aydan *et al.* (1992, 1994,1996) proposed two models for discontinuous rock masses based on the microstructure theory of mechanics (Jones, 1975). Although the first assumption of Singh is the same as that in this approach, this model differs from others. The fundamental differences are as follows:

1   Discontinuities have a finite volume that enables one to model a wide range of discontinuities from joints to faults or fractured zones.
2   The constitutive law of discontinuities is expressed in the conventional sense of mechanics. In other words, the constitutive law is expressed in terms of stresses and strains, and it is uniquely defined.
3   The constitutive law is not restricted to elasticity, and it can be of any kind that can describe the mechanical response of discontinuities.
4   Stress and strain fields of each constituent are related to one another using two concepts: Globally Series and Locally Parallel Model (GSLPM) and Globally Parallel and Locally Series Model (GPLSM) (Fig. 5.29).

5.4.5   Homogenization technique

The homogenization technique was mainly used to obtain the equivalent characteristics of composites (Bakhvalov and Panasenko, 1984; Sanchez-Palencia, 1980) and has been recently applied to soil (Auriault, 1983) and rocks (Fig. 5.30). Assumptions 1 and 3 of the micro-structure model also hold for this technique. Stress and strain fields of constituents are obtained from a perturbation of the displacement field. An influence tensor, which is a gradient of six vectorial functions called characteristic deformation functions for a given representative elementary volume (unit cell), is used to establish relations between the homogenized elasticity tensor and those of its constituents. Except for very simple cases, the equivalent parameters are obtained using a numerical method such as FEM.

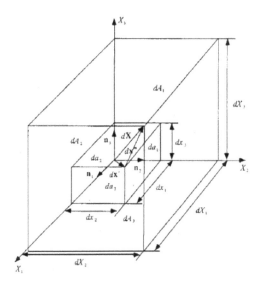

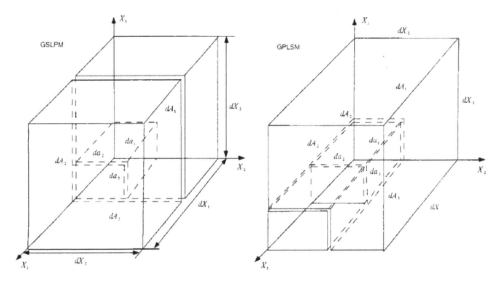

Figure 5.29 Micro-structure model
Source: From Aydan *et al.*, 1992

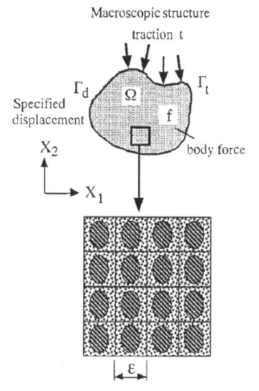

Figure 5.30 Homogenization model

References

Auriault, J.L. (1983) Homogenisation: Application to porous saturated media. *Summer School on Two-Phase Medium Mechanics, Gdansk.*

Aydan, Ö. (1995) The stress state of the earth and the earth's crust due to the gravitational pull. *The 35th US Rock Mechanics Symposium, Lake Tahoe.* pp. 237–243.

Aydan, Ö. (2011) Some issues in tunnelling through rock mass and their possible solutions. *First Asian Tunnelling Conference, ATS11.* pp. 33–44.

Aydan, Ö. & Nawrocki, P. (1998) Rate-dependent deformability and strength characteristics of rocks. *Int. Symp. On the Geotechnics of Hard Soils-Soft Rocks, Napoli, 1.* pp. 403–411.

Aydan, Ö., Tokashiki, N., Seiki, T. & Ito, F. (1992) Deformability and strength of discontinuous rock masses. *Int. Conf. Fractured and Jointed Rock Masses, Lake Tahoe.* pp. 256–263.

Aydan, Ö., Seiki, T., Jeong, G.C. & Tokashiki, N. (1994) Mechanical behaviour of rocks, discontinuities and rock masses. *Proc. of International Symposium Pre-failure Deformation Characteristics of Geomaterials, Sapporo, 2.* pp. 1161–1168.

Aydan, Ö., Tokashiki, N. & Seiki, T. (1996) Micro-structure models for porous rocks to jointed rock masses. *Proc. 3rd Asia-Pacific Conf. on Computational Mechanics, Seoul.*

Aydan, Ö., Daido, M., Tano, H., Nakama, S. & Matsui, H. (2006) The failure mechanism of around horizontal boreholes excavated in sedimentary rock. *50th US Rock Mech. Symp., Paper No. 06-130*, Golden, Colorado.

Aydan, Ö., Tokashiki, N. & Geniş, M. (2012) Some considerations on yield (failure) criteria in rock mechanics ARMA 12–640. *Proc. of 46th US Rock Mechanics/Geomechanics Symposium, Chicago*, 10p, (on CD).

Bakhvalov, N. & Panasenko, G. (1984) *Homogenization: Averaging Processes in Periodic Media*, Kluwer, Dordrecht.

Betournay, M.C., Gorski, B., Labrie, D., Jackson, R. & Gyenge, M. (1991) New considerations in the determining of Hoek and Brown material constants. *Proc. 7th Int. Cong. on Rock Mechanics, Aachen, 1*. pp. 195–200.

Budiansky, B. & O'Connel, R.J. (1976) Elastic moduli of a cracked solid. *International Journal of Solids and Structures*, 12, 81–97.

Donath, F.A. (1964) Strength variation and deformational behavior in anisotropic rock. In: Judd, W.R. (ed.) *State of Stress in the Earth's Crust*. Elsevier, New York. pp. 281–297.

Drucker, D.C. & Prager, W. (1952) Soil mechanics and plastic analysis for limit design. *Quarterly of Applied Mathematics*, 10(2), 157–165.

Farmer, I. (1983) *Engineering Behaviour of Rocks*, 2nd edition, Chapman and Hall, London.

Haimson, B.C. & Fairhurst, C. (1970) Some bit penetration characteristics in pink Tennessee marble. *Proc. 12th US Rock Mechanics Symp.* pp. 547–559.

Hill, R. (1963) Elastic properties of reinforced solids: Some theoretical principles. *Journal of Mechanics and Physics of Solids*, 11, 357–372.

Hirth, G. & Tullis, J. (1994) The brittle-plastic transition in experimentally deformed quartz aggregates. *Journal of Geophysical Research*, 99, 11731–11747.

Hoek, E. & Brown, E.T. (1980) Empirical strength criterion for rock masses. *Journal of Geotechnical Engineering Division, ASCE*, 106(GT9), 1013–1035.

ISRM (2007) The complete ISRM suggested methods for rock characterization, testing and monitoring: 1974–2006. In: Ulusay, R. & Hudson, J.A. (eds.) *Suggested Methods Prepared by the Commission on Testing Methods, International Society for Rock Mechanics*, Compilation Arranged by the ISRM Turkish National Group, Ankara, Turkey.

Jaeger, J.C. & Cook, N.G.W. (1979) *Fundamentals of Rock Mechanics*, 3rd edition. Chapman & Hall, London. pp. 79, 311.

Jones, R.M. (1975) *Mechanics of Composite Materials*. Hemisphere Pub. Co., New York.

Kachanov, L.M. (1958) *The Theory of Creep* (English transl. by A.J. Kennedy). National Lending Library, Boston.

Kawamoto, T., Ichikawa Y. & Kyoya, T. (1988) Deformation and fracturing behaviour of discontinuous rock mass and damage mechanics theory. *International Journal for Numerical and Analytical Methods in Geomechanics*, 12(1), 1–30.

Kyoya, T. (1989) *A Fundamental Study on the Application of Damage Mechanics to the Evaluation of Mechanical Characteristics of Discontinuous Rock Masses*. Ph.D. Thesis, Nagoya University, Nagoya, Japan.

McLamore, R. and Gray, K.E. 1967. The mechanical behaviour of anisotropic sedimentary rocks. *Transactions of the American Society of Mechanical Engineers Series B*, 62–76.

Mirza, U.A. (1978) *Investigation into the Design Criteria for Underground Openings in Rocks Which Exhibit Rheological Behaviour*. PhD thesis, University of Newcastle upon Tyne., Newcastle upon Tyne, UK

Murakami, S. (1985) Anisotropic damage theory and its application to creep crack growth analysis. *Proc. Int. Conf. Constitutive laws for Engineering Materials: Theory and Applications, Elsevier, Amsterdam.* pp. 535–551.

Nasseri, B.M.H., Rao, K.S. & Ramamurthy, T. (2003) Anisotropic strength and deformational behaviour of Himalayan Schists. *International Rock Mechanics and Mining Science*, London, 40(1), 3–23.

Oda, M. (1982) Fabric tensor for discontinuous geological materials. *Soil and Found*ations, Tokyo 22 (4), 96–108.

Oda, M., Suzuki, K. & Maeshibu, T. (1984) Elastic compliance for rock-like materials with random cracks. *Soil and Foundations*, Tokyo, 24(3), 27–24.

Oda, M., Yamabe, T., Ishizuka, Y., Kumasaka, H., Tada, H. & Kimura, K. (1993) Elastic stress and strain in jointed rock masses by means of crack tensor analysis. *Rock Mechanics and Rock Engineering, Vienna*, 26(2), 89–112.

Owen, D.R.J. & Hinton, E. (1980) *Finite Element in Plasticity: Theory and Practice*. Pineridge Press Ltd, Swansea.

Perzyna, P. (1966) Fundamental problems in viscoplasticity. *Advances in Applied Mechanics*, 9(2), 244–368.

Sanchez-Palencia, E. (1980) Non-homogenous media and vibration theory. *Lecture Note in Physics, No. 127, Springer, Berlin*.

Seiki, T. & Aydan, Ö. (2003) Deterioration of Oya Tuff and its mechanical property change as building stone. *Proc. of Int. Symp. on Industrial Minerals and Building Stones, Istanbul, Turkey*. pp. 329–336.

Singh, B. (1973) Continuum characterization of jointed rock masses: Part I*The constitutive equations. *Int. J. Rock Mech. Min Sci.*, 10, 311–335.

Swoboda, G. & Ito, F. (1992) Two-dimensional damage failure propagation of jointed rock mass. *Int. Symp. On Computational Mechanics, Balkema*.

Waversik, W.R. & Fairhurst, C. (1970) A study of brittle rock fracture in laboratory compression experiments. *Int. J. Rock Mech. Min. Sci.*, 7, 561–575.

Wawersik, W.R. (1983) Determination of steady state creep rates an activation parameters for rock salt. In: *High Pressure Testing of Rock, Special Technical Publication of ASTM*, STP86972-91.

Chapter 6

Laboratory and *in-situ* tests

Mechanical, seepage, heat, and diffusion properties related to constitutive laws described in Chapter 5 require tests on rocks and/or rock mass appropriate to physical and environmental conditions. In this chapter, the fundamental principles of available testing techniques used in the field of rock mechanics and rock engineering are explained. It should be noted that the details of each technique may require additional information about equipment and processing, which are explained in the suggested methods (SMs) published by ISRM (Brown, 1981; Hudson and Ulusay, 2007; Ulusay, 2012). Furthermore, size, shape, and environmental conditions are described in related ISRM SMs.

6.1 Laboratory tests on mechanical properties

Laboratory tests on mechanical properties of rocks may be determined from uniaxial tensile and compression tests, triaxial compression experiments, three/four bending experiments, direct shear and Brazilian experiments. Some of these experiments are illustrated in Figure 6.1. The fundamentals of these testing techniques are described in the following subsections.

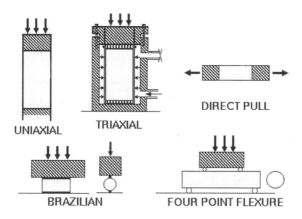

Figure 6.1 Illustration of some testing techniques for determining mechanical properties

6.1.1 Uniaxial compression tests

Specimens from drill cores are prepared by cutting them to the specified length and are thereafter grinded and measured. There are high requirements on the flatness of the end surfaces in order to obtain an even load distribution. The recommended ratio of height/diameter of the specimens is between 2 and 3. Strains and stress in uniaxial compression tests are defined as follows:

Axial strain is calculated from the equation

$$\varepsilon_a = \frac{\Delta l}{l_O} \tag{6.1}$$

where l_O is the original measured axial length, and Δl is the change in measured axial length (defined to be positive for a decrease in length).

Diametrical strain can be determined either by measuring the changes in the diameter of the specimen or by measuring the circumferential strain. In the case of measuring changes in the diameter, the diametric strain is calculated from the equation:

$$\varepsilon_d = \frac{\Delta d}{d_O} \tag{6.2}$$

where d_0 is the original undeformed diameter of the specimen, and Δd is the change in diameter (defined to be negative for an increase in diameter).

In the case of measuring the circumferential strain, ε_c, the circumference is $C = \pi d$; thus the change in circumference is $\Delta C = \pi \Delta d$. Consequently, the circumferential strain ε_c is related to the diametric strain ε_d by:

$$\varepsilon_c = \frac{\Delta C}{C} = \frac{\Delta d}{d_O} \tag{6.3}$$

so that $\varepsilon_c = \varepsilon_d$, where C and d_O are the original circumference and diameter of the specimen, respectively.

The compressive axial stress in the test specimen σ_a is calculated by dividing the compressive load P on the specimen by the initial cross-sectional area A_O.

$$\sigma_a = \frac{P}{A_O} \tag{6.4}$$

where compressive stresses and strains are considered to be positive in this test procedure. For a given stress level, the volumetric strain ε_v is calculated from the equation:

$$\varepsilon_v = \varepsilon_a + 2\varepsilon_d \tag{6.5}$$

The specimens are loaded axially up to failure or any other prescribed level whereby the specimen is deformed and the axial and the radial deformation can be measured using some equipment as shown in Figure 6.2.

There are a tremendous number of studies on the stress and strain distributions induced in uniaxial compression experiments with the consideration of boundary conditions imposed in the experiments. Generally, the stiffness and Poisson's ratio of the platens and specimen are different from each other. Both theoretical and numerical analyses such as FEM indicate that the stress and strain are not uniform within the specimen. Particularly, the distributions are highly nonuniform near the end of the specimen, as shown in Figure 6.3. The nonuniformity

Figure 6.2 A view of experimental setup and instrumentation in uniaxial compression experiment at Nagoya University and University of the Ryukyus

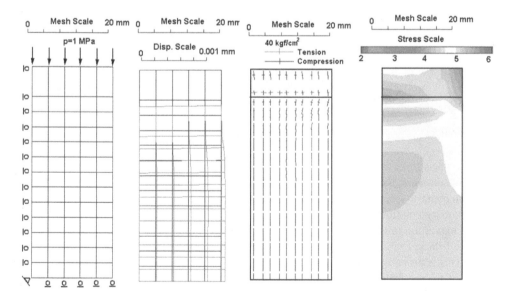

Figure 6.3 Finite element method simulation of axisymmetric rock sample (One-quarter of the specimen is used in view of symmetry.)

of strain–stress distributions strongly depends upon the differences of stiffness and Poisson's ratio of platens and specimen and their geometry. Stress and strain become uniform when the stiffness and Poisson's ratio of platens and specimen are the same, which is not the common case in many experimental studies. An entirely wrong interpretation of nonuniformity stress distributions is caused by the differences of the stiffness and Poisson's ratio as frictional effect. Frictional effect may come into action after relative slip occurs between the platens and the ends of the specimen. In practice, Vaseline oil, Teflon sheets or brush-type platens are used to deal with this issue.

6.1.2 Direct and indirect tensile strength tests (Brazilian tests)

Similar to the uniaxial compression tests, direct tensile tests are used to determine tensile strength and some deformability properties. The definitions of stress and strains are fundamentally the same except for the sign of stresses and strains. The specimens are bonded to the loading platens, or dog-shaped specimens are used to obtain the strain–stress relations.

As the preparation and procedures are quite cumbersome, indirect tensile stress tests are generally preferred. The common procedure is to load a solid or hollow cylindrical specimen under compression, which results in tensile stresses within the samples. A cylindrical specimen is loaded diametrically across the circular cross section. This testing technique is known as the Brazilian tensile strength test. The loading causes tensile stresses perpendicular to the loading direction, which results in a tensile failure. Tensile stress induced in a solid cylinder of rock is theoretically given by the following equation:

$$\sigma_t = \frac{2F}{\pi D t} \tag{6.6}$$

where F, D and t are applied load, diameter and thickness of the rock sample, respectively. The nominal strain of the Brazilian tensile test sample may be given as (see Hondros, 1959; Jaeger and Cook, 1979 for details):

$$\varepsilon_t = 2\left[1 - \frac{\pi}{4}(1-v)\right]\frac{\sigma_t}{E} \quad \text{with} \quad \varepsilon_t = \frac{\delta}{D} \tag{6.7}$$

For most rocks, this formula may be simplified to the following form:

$$\varepsilon_t = 0.82\frac{\sigma_t}{E} \tag{6.8}$$

A plane stress finite element analysis was carried out for the Brazilian test. The properties of the platen were assumed to be those of aluminium with an elastic modulus of 70 GPa. Uniform compressive pressure with an intensity of 20 kgf cm^{-2} was applied on the platens, and boundary conditions are shown in Figure 6.4.

The maximum tensile stress occurs in the vicinity of the center of the sample, and its value is 1.08 kgf cm^{-2}. This is slightly greater than the theoretical estimation of 0.8 kgf cm^{-2}. This is probably due to the slight difference in the application of load boundary conditions. The computed radial displacement of the sample just below the platen was about 0.001 mm, which is almost equal to that estimated from Equation 6.7. Therefore, it is possible to determine the elastic modulus besides the tensile strength of rocks. Furthermore, the strain response in experiments should be similar to those of the uniaxial compression

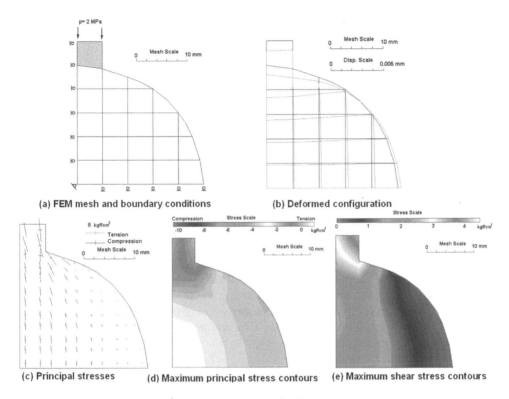

Figure 6.4 Boundary conditions and computed stress distributions

experiments provided that deformability characteristics remain the same under both tension and compression.

Bending tests are also used to determine the tensile strength and deformability of rocks. The maximum tensile stress and maximum flexural strain of rock beam under a three-point bending configuration with a concentrated load (F) and rectangular prismatic shape (b, t) and maximum deflection (δ) may be given in the following form:

$$\sigma_f = \frac{3FL}{2bt^2}, \varepsilon_f = \frac{6\delta t}{L^2}, \frac{\sigma_f}{\varepsilon_f} = \frac{FL^3}{4A\delta}, A = b \cdot t \tag{6.9}$$

Four-point bending is generally recommended due to uniform stress distributions in the area between two applied loads points. Figure 6.5 illustrates stress distribution in a three-point bending test in a photo-elastic test.

6.1.3   Triaxial compression tests

Specimens from drill cores are prepared by cutting them to the specified length and are thereafter ground and measured to obtain the required the flatness of the end surfaces. The recommended ratio of height/diameter of the specimens is between 2 and 3. A membrane is mounted on the surface of the specimen in order to seal the specimen from the

Figure 6.5 Stress distribution in a beam subjected to three-point bending condition

surrounding pressure media. Deformation measurement equipment is mounded on the specimen, and the specimen is inserted into the pressure cell, whereupon the cell is closed and filled with oil. A hydrostatic pressure is applied in the first step. The specimen is then further loaded by increasing the axial load under constant or increasing cell pressure up to failure or any other predefined load level. A test setup and equipment, triaxial cell, and instrumentation used in Nagoya University is shown in Figure 6.6.

6.1.4 Postfailure behavior in uniaxial and triaxial compression tests

The postfailure characteristics of rocks are quite important in rock engineering, when rock failure could not be prevented. For this purpose, servo-control testing devices developed for investigating postfailure characteristics of rocks (e.g. Rummel and Fairhurst, 1970; Waversik and Fairhurst, 1970; Hudson *et al.*, 1972; Kawamoto *et al.*, 1980). The servo-control devices try to provide sufficient support to measure the intrinsic properties of rocks during the postpeak response. Such a support system is provided through the loading system, which utilizes either servo-controlled oil-based jacks or wedge-like solid support. Figure 6.7 shows the principle of the device based on wedge-like solid support concept at Nagoya University, and Figure 6.8 illustrates how wedge-like support system activated during the deformation process.

Figure 6.9(a) shows a general strain–stress responses for a typical rock sample, while Figure 6.9(b) illustrates the brittle and ductile behavior of rocks. The stress drops rapidly when rock exhibits brittle behavior. On the one hand, the stress gradually decreases in the postfailure regime of ductile rocks. Figure 6.10 shows the true response of the triaxial behavior of Ryukyu limestone under different confining pressures.

Figure 6.11 shows views of a granite sample tested by the Nagoya University servo-control testing machine. As noted from Figure 6.11, a macroscopic fracture zone is observed in the final postfailure stage. Figure 6.12 shows an actual image and X-ray CT tomographic image of another granite sample. As noted from the figure, a macroscopic fracture zone and a zone of fine cracks are seen in X-ray CT tomographic image (Aydan *et al.*, 2016c) Figure 6.13 illustrates five different idealized fracturing situations in a given rock sample during the complete strain—stress response (Aydan *et al.*, 1993).

Figure 6.6 Triaxial (a) compression device, (b) cell, and (c) instrumented sample at Nagoya University Rock Mechanics Laboratory

During stages 4 and 5, microscopic fractures coalesces into macroscopic shear bands. It should be noted that there is an argument if the strain–stress response in stages 4 and 5 should be considered a part of constitutive law or not.

6.1.5 Direct shear tests

Direct shear test devices can also be used to obtain the shear strength properties of intact rock or rock discontinuities. There are different types of direct shear test devices. Figure 6.14 shows a shear testing machine named OA-DSTM (Fig. 6.15), designed and built originally in 1991 for direct shear testing under three loading conditions: conventional direct shear loading, direct shear creep loading, and direct shear cyclic loading at Nagoya University (Aydan *et al.*, 1994; Aydan *et al.*, 2016a Both the shear and the normal loads on the shearing plane are designed to be 200 kN, and the system is displacement controlled. Furthermore, the direct shear box was vertical in order to eliminate dead loads on the shearing plane. The normal load is first imposed on the sample, and then shear loading is applied

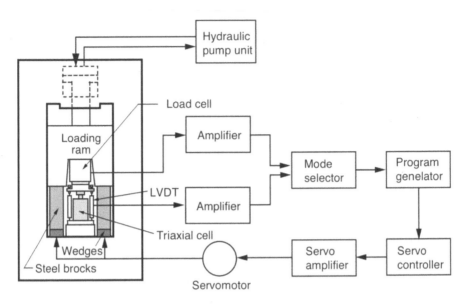

Figure 6.7 Servo-control testing device at Nagoya University

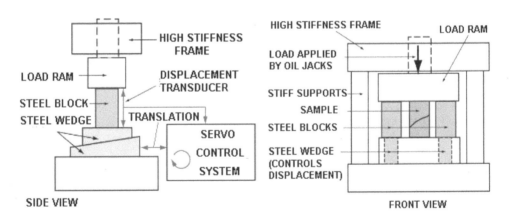

Figure 6.8 The illustration of the wedge-like solid support activation

through vertical jack. However, the system was needed to be upgraded for operational purposes as well as for dealing with dynamic loading conditions. Aydan *et al.* (2016a) have recently upgraded the shear testing machine by adding the dynamic shear loading option.

The size of direct shear samples can be $100 \times 100 \times 100$ mm or $150 \times 75 \times 75$ mm. The original design size was $150 \times 75 \times 75$ mm with the purpose of eliminating the rotational effects on the sample. This can be achieved if the ratio of sample length over sample height is greater than 2. The shear load and displacement and the normal load are directly recorded in computers using the outputs from the system. The outputs are real-time values of shear displacement in millimeters, and shear and normal loads in kilonewtons. Several examples

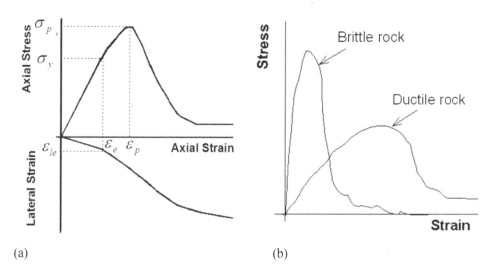

(a)   (b)

Figure 6.9 (a) Complete strain–stress relation of rocks, (b) illustration of brittle and ductile behavior

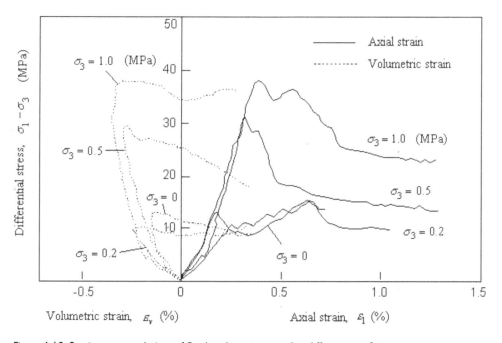

Figure 6.10 Strain–stress relation of Ryukyu limestone under different confining pressures

Figure 6.11 Views of a granite sample before (left) and after testing (right)

Figure 6.12 Actual (left) and X-ray CT (right) images of a granite sample in postfailure stage

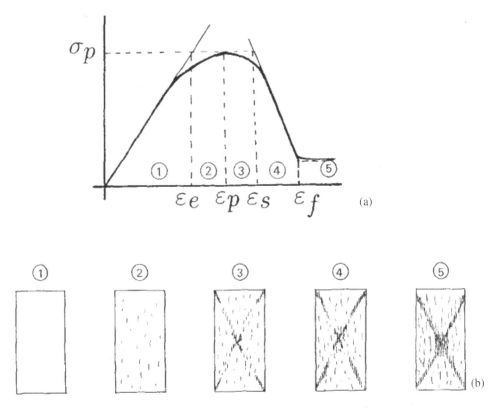

Figure 6.13 Idealized five different fracturing situations in a given rock sample during the complete strain–stress response

Source: From Aydan et al., 1993

Figure 6.14 View of dynamic shear testing machine OA-DSTM.

of various direct shear tests on rock discontinuities and soft rocks are explained. Rock discontinuities are planar (polished and saw-cut of marbles). Soft rocks are sandy Ryukyu limestone (Awa-ishi), coral stone and Oya tuff.

(a) Conventional direct shear testing

Figure 6.15 show the responses measured during the direct shear experiment on the shear response of coral stone (honeycomb-like coral limestone). Once peak load is exceeded, the deformation rate increases as noted in Figure 6.15.

An example of conventional direct shear test results on the polished planar surface of marble is shown in Figure 6.16(a). As seen from the figure, the yielding friction coefficient is about 0.27 (15 degrees), and hardening is observed. When the friction coefficient exceeds 0.52 (27.5 degrees), the relative shear displacement starts to increase. The ultimate friction angle coefficient is about 0.62 (31.8 degrees), while the residual friction coefficient is 0.58 (30.1 degrees).

Figure 6.16(b) shows the view of interface after direct shear testing. As noted from the figure, the contacts on the planar surfaces were quite small despite the fact that the surfaces were polished and planar. Furthermore, striations occurred on the surface parallel to sliding direction, which would be commonly observed on the fault surface and slickenslides

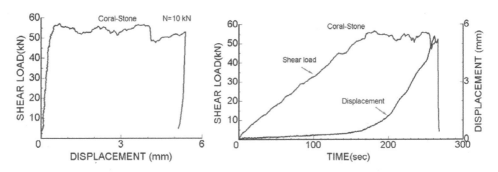

Figure 6.15 Shear displacement–shear load relation of coral stone

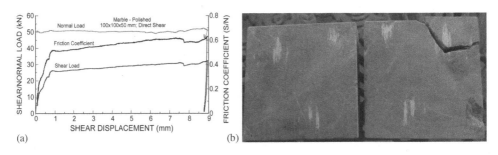

Figure 6.16 (a) Relative displacement, shear, normal and friction coefficient of polished interface between blocks, (b) views of the sheared surfaces

surfaces. In addition, this may also imply that it is practically very difficult to prepare exact planar surfaces that result in full contact of block surfaces.

Tilting tests are carried out on the original blocks of polished and saw-cut surfaces. The results are given in Table 6.1 together with those from obtained direct shear experiment under a normal stress of 5 MPa. From the comparison of the table, the friction angle of polished surfaces seems to be nonrepresentative frictional property of planar discontinuities. On the other hand, the friction angle saw-cut surfaces are closer to the intrinsic friction of planar discontinuities. Nevertheless, it must be noted that the traces of saws on the surfaces would cause the friction angle to be directional (Aydan *et al.*, 1996).

Direct shear test results are plotted together with the results from tilting experiments in Figure 6.17. In the same figure, some failure criteria for rock discontinuities (i.e. Barton and Choubey, 1977; Aydan, 2008; Aydan *et al.*, 1996) are also plotted. It is interesting to note that the shear strength of polished surface under high normal pressure is within the bounds obtained from the friction angle tests determined from tilting tests. The best fit to experimental results is obtained from the failure criterion of Aydan (Aydan, 2008; Aydan *et al.*, 1966). The failure criterion of Barton and Choubey (1977) is close to the upper-bound strength envelope.

Table 6.1 Friction angles of interfaces of marble blocks

Condition	Tilting Test	Direct Shear Test (NS: 5 MPa)			
		Initial	Flow	Peak	Residual
Polished	16–19	15	27	31.8	30.1
Saw-cut	28–35	23	28	31.0	29.0

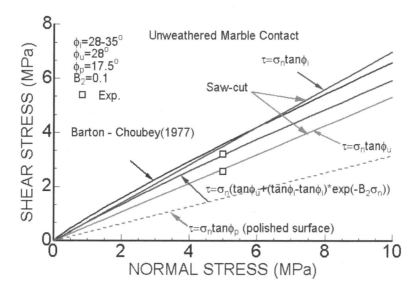

Figure 6.17 Plot of experimental results for polished marble contacts together with some failure criteria for rock discontinuities

(b)   Multistage direct shear testing

A multistage (multistep) direct shear test on a saw-cut surface of sandy Ryukyu limestone sample, which consists of two blocks with dimensions of $150 \times 75 \times 37.55$ mm, was varied out. The initial normal load was about 17 kN and increases to 30, 40, 50, 60 and 70 kN during the experiment. Figure 6.18 shows the shear displacement and shear load responses during the experiment. As noted from the figure, the relative slip occurs between blocks at a constant rate after each increase of normal and shear loads. This experiment is likely to yield shear strength of the interface two blocks under different normal stress levels. Figure 6.19 shows the peak and residual levels of shear stress for each level of normal

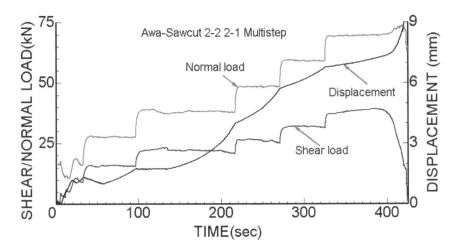

Figure 6.18 Shear stress–shear load response of the interface of sandy limestone blocks during the multistage(step) direct shear experiment.

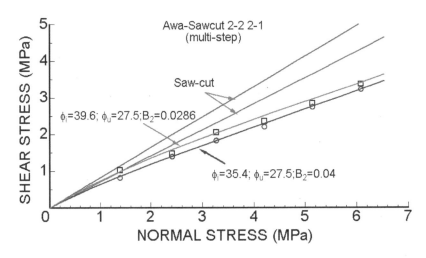

Figure 6.19 Comparison of shear strength envelope for the interface of sandy limestone blocks with experimental results from the multistage(-step) direct shear experiment.

stress increment. Tilting tests were carried out on the same interface, and the apparent friction angles ranged between 35.4 and 39.6 degrees. Tilting test results and direct shear tests are plotted in Figure 6.19 together with shear strength envelopes using the shear strength failure criterion of Aydan (Aydan, 2008; Aydan *et al.*, 1996).

As noted from the figure, the friction angles obtained from tilting tests are very close to the initial part of the shear strength envelopes. However, the friction angle becomes smaller as the normal stress level increases. In other words, the friction angle obtained from tilting tests on saw-cut surfaces cannot be equivalent to the basic friction angle of planar discontinuities and interfaces of rocks. The basic friction angle of the planar interface of sandy limestone blocks is obtained as 27.5 degrees for the range of given normal stress levels.

6.1.6 Tilting tests

Tilting test technique is one of the cheapest techniques to determine the frictional properties of rock discontinuities and interfaces under different environmental conditions (Barton and Choubey, 1977; Aydan *et al.*, 1995; Aydan, 1998). This technique can be used to determine the apparent friction angle of discontinuities (rough or planar) under low stress levels. It definitely gives the maximum apparent friction angle, which would be one of the most important parameters to determine the shear strength criteria of rock discontinuities as well as various contacts. Therefore, the data for determining the parameters of the shear strength criteria for rock discontinuities should utilize both tilting test and direct shear experiment.

(a) Theory of tilting tests

Let us assume that a block is put upon a base block with an inclination α as illustrated in Figure 6.20(a). The dynamic force equilibrium equations for the block can be easily written as follows:

For s-direction

$$W\sin \alpha - S = m\frac{d^2s}{dt^2} \qquad (6.10)$$

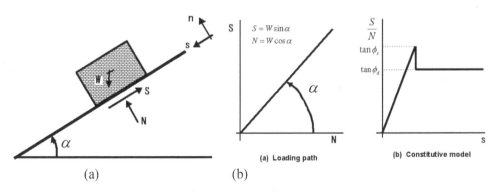

Figure 6.20 (a) Mechanical model for tilting experiments, (b) loading path in tilting experiments and constitutive relation

For *n*-direction

$$W \cos \alpha - N = m \frac{d^2 n}{dt^2} \tag{6.11}$$

Let us further assume that the following frictional laws holds at the initiation and during the motion of the block (Aydan and Ulusay, 2002), as illustrated in Figure 6.20(b):

At initiation of sliding

$$\frac{S}{N} = \tan \phi_s \tag{6.12}$$

During motion

$$\frac{S}{N} = \tan \phi_d \tag{6.13}$$

At the initiation of sliding, the inertia terms are zero so that the following relation is obtained:

$$\tan \alpha = \tan \phi_s \tag{6.14}$$

The preceding relation implies that the angle of inclination (rotation) at the initiation of sliding should correspond to the static friction angle of the discontinuity.

If the normal inertia term is negligible during the motion and the frictional resistance is reduced to dynamic friction instantaneously, one can easily obtain the following relations for the motion of the block:

$$\frac{d^2 s}{dt^2} = A \tag{6.15}$$

where $A = g(\sin \alpha - \cos \alpha \tan \phi_d)$.

The integration of differential Equation (6.15) will yield the following:

$$s = A \frac{t^2}{2} + c_1 t + c_2 \tag{6.16}$$

Since the following holds at the initiation of sliding:

$$s = 0 \quad \text{and} \quad v = 0 \quad \text{at} \quad t = T_s \tag{6.17}$$

Equation (6.16) takes the following form:

$$s = \frac{A}{2}(t - T_s)^2 \tag{6.18}$$

Coefficient A can be obtained either from a given displacement s_n at a given time t_n with the condition, that is:

$$t_n > T_s$$

$$A = 2 \frac{s_n}{(t_n - T_s)^2} \tag{6.19}$$

or from the application of the least squares technique to measured displacement response as follows:

$$A = 2 \frac{\sum_{i=1}^{n} s_i(t_i - T_s)^2}{\sum_{i=1}^{n} (t_i - T_s)^4} \tag{6.20}$$

Once constant A is determined, the dynamic friction angle is obtained from the following relation:

$$\phi_d = \tan^{-1}\left(\tan\alpha - \frac{1}{\cos\alpha}\frac{A}{g}\right) \tag{6.21}$$

(b)   Tilting device and setup

An experimental device consists of a tilting device operated manually. During experiments, the displacement of the block and rotation of the base are measured through laser displacement transducers produced by KEYENCE, while the acceleration responses parallel and perpendicular to the shear movement are measured by a three-component accelerometer (Tokyo Sokki) attached to the upper block and WE7000 (Yokogawa) data acquisition system. The measured displacement and accelerations are recorded onto laptop computers. The weight of the accelerometer is about 98 gf. Figure 6.21 shows the experimental setup.

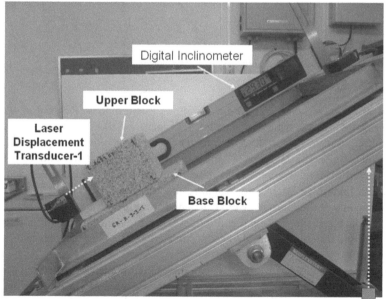

Figure 6.21 View of experimental setup for tilting device

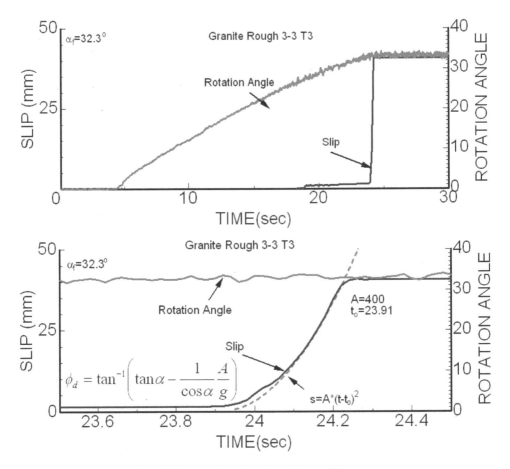

Figure 6.22 Responses of rough discontinuity of granite during a tilting test

A series of tilting tests are carried out on some discontinuities. Responses of some of these experiments are described as examples (Aydan, 2019). The measured responses during a tilting test on a rough discontinuity plane is shown in Figure 6.22 as an example. Figure 6.23 shows views of the tilting test on rough discontinuity plane of granite. As noted from the responses of rotation angle, relative displacement, and acceleration shown in Figure 6.23, fairly consistent results are observed. The static and dynamic friction coefficients of the interface were calculated from measured displacement response and weight of the upper block using the tilting testing equipment shown in Figure 6.22. The static and dynamic friction coefficients were estimated at 32.3–37.6 degrees and 30.3–35.6 degrees, respectively.

Similarly experimental results on saw-cut discontinuity planes of Ryukyu limestone samples are shown in Figure 6.24, which shows responses measured during a tilting experiment on a saw-cut plane of Ryukyu limestone. The static and dynamic friction coefficients of the interface were calculated from measured displacement responses explained in the previous section, and they were estimated at 28.8–29.6 degrees and 24.3–29.2 degrees, respectively.

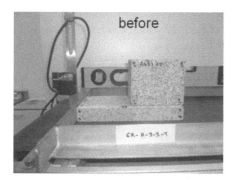
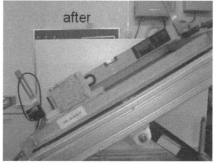

Figure 6.23 Views of tilting experiment on rough discontinuity plane of granite

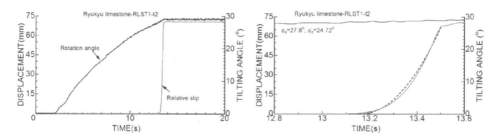

Figure 6.24 Responses of saw-cut discontinuity planes of Ryukyu limestone samples during a tilting test

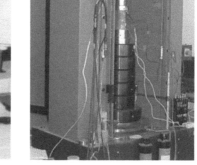

(a) Cantilever type creep testing apparatus   (b) Servo-controlled testing rig

Figure 6.25 Views of some creep testing apparatuses

6.1.7 Experimental techniques for creep tests

The methods for creep tests described herein are concerned with the creep characteristics of rocks under the indirect tensile stress regime of Brazilian test, uniaxial and triaxial compression tests and direct shear tests with the consideration of available creep testing techniques used in the rock mechanics field as well as other disciplines of engineering under laboratory conditions.

(a) Apparatuses

Apparatuses for creep tests can be of the cantilever type or load/displacement-controlled type (Fig. 6.26). Although the details of each testing machine may differ, the required features of apparatuses for creep tests are described herein.

Cantilever-type apparatus has been used in creep tests since early times. It is, practically, the most suitable apparatus for creep tests, as the load level can be easily kept constant in time-space (Fig. 6.26(a)). The severest restrictions of this type of apparatus are the level of applicable load, which depends upon the length of the cantilever arm and its oscillations

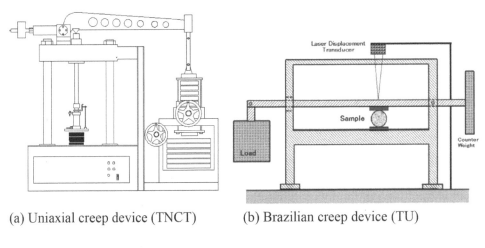

(a) Uniaxial creep device (TNCT)　　(b) Brazilian creep device (TU)

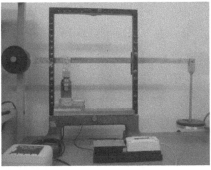

(c) Impression creep test device (TU)

Figure 6.26 Schematic illustration of cantilever type apparatuses for creep tests

during the application of the load. Cantilever-type apparatus utilizing a multiarm lever overcomes load limit restrictions, and up to 500 kN loads can be applied to samples. The oscillation problem is also technically dealt with.

Load is applied onto samples by attaching deadweights to the lever, which may be done manually for low-stress creep tests or mechanically for high-stress creep tests. In triaxial experiments, special load cells are required, and the confining pressure is generally provided through oil pressure. The utmost care must be taken for keeping the confining pressure constant with the consideration of continuous power supply for the compressor of the confining pressure system.

(b) Brazilian creep tests

The loading jigs and procedure used in the suggested method for Brazilian tests by ISRM should be followed unless the size of the samples differs from the conventional size (Fig. 6.26(b)). The displacement should be measured continuously or periodically, as suggested in the SM. The load application rate may be higher than that used in the SM. Once the load reaches the designated load level, it should be kept constant thereafter. If experiments are required to be carried out under a saturated condition, the jigs and sample should be put into a water-filled special cell.

(c) Uniaxial compression creep tests

Displacements are measured continuously or periodically as suggested in the SM. The load application rate may be higher than that used in the SM. Once the load reaches the designated load level, it should be kept constant thereafter. If experiments have to be carried out under saturated condition, the sample should be put into a water-filled special cell.

(d) Triaxial compression creep tests

The displacement is measured continuously or periodically as suggested in the SM. The load application rate may be higher than that used in the SM. Once the load reaches the designated load level, it should be kept constant thereafter. If experiments are required to be carried out under saturated condition, the sample should be put into a water-filled special cell.

(e) Impression creep test as an index test

The impression creep test technique utilizes the indenter, which is a cylinder with a flat end. The indenter makes a shallow impression on the surface of the specimen, and it is therefore named impression creep. There may be two different loading schemes during this experiment, namely, direct application of the deadweight (Fig. 6.27) or load by a cantilever frame (Fig. 6.26(c)). The potential use of this technique for the creep characteristics of rocks was explored by (Aydan *et al.*, 2016; Rassouli *et al.*, 2010). The critical issue with this technique is the definition of strain and stress, which can be associated with conventional creep experiments.

(f) Direct shear test device

The servo-control shear testing device shown in Figure 6.14 can be also used for creep tests on rock discontinuities.

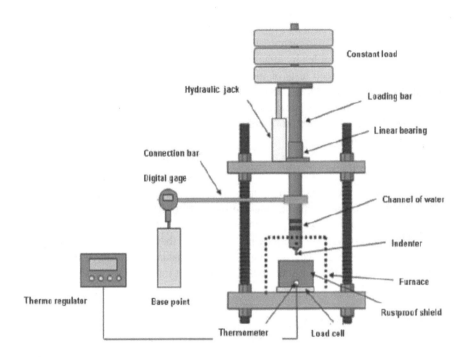

Figure 6.27 Impression creep apparatus utilizing deadweight

Source: Rassouli *et al.*, 2010

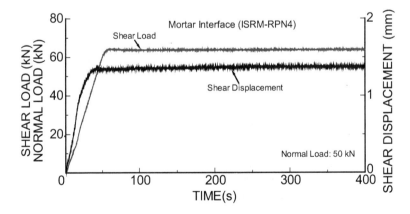

Figure 6.28 Direct shear creep test on a mortar interface with a surface roughness profile number 4

An example of direct creep test on mortar sample having a surface roughness profile number RPN4 of ISRM is shown in Figure 6.28. As noted from experimental responses, the testing device can keep shear load constant on the sample, which is the major problem when servo-control testing machines are used. Although the duration of the test is short, the creep behavior of the interface indicates an almost linear response.

6.2 *In-situ* mechanical tests

6.2.1 *Conventional mechanical tests*

It is generally expensive to carry out experiments on large rock samples *in-situ* if rock mass is considered to be the equivalent continuum. Such experiments were carried out during the construction of Kurobe Dam in Japan (Nose, 1962), and the *in-situ* testing techniques are illustrated in Figure 6.29. The diameter of plate-bearing tests generally ranges between 30 and 60 cm. Shear strength samples are generally about 60 cm long and 30–40 cm high. Four tests are carried out to determine the peak cohesion and friction angle of rock masses. In addition, tests are repeated on sheared samples to determine residual strength parameters. The common size of triaxial samples is about 100 cm. However, the largest size of the triaxial test at Kurobe dam was 280 cm. Historically, uniaxial compression tests on rock masses were probably first undertaken in South Africa (i.e. Bieniawski, 1974; Van Heerden, 1975) using coal pillars. However, the first triaxial compressive strength tests were undertaken at the Kurobe Dam site by Kansai Electric Power Company (Nose, 1962). However, triaxial compression tests are not carried out due to their high cost and the huge difference between strength values obtained from triaxial compression tests and *in-situ* shear strength tests as seen in the Kurobe Dam project. This problem was pointed out by Hibino (2007), who was actively involved in the large powerhouse and dam construction projects. In addition, the natural underground openings and steep cliffs associated with Ryukyu limestone present some stability problems to the superstructures on the ground surface. Figure 6.30 shows an *in-situ* plate-loading test and rock shear test at the construction site of Minami-Daitojima fishing port in Ryukyu archipelago.

6.2.2 *In-situ creep test*

Results of an *in-situ* creep test method are used to predict time-dependent deformation characteristics of rock mass resulting from loading. This test method may be useful in structural design analysis where loading is applied over an extensive period. This test method is normally performed at ambient temperature, but equipment can be modified or substituted for operations at other temperatures. There are applications of this test

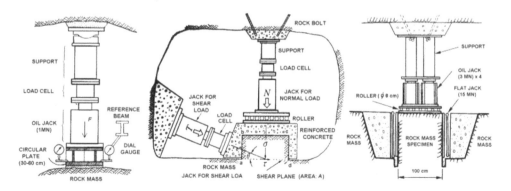

Figure 6.29 Illustration of *in-situ* testing techniques in Japan

Source: Aydan *et al.*, 2014, based on the original drawings by Hibino, 2007

Figure 6.30 Views of *in-situ* experiments on rock mass in Minami Daitojima fishing port, Japan: (a) plate loading test, (b) rock shear test

Source: Tokashiki and Aydan, 2012

technique in pillars of rock salt mines. *In-situ* creep tests are generally plate-bearing tests, with the direct shear creep test using the setups shown in Figure 6.29.

(a) Plate-bearing creep tests

The diameter of platens used in plate-bearing tests is 300 mm, and the maximum load is about 500 kN. The maximum nominal pressure is about 7.2 MPa. The deformation modulus is obtained from the following relation based on Boussinesq's solution:

$$E_0 = \frac{1-v^2}{D} \cdot \frac{F}{\delta_0} \tag{6.22}$$

where E, v, δ_0, D and F are deformation modulus, Poisson's ratio, instantaneous settlement, diameter of platen and applied load. Poisson's ratio is generally assumed to be 0.2 or 0.25. The total displacement is the sum of initial displacement and delayed creep displacement, given as:

$$\delta_t = \delta_0 + \delta_c \tag{6.23}$$

The creep displacement is given as a fraction of the total displacement using a five-element generalized Voigt-Kelvin model (Fig. 6.31):

$$\delta_t = \delta_0 \left(1 + \frac{E_1}{E_0} \left(1 - \exp(-\frac{E_1}{\eta_1}t) \right) + \frac{E_2}{E_0} \left(1 - \exp(-\frac{E_2}{\eta_2}t) \right) \right) \tag{6.24}$$

Thus, the creep displacement would be given as:

$$\delta_c = \frac{1-v^2}{D} \cdot \frac{F}{E_0} \left[\frac{E_1}{E_0} \left(1 - \exp(-\frac{E_1}{\eta_1}t) \right) + \frac{E_2}{E_0} \left(1 - \exp(-\frac{E_2}{\eta_2}t) \right) \right] \tag{6.25}$$

Figure 6.32 shows examples of plate-bearing creep tests on a rhyolite foundation of a dam site in Central Japan. Rock mass classified as CL and CM in the rock mass classification

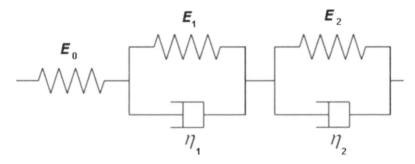

Figure 6.31 Five-element generalized Voigt-Kelvin model

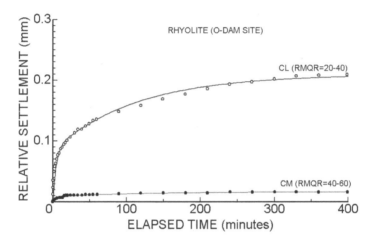

Figure 6.32 Creep displacement of rhyolite rock mass at a dam site in Central Japan

Table 6.2 Values of parameters for creep responses measured in plate-bearing test

Rock Class		δ_c/δ_e	E_0 (MPa)	E_1/E_0	E_2/E_0	E_1/η_1 (1 min^{-1})	E_2/η_2 (1 min^{-1})
DENKEN	RMQR						
CL	20–40	0.08	534	0.4	0.6	0.238	0.009
CM	40–60	0.03	3332	0.5173	0.4828	0.300	0.008

system (DENKEN) of the Central Research Institute of Electric Power Companies of Japan. The values of parameters for creep responses shown in Figure 6.32 are listed in Table 6.2.

(b)   Direct rock shear creep tests

In-situ direct shear rock test setup shown in Figure 6.29(b) is utilized for creep tests. The shearing area is 600×600 mm, and the height of the sample is 300 mm. Direct shear creep experiments are done in two stages. The first stage is called primary creep stage, and the specimen is loaded at a level of one-third of the ultimate peak shear strength at a given normal load. The duration of the primary creep stage is generally more than 90 minutes. The second stage is called the secondary creep stage, and the specimen is loaded at a level of two-thirds of the ultimate peak shear strength at a given normal load. The duration of the primary creep stage is generally more than 120 minutes. However, the duration of the creep tests may be several days to months depending on the importance of the structure. A generalized Voigt-Kelvin model having three elements is generally used as a model for the primary and secondary creep stages:

$$w_t = w_e + w_c = w_e(1 + \alpha(1 - \exp(-\beta t))); \alpha = \frac{K_e}{K_v}; \beta = \frac{K_v}{\eta_v} \tag{6.26}$$

where τ, K_e, K_v and η_v are applied shear stress, Hookean stiffness, Kelvinean stiffness and Kelvinean viscosity.

Figure 6.33 shows examples of responses during direct shear creep tests on a rhyolite foundation of a dam site in Central Japan. Rock mass is classified as CM in the rock mass classification system (DENKEN) of the Central Research Institute of Electric Power Companies of Japan. The values of parameters for creep responses shown in Figure 6.33 are listed in Table 6.3.

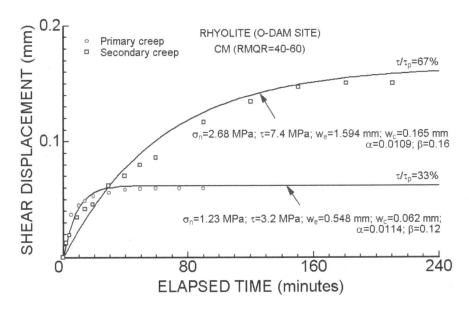

Figure 6.33 Direct rock shear creep displacement of rhyolite rock mass at a dam site in Central Japan

Table 6.3 Values of parameters for creep responses measured in direct shear test

Stages	τ (MPa)	αw_e(mm)	K_e MPa mm^{-1}	w_e(mm)	$\alpha = \frac{K_e}{K_v}$	$\beta = \frac{K_v}{\eta_v}(1\ min^{-1})$
Primary creep	3.2	0.062	5.839	0.548	0.114	0.12
Secondary Creep	7.4	0.165	4.888	1.594	0.109	0.16

6.3 Thermal properties of rocks and their measurements

Thermal properties such as specific heat, heating or cooling coefficient and thermal conductivity are important to assess the heat transport through solids as noted from Equation (4.1). There are many techniques to measure thermal properties such as specific heat coefficient, thermal conductivity, thermal diffusion and thermal expansion coefficient. The details of such techniques can be found in various publications and textbooks (i.e. Clark, 1966; Somerton, 1992). Specific heat coefficient is commonly measured using the calorimeter tests.

The earlier and common technique for thermal conductivity measurement is the divided bar technique (Birch, 1950) based on the steady-state heat flow assumption, and it is illustrated in Figure 6.34. This technique utilizes reference materials with well-known thermal properties.

There are also techniques for measuring thermal conductivity utilizing transient heat flow (i.e. Carslaw and Jaeger, 1959; Popov *et al.*, 1999, 2016; Sass *et al.*, 1984). These techniques utilize line or plane sources, and temperature variations are measured by either contact sensor or infrared camera. These techniques utilize the analytical solutions developed by Carslaw and Jaeger (1959)

An experimental technique using a device similar to a calorimeter-type apparatus is described in this section in order to measure thermal properties of rock materials from a single experiment, and its applications are given. Let us consider a solid (or group of solids) is enveloped by fluid (i.e. water (w)) as illustrated in Figure 6.35. It is assumed that solid and fluid have different thermal properties and temperature.

For theoretical modeling, the following parameters are defined as follows: Q: heat; ρ: density; k: thermal conductivity; c: specific heat coefficient; T: temperature; m: mass;

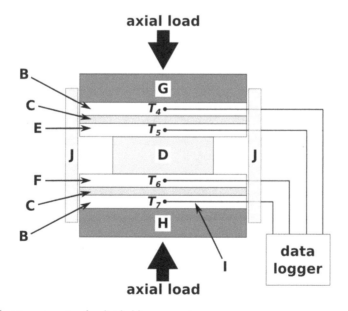

Figure 6.34 Key components of a divided-bar apparatus

Source: Popov *et al.*, 2016A – pivot point, B – brass disks, C – reference material, D – rock specimen, E – hot plate, F – cold plate, G – heat source (concealed Peltier device), H – heat sink, I – holes for the insertion of temperature sensors, J – thermal insulation

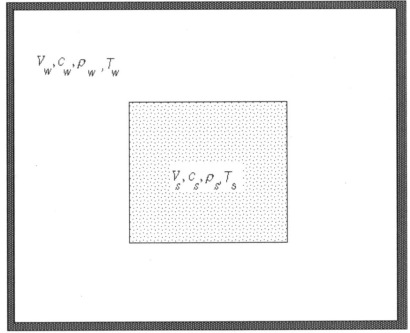

a) Physical Model

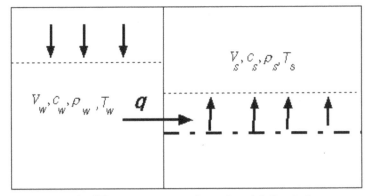

b) Thermo-mechanical Model

Figure 6.35 Illustration of physical and thermomechanical model

h:cooling coefficient; V: volume; A_s: surface area of solid; λ: thermal expansion coefficient. The heat of a body is given in the following form:

$$Q = m \cdot c \cdot T = \rho V \cdot c \cdot T \tag{6.27}$$

Its unit is the joule ($J = N \cdot m$).

Assuming that mass and specific heat coefficient are constant, the heat rate (heat flux) is given in the following form:

$$\frac{dQ}{dt} = q = m \cdot c \cdot \frac{dT}{dt} \tag{6.28}$$

The Newton cooling law is written in the following form:

$$q = h \cdot A_s \cdot \Delta T \tag{6.29}$$

where ΔT is the temperature difference between solid and enveloping fluid, and its unit is the watt $(W = J/s)$.

In this particular model, the temperature of the surrounding fluid is assumed to be higher than the solid enveloped by the fluid. Furthermore, there is no heat flow from the system outward. In other words, it is thermally isolated. The heat flux from fluid can be given as:

$$q_w = -\rho_w \cdot c_w \cdot V_w \frac{dT_w}{dt} \tag{6.30}$$

The heat from fluid into solid should be equal to the use of the Newton cooling law:

$$-\rho_w \cdot c_w V_w \frac{\partial T_w}{\partial t} = h \cdot A_s (T_w - T_s) \tag{6.31}$$

Similarly, the heat change of the solid should be equal to that supplied from fluid as given by:

$$\rho_s \cdot c_s \cdot V_s \frac{\partial T_s}{\partial t} = h \cdot A_s (T_w - T_s) \tag{6.32}$$

It should be noted that the sign of heat flux is a plus sign (+). Rewriting Equation (6.31) yields the following:

$$T_s = T_w + \frac{\rho_w c_w v_w}{h \cdot A_S} \cdot \frac{\partial T_w}{\partial t} \tag{6.33}$$

If the derivation of Equation (6.32) with respect to time is inserted into Equation (6.30), one easily gets the following:

$$\frac{\partial^2 T_w}{\partial t^2} + \alpha \frac{\partial T_w}{\partial t} = 0 \tag{6.34}$$

where

$$\alpha = h \cdot A_s \frac{\rho_w \cdot c_w \cdot V_w + \rho_s \cdot c_s \cdot V_s}{\rho_w \cdot c_w \cdot V_w \cdot \rho_s \cdot c_s \cdot V_s} \tag{6.35}$$

The solution of Equation (6.34) is obtained as follows:

$$T_w = C_1 + C_2 e^{-\alpha t} \tag{6.36}$$

Integral coefficients C_1 and C_2 of Equation (6.36) are obtained from the following conditions:

$$T_w = T_i \quad \text{at} \quad t = 0 \quad \text{and} \quad T_1 = T_f \quad \text{at} \quad t = \infty \tag{6.37}$$

as

$$C_1 = T_f, C_2 = T_i - T_f \tag{6.38}$$

Using integral constants given by Equation (6.38), Equation (6.32) becomes:

$$T_w = T_f + (T_i - T_f)e^{-\alpha t}, \frac{\partial T_w}{\partial t} = -\alpha(T_i - T_f)e^{-\alpha t} \tag{6.39}$$

The average temperature of solid is obtained by inserting Equation (6.39) into Equation (6.33) as:

$$T_s = T_f - \frac{\rho_w \cdot c_w \cdot V_w}{\rho_s \cdot C_s \cdot V_s}(T_i - T_f)e^{-\alpha t} \tag{6.40}$$

As $T_s = T_o$ at time $t = 0$, Equation (6.40) can be rewritten as:

$$\frac{T_f - T_o}{T_i - T_f} = \frac{\rho_w \cdot c_w \cdot V_w}{\rho_s \cdot c_s \cdot V_s} \tag{6.41}$$

Inserting Equation (6.41) into Equation (6.40) yields:

$$T_s = T_f - (T_f - T_o)e^{-\alpha t} \tag{6.42}$$

The temperature difference between the solid and enveloping fluid can be obtained from Equations (6.40) and (6.42) as:

$$\Delta T_{ws} = (T_w - T_s) = (T_i - T_o)e^{-\alpha t} \tag{6.43}$$

Therefore, if the values of T_o, T_f, T_i, ρ_w, ρ_s, V_w, V_s, c_w are known, the specific heat coefficient of solid can be easily obtained. For example, the specific heat coefficient of water is 4.1783–4.2174 J g/K^{-1} for a temperature range of 0–90°C. As the thermal properties of water remain almost constant for the given temperature range, the water would be used as fluid in the experimental setup. After obtaining the specific heat coefficient, the coefficient α is obtained from Equation (4.34), (4.36) or (4.37) using the curve-fitting technique to experimental response. Then using the value of coefficient α, the value of cooling coefficient is obtained from Equation (4.30).

For determining the thermal conductivity coefficient(k), the following approach is used. Fourier law may be written for a one-dimensional situation as:

$$q = -kA\frac{\partial T}{\partial x} \tag{6.44}$$

Assuming the specimen has a length (L) and using the Newton's cooling law, we may write the following relationship:

$$hA\Delta T = kA\frac{\Delta T}{L} \tag{6.45}$$

Equation (6.44) can be rewritten, and the following relation holds between cooling coefficient and thermal conductivity:

$$k = h \cdot L \tag{6.46}$$

The characteristic length of a solid sample can be obtained from the volume of the solid from the following relationship:

$$L = \sqrt[3]{V_s} \tag{6.47}$$

Linear thermal expansion coefficient (λ) is defined as:

$$\lambda = \frac{1}{L} \cdot \frac{dL}{dT} \tag{6.48}$$

where L is the length of sample. $\frac{dL}{dT}$ is the variation of length of sample with respect to temperature variation, and it is determined under the unstrained condition or 100 gf load on the sample. If the variation of length of sample at the equilibrium state with respect to the initial length before the commencement of the experiment is measured, it is straightforward to obtain the linear expansion coefficient. Similarly, width or diametrical changes can be also measured, and thermal expansion coefficients can be evaluated from the variation of side length or diameter for a given temperature difference.

The technique described in this subsection is unique and quite practical considering the labor required in other techniques. The device for determining the thermal properties of geo-materials consists of a thermostat cell equipped with temperature sensors. The fundamental features of this device are illustrated in Figure 6.36. In the experiments, the temperature of sample, water, air and thermostat are measured. The method utilizes the thermal properties of water, whose properties remain to be the same up to 90°C, to infer the thermal properties of geo-material substances. If the continuous measurements of temperatures are available, one can easily infer the thermal properties from the following equations as follows.

Specific heat of geo-material

$$c_s = \frac{\rho_w \cdot c_w \cdot V_w}{\rho_s \cdot V_s} \frac{T_i - T_f}{T_f - T_o} \tag{6.49}$$

where ρ_w is the density of water, c_w is the specific heat coefficient of water, V_w is the volume of water, ρ_s is the density of sample, c_s is the specific heat coefficient of sample, V_s is the volume of sample, T_i is the initial temperature of water, T_o is the initial temperature of sample, and T_f is the equilibrium temperature.

The heat conduction coefficient (α) is obtained from fitting experimental results to the following equation:

$$\Delta T_{ws} = (T_w - T_s) = (T_i - T_o)e^{-\alpha t} \tag{6.50}$$

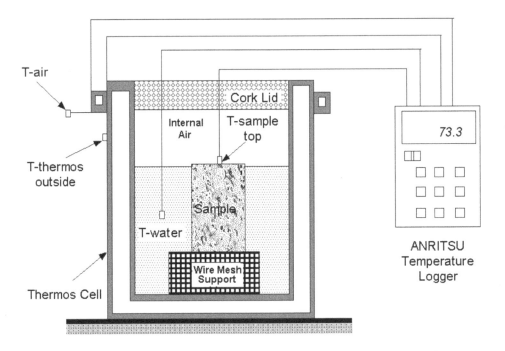

Figure 6.36 Illustration of experimental setup

If the heat conduction coefficient (α) is determined, then Newton's cooling coefficient is determined from the following equation:

$$h = \frac{\alpha}{A_s} \cdot \frac{\rho_w \cdot c_w \cdot V_w \cdot \rho_s \cdot c_s \cdot V_s}{\rho_w \cdot c_w \cdot V_w + \rho_s \cdot c_s \cdot V_s} \tag{6.51}$$

Finally, thermal conductivity coefficient is obtained from the following equation:

$$k = h \cdot L \tag{6.52}$$

where L is the characteristics sample side length.

If the sample temperature can be measured, it will be very easy to determine the specific heat coefficient of the sample and subsequent properties. This is possible for granular materials since the temperature sensor can be embedded in the center of the sample. However, it is quite difficult to determine the equilibrium temperature T_f for solid samples. Therefore, the following procedure is followed for this purpose:

- Step 1: Determine the heat conduction coefficient (α).
- Step 2: Plot the following equation in time space:

$$T_s = T_w - (T_i - T_o)e^{-\alpha t} \tag{6.53}$$

- Step 3: Determine the peak value from Equation (6.52), and assign it as equilibrium temperature T_f.
- Step 4: Then proceed to determine the rest of thermal properties using the procedure previously described.

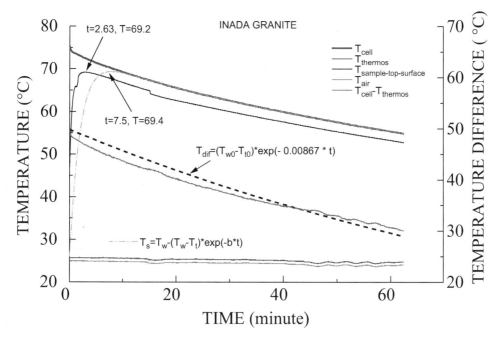

Figure 6.37 Application of the procedure to Inada granite sample

Figure 6.37 shows the application of the method to a cylindrical Inada granite sample. The temperature of the sample at the top was also measured. As noted in the figure, the temperature at the top of the sample achieves the peak value before the computed response. Since the temperature of the sample is averaged over the total volume of the sample at the proposed temperature, the computed sample temperature achieves its peak value later then that at the top surface of the sample.

Thermal conductivity could be measured using the TK04 system described by Blum (1997). This system employs a single-needle probe (Von Herzen and Maxwell, 1959), heated continuously, in a half-space configuration for hard rock. The needle probe is a thin metal tube that contains a thermistor and a heater wire. The needle is assumed to be approximately an infinitely long, continuous medium; the temperature near the line source is measured as a function of time. If it is assumed that the sediment or rock sample to be measured can be represented as a solid in a fluid medium, it is then possible to determine a relationship between thermal diffusivity and thermal conductivity. With this assumption, the change in temperature of the probe as a function of time is given to a good approximation by (Von Herzen and Maxwell, 1959):

$$T(t) = \frac{q}{4k} \ln\left(4 \frac{\alpha_t}{Ba^2}\right) \tag{6.54}$$

where T = temperature (°C), q = heat input per unit time per unit length (W m^{-1}), k = thermal conductivity of the sediment or rock sample (W m·°C^{-1}), t = time after the initiation of the heat (s), α= thermal diffusivity of the sample (m^2 s^{-1}), B = a constant (1.7811), and a = the probe radius (m).

This relationship is valid when t is large compared with a^2/α. A plot of T *versus* $\ln(t)$ yields a straight line, the slope of which determines k (Von Herzen and Maxwell, 1959).

6.4   Tests for seepage parameters

6.4.1   Falling head tests

When rock is quite permeable, falling head tests, which utilize the deadweight of fluid, are also used for determining the permeability of rocks and discontinuities. In this subsection, analytical solutions for falling head tests for longitudinal flow and radial flow conditions are derived.

(a)   Longitudinal falling head test method

Experimental setup used for this kind of test is shown in Figure 6.38 (Aydan *et al.*, 1997). As seen from the figure, two manometers having cross sections a are assumed to be attached to both ends of the sample. During a test, the change of pressure and velocity of flow can be measured through these manometers. The level h_2 of water at the lower tank is assumed to be constant in the following formulation. When an experiment starts, flow rate inside the pipe can be given as:

$$v_p = -a \frac{\partial h_1}{\partial t} \tag{6.55}$$

where h_1 is the level of water inside the manometer (1). At a given time, flow rate through the cross-section area A of the specimen is given by:

$$v_t = \bar{v}A \tag{6.56}$$

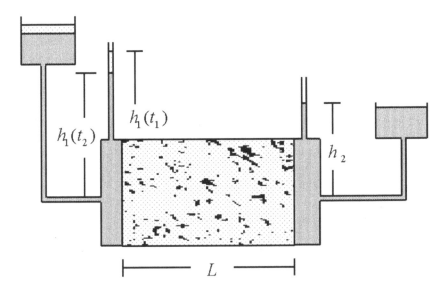

Figure 6.38 Illustration of longitudinal falling head test

It is assumed that flow rate through the specimen should be equal to the flow rate of the pipe. Then the pressure gradient in the specimen can be given in the following form:

$$\frac{\partial p}{\partial x} \approx -\rho g \frac{(h_1 - h_2)}{L} \tag{6.57}$$

where ρ is density, and g is gravitational acceleration. Substituting Equation (6.56) into Equation (6.55) and equalizing the resulting equation to Equation (6.54) yields the following differential equation for the change of water height h_1:

$$\frac{\partial h_1}{h_1 - h_2} = -\frac{kA\rho g}{La\eta} \partial t \tag{6.58}$$

where L is sample length. Solution of the preceding differential equation is:

$$h_1 = h_2 + Ce^{-\alpha t} \tag{6.59}$$

where

$$\alpha = \frac{kA}{La} \frac{\rho g}{\eta}$$

If initial conditions are given by

$$h_1 = h_{10} \quad \text{at} \quad t = 0$$

where h_{10} is water height at manometer 1 at $t = 0$. Thus the integration coefficient C is obtained as follows:

$$C = h_{10} - h_2 \tag{6.60}$$

Inserting the preceding integration coefficient in Equation (6.58) yields the following:

$$-\alpha t = \ln(\frac{\Delta h}{\Delta h_o}) \tag{6.61}$$

where $\Delta h = h_1 - h_2, \Delta h_o = h_{10} - h_2$.

If α is substituted in the preceding equation, the following expression for permeability is obtained:

$$k = \frac{La}{A} \frac{\ln(\frac{h_{10} - h_2}{h_1 - h_2})}{t} \frac{\eta}{\rho g} \tag{6.62}$$

(b) Radial falling head test method

Experimental setup used for this of kind test is shown in Figure 6.39 (Aydan *et al.*, 1997). As seen from the figure, a manometer is placed on the top of the cylindrical hole drilled in the middle of test specimen. The cross-section area of this manometer is denoted by A_h. During the test, the change of pressure and velocity of flow can be measured with this manometer. The level h_2 of water at the outer container is assumed to be constant. When experiment starts, flow rate inside the manometer can be given as:

$$q = -\rho g A_h \frac{\partial h_1}{\partial t} \tag{6.63}$$

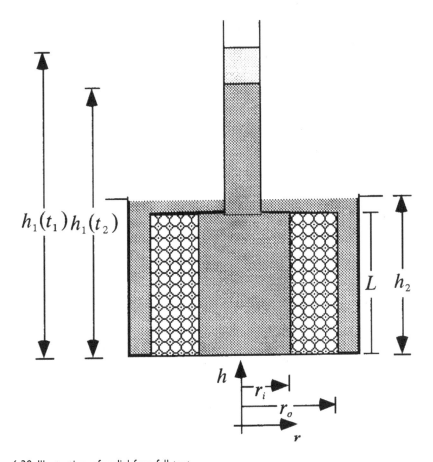

Figure 6.39 Illustration of radial free-fall test

where h_1 is the level of water inside the manometer. At a given time, the flow rate through a cross-section area of hole (A_p) inside the test specimen is given by:

$$v_t = \bar{v}A_p \tag{6.64}$$

It is assumed that flow rate through the hole perimetry should be equal to the flow rate of the pipe. The pressure gradient in the specimen may be given in the following form:

$$\frac{\partial p}{\partial r} \approx -\frac{\partial}{\partial r}(\rho g(h_1 - h_2)) = -\rho g\frac{\partial(h_1 - h_2)}{\partial r} = -\rho g\frac{(h_1 - h_2)}{r\ln(r_o/r_i)} \tag{6.65}$$

Substituting Equation (6.65) into Equation (6.64) and equalizing the resulting equation to Equation (6.63) yields the following differential equation for the change of water height h_1:

$$\frac{\partial h_1}{h_1 - h_2} = -\frac{kA_p}{\eta A_h}\frac{1}{r_i\ln(r_o/r_i)}\partial t \tag{6.66}$$

Solution of the preceding differential equation is:

$$h_1 = h_2 + Ce^{-\alpha t} \tag{6.67}$$

where

$$\alpha = \frac{k}{r_i \ln(r_o/r_i)} \frac{A_p}{A_h} \frac{1}{\eta}$$

Introducing the following initial conditions:

$$h_1 = h_{10} at \quad t = 0$$

yields the integration constant C as:

$$C = h_{10} - h_2 \tag{6.68}$$

If integration constant is inserted into Equation (6.67), the following equation is obtained:

$$-\alpha t = \ln\left(\frac{h_1 - h_2}{h_{10} - h_2}\right) \tag{6.69}$$

If α is substituted into the preceding equation, the following expression for permeability is obtained:

$$k = \eta r_i \ln(r_o/r_i) \frac{A_h}{A_p} \ln\left(\frac{h_{10} - h_2}{h_1 - h_2}\right) \frac{1}{t} \tag{6.70}$$

(c) Transient pulse test method

(I) LONGITUDINAL FLOW TESTS

Brace *et al.* (1968) proposed a transient pulse method for longitudinal flow tests. In this method, the following assumptions are made (Aydan, 1998):

- Fluid flow obeys Darcy's law.
- The change of fluid density inside pores with respect time is negligible.
- The volume of reservoirs (V_1, V_2) is constant.
- The relation between pressure and volumetric strain of fluid is linear.

Permeability is obtained from pressure changes, which are applied to the ends of a specimen, with respect to time (Fig. 6.40). During experiments, flow rate is not measured. The volumetric strain of fluid inside reservoirs V_1 ve V_2 can be written as follows:

$$\varepsilon_V^1 \approx \frac{\Delta V_1}{V_1}, \quad \varepsilon_V^2 \approx \frac{\Delta V_2}{V_2} \tag{6.71}$$

Similarly, for volumetric strain rate of fluid, the following relations can also be written as:

$$\dot{\varepsilon}_V^1 \approx \frac{\Delta \dot{V}_1}{V_1}, \quad \dot{\varepsilon}_V^2 \approx \frac{\Delta \dot{V}_2}{V_2} \tag{6.72}$$

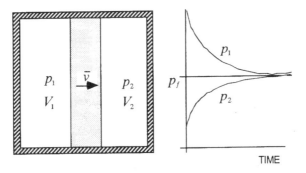

Figure 6.40 Illustration of transient pulse-method for longitudinal flow

or

$$\Delta \dot{V}_1 = \dot{\varepsilon}_V^1 V_1, \quad \Delta \dot{V}_2 = \dot{\varepsilon}_V^2 V_2 \tag{6.73}$$

If the following relations exist between the volumetric strain of fluid and pressure in reservoirs:

$$\varepsilon_V^1 = -c_f p_1, \quad \varepsilon_V^2 = -c_f p_2 \tag{6.74}$$

and, the compressibility coefficient (c_f) is constant, for volumetric strain rate, the following relations can be also written:

$$\dot{\varepsilon}_V^1 = -c_f \dot{p}_1, \quad \dot{\varepsilon}_V^2 = -c_f \dot{p}_2 \tag{6.75}$$

Flow rates may be defined as:

$$v_{t_1} = \Delta \dot{V}_1, \quad v_{t_2} = \Delta \dot{V}_2 \tag{6.76}$$

Using Equations (6.72), (6.75) and (6.76), flow rates can be rewritten in the following form:

$$v_{t_1} = -c_f V_1 \frac{\partial p_1}{\partial t}, \quad v_{t_2} = -c_f V_2 \frac{\partial p_2}{\partial t} \tag{6.77}$$

Introducing the following boundary conditions:

$$p = p_1 \quad at \quad x = 0, \quad p = p_2 \quad at \quad x = L$$

and using Darcy law, the following relations can be obtained for flow rates:

$$v_{t_1} = -\frac{kA}{\eta} \left(\frac{dp_1}{dx} \right)_{x=0}, \quad v_{t_2} = -\frac{kA}{\eta} \left(\frac{dp_2}{dx} \right)_{x=L} \tag{6.78}$$

where A is the cross-section area of the sample, and L is the length of the sample. Pressure gradients in the preceding equations are as follows:

$$\frac{dp_1}{dx} \approx -\frac{(p_1 - p_2)}{L}, \quad \frac{dp_2}{dx} \approx -\frac{(p_2 - p_1)}{L}$$

Inserting the preceding equation into Equation (6.78) and equating the resulting equation to Equation (6.77) yields the following set of equations:

$$\frac{\partial p_1}{\partial t} = -\beta \frac{1}{V_1}(p_1 - p_2)$$ (6.79)

$$\frac{\partial p_2}{\partial t} = \beta \frac{1}{V_2}(p_1 - p_2)$$ (6.80)

where

$$\beta = \frac{kA}{c_f \eta L}$$

Brace *et al.* (1968) solved similar equations by using the Laplace transformation technique. Herein, the method of elimination will be used for solving the preceding set of equations (Kreyszig, 1983). Equation (6.79) can be rearranged as follows:

$$p_2 = p_1 + \frac{V_1}{\beta}\frac{\partial p_1}{\partial t}$$ (6.81)

Taking the time derivative of these equations, the following expression is obtained:

$$\frac{\partial p_2}{\partial t} = \frac{\partial p_1}{\partial t} + \frac{V_1}{\beta}\frac{\partial^2 p_1}{\partial t^2}$$ (6.82)

Substituting Equations (6.81) and (6.82) into Equation (6.80) and rearranging the resulting equation yields the following homogeneous differential equation:

$$\frac{\partial^2 p_1}{\partial t^2} + \alpha \frac{\partial p_1}{\partial t} = 0$$ (6.83)

where

$$\alpha = \beta \frac{V_1 + V_2}{V_1 V_2}$$

The general solution of this differential equation is:

$$p_1 = C_1 + C_2 e^{-\alpha t}$$ (6.84)

Introducing the following initial conditions:

$$p_1 = p_i \quad at \quad t = 0, \qquad p_1 = p_f \quad at \quad t = \infty$$

where p_i is the applied initial pressure at Reservoir 1 (V_1), p_f is final pressure at the end of the test, yielding the integration constants C_1 and C_2 as:

$$C_1 = p_f, \quad C_2 = p_i - p_f$$ (6.85)

Inserting these integration constants into Equation (6.84 (3.51)) gives the following equation:

$$p_1 = p_f + (p_i - p_f)e^{-\alpha t}$$ (6.86)

Taking the time derivative of the preceding equation:

$$\frac{\partial p_1}{\partial t} = -(p_i - p_f)\beta \frac{V_1 + V_2}{V_1 V_2} e^{-\alpha t} \tag{6.87}$$

Substituting Equations (6.86) and (6.87) into Equation (6.79) and rearranging yields the following equation:

$$p_2 = p_f - (p_i - p_f)\frac{V_1}{V_2} e^{-\alpha t} \tag{6.88}$$

For the following initial condition for P_2:

$$p_2 = p_0 \quad at \quad t = 0$$

Equation (6.87) takes the following form:

$$(p_i - p_f) = (p_f - p_0)\frac{V_2}{V_1} \tag{6.89}$$

The preceding equation can be rewritten in a different way for $p_i - p_0$ as follows:

$$(p_i - p_f) = (p_i - p_0)\frac{V_2}{V_1 + V_2} \tag{6.90}$$

Inserting this equation into Equation (6.85) and rearranging yields the following:

$$-\alpha t = \ln(\frac{p_i - p_f}{p_i - p_0}\frac{V_1 + V_2}{V_2}) \tag{6.91}$$

where

$$\alpha = \frac{kA}{c_f L\eta}\frac{V_1 + V_2}{V_1 V_2}$$

From the preceding equations, one gets the following equation to compute permeability:

$$k = \frac{\eta c_f L}{A}\frac{V_1 V_2}{V_1 + V_2}\ln(\frac{\Delta p_o}{\Delta p}\frac{V_2}{V_1 + V_2})\frac{1}{t} \tag{6.92}$$

where $\Delta p = p_1 - p_f, \Delta p_o = p_i - p_o$. When gas is used as a permeation fluid, p_1 and p_2 are replaced with $U_1 \, (= p_1^2)$ and $U_2(= p_2^2)$, and permeability can be calculated using the same relation previously given.

If the volume of Reservoir 2 (V_2) is much greater than the volume of Reservoir 1 (V_1), ($V_2?V_1$) (for instance, outer side of specimen is open to air) p_0 ve p_f given in the preceding equation will be equal to atmospheric pressure (p_a). For this particular case, Equation (6.91) takes the following form:

$$k = \frac{\eta c_f L V_1}{A}\ln(\frac{p_i - p_a}{p_1 - p_a})\frac{1}{t} \tag{6.93}$$

For different values of α, the relations between the normalized pressure change and time and natural logarithm of the normalized pressure change and time for transient pulse tests were computed and are shown in Figure 6.41. As seen in Figure 6.41, there is a linear

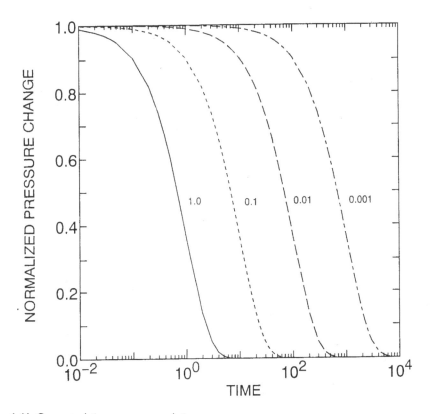

Figure 6.41 Computed time–pressure relations

relation between the natural logarithm of normalized pressure change and time. In both figures, time is taken as a unitless parameter. However, unit depends on the description of the problem. For instance, it can be year, day or second.

Theoretical and experimental curves for a transient pulse test on a rock salt (halite) specimen are shown in Figure 6.42. As seen from this figure, there is a slight difference between the curve obtained from the test and the curve from the theory, particularly in the initial stages. In order to obtain permeability value, the linear part of normalized pressure change and time relation is generally used to compute permeability for the selected range.

(2) RADIAL TRANSIENT PULSE METHOD

The transient pulse method is also extended to radial flow by Aydan *et al.* (1997) This method is fundamentally very similar to that for longitudinal flow (Fig. 6.43). The only differences are associated with the pressure gradient and surface area at the inner and outer radii.

Volumetric strain of fluid inside reservoirs V_1 ve V_2 can be written as follows:

$$\varepsilon_V^1 \approx \frac{\Delta V_1}{V_1}, \quad \varepsilon_V^2 \approx \frac{\Delta V_2}{V_2} \tag{6.94}$$

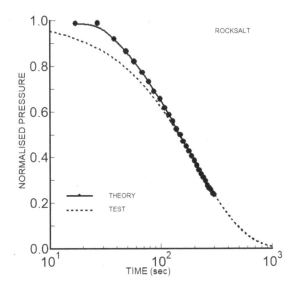

Figure 6.42 An experimental result on rock salt
Source: From Aydan and Üçpirti, 1997

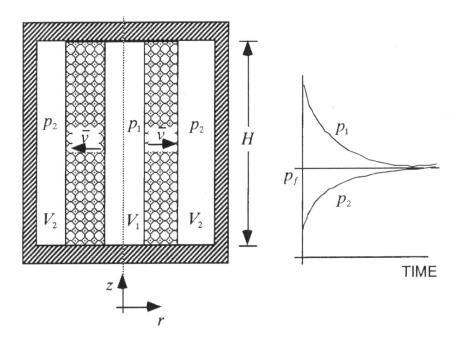

Figure 6.43 Illustration of the radial transient pulse test

Similarly, for volumetric strain rate of fluid, the following relations can also be written as:

$$\dot{\varepsilon}_V^1 \approx \frac{\Delta \dot{V}_1}{V_1}, \quad \dot{\varepsilon}_V^2 \approx \frac{\Delta \dot{V}_2}{V_2} \tag{6.95}$$

or

$$\Delta \dot{V}_1 = \dot{\varepsilon}_V^1 V_1, \quad \Delta \dot{V}_2 = \dot{\varepsilon}_V^2 V_2 \tag{6.96}$$

If the following relation exists between volumetric strain of fluid and pressure:

$$\varepsilon_V^1 = -c_f p_1, \quad \varepsilon_V^2 = -c_f p_2 \tag{6.97}$$

and, compressibility coefficient (c_f) is constant, for volumetric strain rate, the following relation can be also written:

$$\dot{\varepsilon}_V^1 = -c_f \dot{p}_1, \quad \dot{\varepsilon}_V^2 = -c_f \dot{p}_2 \tag{6.98}$$

Flow rate may be given as:

$$v_{t_1} = \Delta \dot{V}_1, \quad v_{t_2} = \Delta \dot{V}_2 \tag{6.99}$$

Using Equations (6.96), (6.98) and (6.99), flow rate can be rewritten in the following form:

$$v_{t_1} = -c_f V_1 \frac{\partial p_1}{\partial t}, \quad v_{t_2} = -c_f V_2 \frac{\partial p_2}{\partial t} \tag{6.100}$$

Introducing the following boundary conditions:

$$p = p_1 \quad at \quad r = r_1, \quad\quad p = p_2 \quad at \quad r = r_2$$

and using Darcy's law, then the following relation can be obtained for flow rate:

$$v_{t_1} = -\frac{kA_{p1}}{\eta}\left(\frac{dp_1}{dr}\right)_{r=r_1}, \quad v_{t_2} = -\frac{kA_{p2}}{\eta}\left(\frac{dp_2}{dr}\right)_{r=r_2} \tag{6.101}$$

where A_{p1} is the surface area of the pressure injection hole, and A_{p2} is the area of the pressure release surface. Pressure gradients in the preceding equations are as follows:

$$\frac{dp_1}{dr} \approx -\frac{1}{r_1}\frac{(p_1 - p_2)}{\ln(r_2/r_1)}, \quad \frac{dp_2}{dr} \approx -\frac{1}{r_2}\frac{(p_2 - p_1)}{\ln(r_2/r_1)} \tag{6.102}$$

Inserting the preceding equation into Equation (6.101) and equalizing the resulting equation to Equation (6.100 (3.67)) yields the following set of equations:

$$\frac{\partial p_1}{\partial t} = -\beta \frac{A_{p1}}{V_1 r_1}\frac{(p_1 - p_2)}{\ln(r_2/r_1)} \tag{6.103}$$

$$\frac{\partial p_2}{\partial t} = \beta \frac{A_{p2}}{V_2 r_2}\frac{(p_1 - p_2)}{\ln(r_2/r_1)} \tag{6.104}$$

where

$$\beta = \frac{k}{c_f \eta}$$

Equation (6.102 (3.70)) can be rearranged as follows:

$$p_2 = p_1 + \frac{V_1 r_1 \ln(r_2/r_1)}{\beta A_{p1}} \frac{\partial p_1}{\partial t} \qquad (6.105)$$

Taking the time derivative of the preceding equation, the following expression is obtained:

$$\frac{\partial p_2}{\partial t} = \frac{\partial p_1}{\partial t} + \frac{V_1 r_1 \ln(r_2/r_1)}{\beta A_{p1}} \frac{\partial^2 p_1}{\partial t^2} \qquad (6.106)$$

Substituting Equations (6.105) and (6.106) into Equation (6.104) and rearranging the resulting equation yields the following homogeneous differential equation:

$$\frac{\partial^2 p_1}{\partial t^2} + \alpha \frac{\partial p_1}{\partial t} = 0 \qquad (6.107)$$

where

$$\alpha = \beta \frac{V_2 r_2 A_{p1} + V_1 r_1 A_{p2}}{\ln(r_2/r_1) V_1 V_2 r_2 r_1}$$

General solution of this differential equation is:

$$p_1 = C_1 + C_2 e^{-\alpha t} \qquad (6.108)$$

Introducing the following initial conditions:

$$p_1 = p_i \quad at \quad t = 0, \qquad p_1 = p_f \quad at \quad t = \infty$$

yields the integration constants C_1 and C_2 as:

$$C_1 = p_f, \quad C_2 = p_i - p_f \qquad (6.109)$$

Inserting these integration constants into Equation (6.108) gives the following equation:

$$p_1 = p_f + (p_i - p_f) e^{-\alpha t} \qquad (6.110)$$

Taking the time derivative of the preceding equation:

$$\frac{\partial p_1}{\partial t} = -(p_i - p_f) \beta \frac{V_2 r_2 A_{p1} + V_1 r_1 A_{p2}}{V_2 r_2 V_1 r_1 \ln(r_2/r_1)} e^{-\alpha t} \qquad (6.111)$$

Substituting Equations (6.110) and (6.111) into Equation (6.103) and rearranging yields the following equation:

$$p_2 = p_f - (p_i - p_f)\frac{V_1 r_1 A_{p2}}{V_2 r_2 A_{p1}} e^{-\alpha t} \tag{6.112}$$

For the following initial condition for p_2:

$$p_2 = p_0 \quad at \quad t = 0$$

Equation (6.111) takes the following form:

$$(p_i - p_f) = (p_f - p_0)\frac{V_2 r_2 A_{p1}}{V_1 r_1 A_{p2}} \tag{6.113}$$

The preceding equation can be rewritten in a different way for $p_i - p_0$ as follows:

$$(p_i - p_f) = (p_i - p_0)\frac{V_2 r_2 A_{p1}}{V_1 r_1 A_{p2} + V_2 r_2 A_{p1}} \tag{6.114}$$

Inserting this equation into Equation (6.110), and rearranging yields the following:

$$-\alpha t = \ln(\frac{p_1 - p_f}{p_i - p_0}\frac{V_1 r_1 A_{p2} + V_2 r_2 A_{p1}}{V_2 r_2 A_{p1}}) \tag{6.115}$$

where

$$\alpha = \frac{k}{c_f \eta}\frac{V_2 r_2 A_{p1} + V_1 r_1 A_{p2}}{V_2 r_2 V_1 r_1 \ln(r_2/r_1)}$$

From the preceding equations, one can use the following equation to compute permeability:

$$k = \eta c_f \frac{V_2 r_2 V_1 r_1 \ln(r_2/r_1)}{V_2 r_2 A_{p1} + V_1 r_1 A_{p2}} \ln(\frac{\Delta p_o}{\Delta p}\frac{V_2 r_2 A_{p1}}{V_2 r_2 A_{p1} + V_1 r_1 A_{p2}})\frac{1}{t} \tag{6.116}$$

When gas is used as a permeation fluid, p_1 and p_2 are replaced with U_1 $(= p_1^2)$ and $U_2 (= p_2^2)$, and permeability can be calculated using the same relation as previously given.

If the volume of reservoir 2 (V_2) is much greater than the volume of reservoir 1 (V_1), $(V_2 ? V_1)$ (for instance, the outer side of specimen is open to air) p_0 ve p_f given in the preceding equation will be equal to atmospheric pressure (p_a). For this particular case, Equation (6.116 (3.83)) becomes:

$$k = \eta c_f \frac{V_1 r_1 \ln(r_2/r_1)}{A_{p1}} \ln(\frac{p_i - p_a}{p_1 - p_a})\frac{1}{t} \tag{6.117}$$

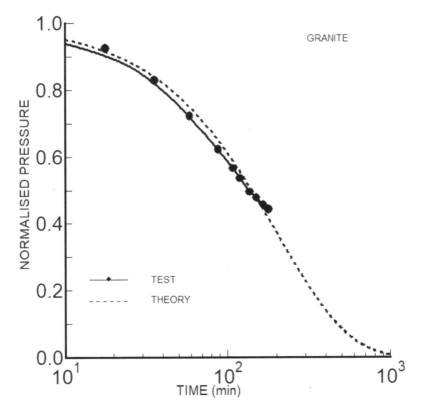

Figure 6.44 An experimental result on granite

Theoretical and experimental curves for a transient pulse test on the granite specimen are shown in Figure 6.44. As seen from this figure, there is a slight difference between the curve obtained from the test and the curve from the theory, particularly in the initial stages. In order to obtain permeability value, the linear part of normalized pressure change and time relation is generally used to compute permeability for the selected range.

References

Aydan, Ö. (1998) Finite element analysis of transient pulse method tests for permeability measurements. *The 4th European Conf. on Numerical Methods in Geotechnical Engineering-NUMGE98, Udine.* pp. 719–727.

Aydan, Ö. (2016a) *Time Dependency in Rock Mechanics and Rock Engineering.* CRC Press, Taylor and Francis Group. p. 241.

Aydan, Ö. (2016b) *Considerations on Friction Angles of Planar Rock Surfaces with Different Surface Morphologies from Tilting and Direct Shear Tests.* ARMS2016, Bali.

Aydan, Ö. (2008) New directions of rock mechanics and rock engineering: Geomechanics and Geoengineering. 5th Asian Rock Mechanics Symposium (ARMS5), Tehran, 3–21.

Aydan, Ö. & Üçpırtı, H. (1997) The theory of permeability measurement by transient pulse test and experiments. *Journal. of the School of Marine Science and Technology,* Tokai University, 43, 45–66.

Aydan, Ö. & Ulusay, R. (2002) Back analysis of a seismically induced highway embankment during the 1999 Düzce earthquake. *Environmental Geology*, 42, 621–631.

Aydan, Ö., Akagi, T. & Kawamoto, T. (1993) Squeezing potential of rocks around tunnels; theory and prediction. *Rock Mechanics and Rock Engineering*, 26(2), 137–163.

Aydan, Ö., Ulusay, R. & Tokashiki, N. (2014) A new rock mass quality rating system: Rock Mass Quality Rating (RMQR) and its application to the estimation of geomechanical characteristics of rock masses. *Rock Mech Rock Eng*, 47, 1255–1276.

Aydan, Ö., Akagi, T., Okuda, H. & Kawamoto, T. (1994) The cyclic shear behaviour of interfaces of rock anchors and its effect on the long term behaviour of rock anchors. *Int. Symp. on New Developments in Rock Mechanics and Rock Engineering, Shenyang*. pp. 15–22.

Aydan, Ö., Shimizu, Y. & Kawamoto, T. (1995) A portable system for in-situ characterization of surface morphology and frictional properties of rock discontinuities. *Field Measurements in Geomechanics, 4th International Symposium*. Bergamo, pp. 463–470.

Aydan, Ö., Üçpirti, H. & Turk, N. (1997) Theory of laboratory methods for measuring permeability of rocks and tests (in Turkish). *Bulletin of Rock Mechanics*, Ankara, 13, 19–36.

Aydan, Ö., Ito, T. & Rassouli, F. (2016a) Chapter 11: Tests on creep characteristics of rocks. CRC Press., London, pp. 333–364.

Aydan, Ö., Tokashiki, N., Tomiyama, J., Iwata, N., Adachi, K. & Takahashi, Y. (2016b) *The Development of a Servo-control Testing Machine for Dynamic Shear Testing of Rock Discontinuities and Soft Rocks*. EUROCK2016, Ürgüp. pp. 791–796.

Aydan, Ö., Tokashiki, N. & Edahiro, M. (2016c) Utilization of X-Ray CT Scanning technique in Rock Mechanics Applications. ARMS2016, Bali.

Aydan, Ö., Ohta, Y., Kiyota, R. & Iwata, N. (2019) The evaluation of static and dynamic frictional properties of rock discontinuities from tilting and stick-slip tests. *46th Rock Mechanics Symposium of Japan*. Tokyo, pp. 105–110.

Barton, N.R. & Choubey, V. (1977) The shear strength of rock joints in theory and practice. *Rock Mechanics*, 10, 1–54.

Bieniawski, Z.T. (1974) Geomechanics classification of rock masses and its application in tunnelling. *Third Int. Congress on Rock Mechanics, ISRM*, Denver, IIA, 27–32.

Birch, F. (1950) Flow of heat in the Front Range, Colorado, *Geological Society of America Bulletin*, 61, 567–630.

Blum, P. (1997) Physical properties handbook. *ODP Technical Note*, 26.

Brace, W.F., Walsh, J.B. & Frangos, W.T. (1968) Permeability of granite under high pressure. *Journal of Geophysical Research*, 73(6), 2225–2236.

Brown, E.T. (1981) *Suggested Methods for Rock Characterization, Testing, Monitoring*. Pergamon Press, Oxford.

Carslaw, H. & Jaeger, J. (1959) *Conduction of Heat in Solids*. 2nd edition. Oxford University Press, Oxford, 510pp.

Clark, S.P., Jr. (1966) Handbook of physical constants. *Geological Society of America Memoir*, 97, 587p.

Hibino, S., 2007. Necessary Knowledge of Rock Mass for Engineers (Gijutsusha ni hitsuyo na ganban no chishiki), Kajima Pub. Co., Tokyo (in Japanese).

Hondros, G. (1959) The evaluation of Poisson's ratio and the modulus of materials of low tensile resistance by the Brazilian (indirect tensile) tests with particular reference to concrete. *Australian Journal of Applied Sciences*, 10, 243–268.

Hudson, J.A. & Ulusay, R. (2007) *The Complete ISRM Suggested Methods for Rock Characterization, Testing, Monitoring, 1974–2006*, ISRM. Ankara.

Hudson, J.A., Crouch, S.L. & Fairhurst, C. (1972) Soft, stiff and servo-controlled testing machines: A review with reference to rock failure. *Engineering Geology*, 6, 155–189.

Jaeger, J.C. & Cook, N.G.W. (1979) *Fundamentals of Rock Mechanics*, 3rd edition. Chapman & Hall, London. pp. 79, 311.

Kawamoto, T., Tokashiki, N. & Ishizuka, Y. (1980) On uniaxial compression test of rock-like materials using a new type of high stiff testing machine Japan. *Material Science Journal (Zairyo)*, 30 (322), 517–523.

Kreyszig, E. (1983) *Advanced Engineering Mathematics*. John *Wiley* & Sons, New York.

Popov, Y., Pribnow, D., Sass, J., Williams, C. & Burkhardt, H. (1999) ISRM Suggested Methods for determining thermal properties of rock samples. Characterisation of rock thermal conductivity by high-resolution optical scanning. *Geothermics*, 28, 253–276.

Popov, Y., Beardsmore, G., Clauser, C. & Roy, S. (2016) ISRM suggested methods for determining thermal properties of rocks from laboratory tests at atmospheric pressure. *Rock Mechanics and Rock Engineering*.

Rassouli, F.S., Moosavi, M., & Mehranpour, M.H. (2010) The effects of different boundary conditions on creep behavior of soft rocks. *The 44th U.S. Rock mechanics Symposium & 5th U.S. Canada symposium*, Salt Lake City, Utah.

Rummel, F. & Fairhurst, G. (1970) Determination of the post-failure behaviour of brittle rock using a servo-controlled testing machine. *Rock Mechanics*, 2, 189–204.

Sass, J., Stone, C. & Munroe, R. (1984) Thermal conductivity determinations on solid rock – A comparison between a steady-state divided-bar apparatus and a commercial transient line-source device. *Journal of Volcanology and Geothermal Research*, 20(1–2), 145–153.

Somerton, W.H. (1992) *Thermal properties and temperature-related behavior of rock/fluid systems*. Developments in petroleum science, 37. Elsevier Science Publishers B.V., Amsterdam, Netherlands, 257p.

Tokashiki, N. & Aydan, Ö. (2012). Estimation of Rockmass Properties of Ryukyu Limestone. Asian Rock Mechanics Symposium, Seoul, 725–734.

Ulusay, R. (2012) *The ISRM Suggested Methods for Rock Characterization, Testing, Monitoring, 2007–2014*. Springer, Vienna.

Van Heerden, W.L. 1975. In-situ complete stress-strain characteristics of large coal specimens. *J.S. Afr. Min. Metall.*, 75, 207–217.

Von Herzen, R.P. & Maxwell, A.E. (1959) The measurement of thermal conductivity of deep-sea sediments by a needle-probe method. *Journal Geophysical Research*, 69, 1557–1563.

Waversik, W.R. & Fairhurst, C. (1970) A study of brittle rock fracture in laboratory compression experiments. *International Journal Rock Mechanics and Mining Science*, 7, 561–575.

Chapter 7

In-situ stress estimation, measurement and inference methods

The stress state of the Earth is of paramount importance in geomechanics and geophysics. Particularly, the virgin stress state in the Earth's crust is of great interest in mining and civil engineering since the stability of excavations is very much influenced by that. Geophysicians are also concerned with the stress state of the crust in association with understanding the earthquake mechanism and predicting earthquakes.

Many *in-situ* stress inference techniques are classified into broadly direct or indirect techniques (Amadei and Stephansson, 1997). Direct techniques are generally costly, and they are only utilized for some important structures. The direct techniques utilize boreholes and assume that the surrounding rock behave elastically. However, the acoustic emission (AE) method can be used as a direct stress measurement method as it is less costly and it can be performed under well controlled conditions in laboratory once sampling is done.

Indirect stress inference techniques utilizing borehole breakouts, fault striations and earthquake focal mechanism solutions are also proposed and used (Zoback and Healy, 1992; Angellier, 1984; Aydan, 2000a; Aydan and Kim, 2002). Recently a new stress inference technique utilizing damage zone around blast holes was proposed by Aydan (2012), and it was applied to several sites in Japan and Turkey.

7.1 *In-situ* stress estimation methods

7.1.1 Empirical approaches

The empirical approaches to estimate the crustal stresses are based on some empirical formulas utilizing stress measurements in various engineering projects and some earthquake prediction studies. The measurements of *in-situ* stresses in various engineering projects were carried out in South Africa (Hast, 1969) first and later in other countries. Measurements indicated that horizontal stresses could be several times vertical stresses in the shields such as the Canadian or Scandinavian shields (Herget, 1986; Stephanson *et al.*, 1986). Furthermore, many measurements in association with the excavations in mining and civil engineering fields and earthquake prediction projects stress measurements in the Earth crust have been undertaken over last 50 years. However, most of the measurements are restricted to a depth below 5000 m. Brown and Hoek (1978) proposed some empirical relations, given here, to estimate the ratio of the horizontal stress to normalized vertical stress.

$$\lambda = 1.0 + \frac{2}{h} \tag{7.1}$$

7.1.2 Analytical approaches

Many proposals for the stress state of the Earth have been made in this century, and these proposals may be classified on the basis of their main characteristics.

7.1.2.1 Approaches assuming the Earth in liquid state

Jeffreys and Bullen (1940) considered that the stress state of the Earth was hydrostatic, and they calculated the pressure of the Earth by considering the variation of density and gravitational acceleration. Anderson and Hart (1976) modified this approach by considering recent findings.

Nadai (1950) also derived the following formula by assuming that the density of the Earth is constant and that the gravitational acceleration varies linearly with depth:

$$\sigma_r = \sigma_\theta, \; \sigma_r = \frac{\rho_m g_o R_o}{2}\left(1 - \left(\frac{r}{R_0}\right)^2\right) \tag{7.2}$$

where σ_r is radial stress, σ_θ is tangential stress, R_0 is the radius of the Earth, r is radial distance, ρ_m is the mean density of the Earth, and g_0 is gravitational acceleration at the Earth's surface.

7.1.2.2 Approaches assuming the Earth in solid state

Terzaghi and Richart (1952) considered a vertical column and formulated the stress state of the Earth by assuming that lateral strains are zero and the medium is elastic as:

$$\sigma_\theta = \frac{v}{1-v}\sigma_r, \sigma_r = \rho g H \tag{7.3}$$

where v is Poisson's ratio, H is the depth from ground surface, ρ is density, and g is gravitational acceleration.

Salustowicz (1968) derived the following formula by assuming that the density of the Earth is constant and that it consists of a homogeneous elastic material together with the consideration of a linearly decreasing gravitational acceleration:

$$\sigma_r = \rho_m g_o R_o \alpha\left(1 - \left(\frac{r}{R_0}\right)^2\right), \tag{7.4}$$

$$\sigma_\theta = \rho_m g_o R_o \alpha\left(1 - \beta\left(\frac{r}{R_0}\right)^2\right), \tag{7.5}$$

where

$$\alpha = \frac{3-v}{10(1-v)}, \beta = \frac{1+3v}{(3-v)}$$

7.1.2.3 Approaches based on the assumption that the crust and mantle are in solid state and the core is in liquid state

Aydan (Aydan, 1993; Aydan and Kawamoto, 1994) proposed two models, namely:

- Two-layered model (TLM)
- Multilayer model (MLM)

Two-layered model (TLM) (Fig. 7.1(a)): By considering that the gravitational acceleration remains constant up to the interface between the lower mantle and the outer core and thereafter decreases linearly (Fowler, 1990), the Earth was modeled as a simple spherical body consisting of a core and a mantle. By introducing the averaged values of physical and Lamé's constants for each zone, the following formula were developed for radial and tangential stresses in the mantle and the core:

Mantle

$$\sigma_r^* = \frac{R_i^3 R_o^3}{R_o^3 - R_i^3}\left[Pi\left(\frac{1}{r^3} - \frac{1}{R_o^3}\right) + \frac{\lambda^* + \mu^*}{\lambda^* + 2\mu^*}\rho_o g_o\left(\frac{R_o^4 - R_i^4}{R_o^3 R_i^3} - \frac{R_o - R_i}{r^3}\right)\right] - \frac{\lambda^* + \mu^*}{\lambda^* + 2\mu^*}\rho_o g_o r$$

(7.6)

$$\sigma_\theta^* = \frac{R_i^3 R_o^3}{R_o^3 - R_i^3}\left[-Pi\left(\frac{1}{2r^3} + \frac{1}{R_o^3}\right) + \frac{\lambda^* + \mu^*}{\lambda^* + 2\mu^*}\rho_o g_o\left(\frac{R_o^4 - R_i^4}{R_o^3 R_i^3} + \frac{R_o - R_i}{2r^3}\right)\right] - \frac{2\lambda^* + \mu^*}{\lambda^* + 2\mu^*}\rho_o g_o\frac{r}{2}$$

(7.7)

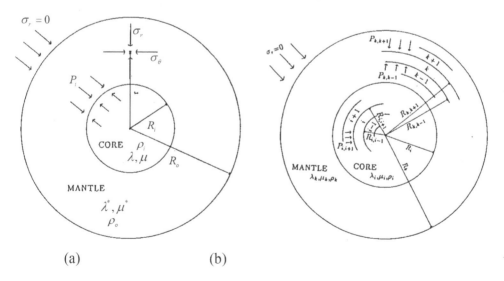

(a)   (b)

Figure 7.1 (a) Double-layer model, and (b) multilayer model

Core

$$\sigma_r = p_i + \frac{\rho_i g_o R_i}{2}\left(1 - \frac{r^2}{R_i^2}\right), \sigma_\theta = \sigma_r \tag{7.8}$$

where λ, μ, λ^*, μ^* are average Lamé's constants for the outer and inner zones, respectively, and R_i is the radius of the inner zone:

$$P_i = \frac{\frac{R_i^3 R_o^3}{R_o^3 - R_i^3}\frac{\lambda^* + \mu^*}{\lambda^* + 2\mu^*}\rho_o g_o\left[\frac{1}{4\mu^*}\frac{R_o - R_i}{R_i^2} - \frac{R_i}{3\lambda^* + 2\mu^*}\frac{R_o^4 - R_i^4}{R_o^3 R_i^3}\right] - \frac{\rho_o g_o}{\lambda^* + 2\mu^*}\frac{R_i^2}{4} - \frac{\rho_i g_o}{\lambda}\frac{R_i^2}{15}}{\frac{R_i^3 R_o^3}{R_o^3 - R_i^3}\left[\frac{1}{4\mu^*}\frac{1}{R_i^2} + \frac{1}{3\lambda^* + 2\mu^*}\frac{R_i}{R_o^3}\right] + \frac{R_i}{3\lambda}} \tag{7.9}$$

Multilayered model (MLM) (Fig. 7.1(b)): Since physical and Lamé's constants vary with depth, the preceding approach was extended to simulate a multilayered structure of the Earth in order to have a better solution for its stress state. The solutions for each layer were essentially similar to those for the two-layered model. The displacement for a layer numbered i in the core is obtained by assuming the Earth consists of n layers having constant physical and mechanical parameters in each respective layer as:

$$u^i = A_1^i r + A_2^i \frac{1}{r^2} - \frac{\rho_i g_o}{10\lambda_i}\frac{r^3}{R_{mci}} \tag{7.10}$$

By introducing the following conditions at the interface between layer i and layer $i + 1$:

$$\sigma_r = P_{i,i+1} \quad \text{at} \quad r = R_{i,i+1}$$

the integration constants A_1^i and A_2^i are obtained as

$$A_1^i = \frac{1}{3\lambda_i}\left[P_{i,i+1} + \rho_i g_o \frac{R_{i,i+1}^2}{2R_{mc}}\right], A_2^i = 0$$

where R_{mc} is the radius of the interface between the mantle and the core.

Similarly, for a typical layer numbered k in the mantle, displacement u^k can be obtained in the following form:

$$u^k = A_1^k r + A_2^k \frac{1}{r^2} - \frac{\rho_k g_o}{\lambda^k + 2\mu^k}\frac{r^2}{4} \tag{7.11}$$

By introducing the following conditions at the interface between layer k and layer $k + 1$ and the interface between layer k and layer $k - 1$:

$$\sigma_r = P_{k,k-1} \text{ at } r = R_{k,k-1}$$

$$\sigma_r = P_{k,k+1} \text{ at } r = R_{k,k+1}$$

the integration constants A_1^k and A_2^k can be obtained as:

$$A_1^k = \frac{1}{3\lambda^k + 2\mu^k} \frac{R_{k,k-1}^3 R_{k,k+1}^3}{R_{k,k+1}^3 - R_{k,k-1}^3} \left[\frac{\lambda^k + \mu^k}{\lambda^k + 2\mu^k} \rho_k g_o \left(\frac{R_{k,k+1}^4 - R_{k,k+1}^4}{R_{k,k-1}^3 R_{k,k-1}^3} \right) + \frac{P_{k,k+1} R_{k,k+1}^3 - P_{k,k-1} R_{k,k-1}^3}{R_{k,k-1}^3 R_{k,k+1}^3} \right]$$

$$A_2^k = \frac{1}{4\mu^k} \frac{R_{k,k-1}^3 R_{k,k+1}^3}{R_{k,k+1}^3 - R_{k,k-1}^3} \left[\frac{\lambda^k + \mu^k}{\lambda^k + 2\mu^k} \rho_k g_o \left(\frac{R_{k,k+1}^4 - R_{k,k+1}^4}{R_{k,k-1}^3 R_{k,k-1}^3} \right) + (p_{k,k+1} - P_{k,k-1}) \right]$$

Introducing the continuity condition of displacement, that was, $u_i - u_{i+1} = 0$ at each interface $r = R_{i,i+1}$ yielded a linear equation system for interlayer pressures $P_{1,2}, \dots P_{i,i-1}, P_{i,i+1}, P_{n-1,n}$. The solution of this equation system together with the condition of $\sigma_r = 0$ at the Earth's surface gives the pressures on both sides of layers, from which the displacement and stresses of each layer could be easily obtained.

7.1.2.4 Approaches assuming that the crust is in solid state and plastic

Possible stress state in the crust associated with normal and thrust faulting were first discussed by Hubbert (1951) and Anderson (1951). This approach assumes that the crust is in plastic state and that the materials obey a Mohr-Coulomb–type yield criterion. The stress states associated with various faulting regimes are summarized and discussed in a textbook by Jaeger and Cook (1979) as follows:

Normal faulting regime

$$\sigma_v = q\sigma_h + \sigma_c, q = \frac{1 + \sin\phi}{1 - \sin\phi} \tag{7.12}$$

Thrust faulting regime

$$\sigma_H = q\sigma_v + \sigma_c \tag{7.13}$$

Strike-slip faulting regime

$$\sigma_H = q\sigma_h + \sigma_c \tag{7.14}$$

where $\sigma_v = \rho g H$, ϕ is the friction angle, σ_c is uniaxial strength of crust, σ_H is the maximum horizontal stress, and σ_h is the minimum horizontal stress.

As noted from the preceding equations, one of the stress components is always indeterminate. Parameters q and σ_c may be temperature and/or confining pressure dependent. This fact is also considered in some publications (i.e. Brace and Kohlstedt, 1980; McGarr, 1980; Shimada, 1993).

7.1.2.5 Comparisons

The approaches summarized in this subsection are compared with one another, and their main characteristics are discussed herein. Figure 7.2 shows plots of the distributions of

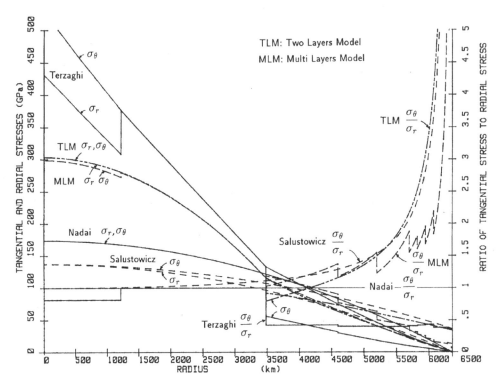

Figure 7.2 Comparison of various approaches

radial and tangential stresses and the ratio of tangential stress to radial stress in the Earth obtained from various models. Nadai's and Salustowicz's models underestimate the radial stress distributions. While the lateral stress coefficient remains 1 for Nadai's model, it is remarkable to note that the coefficients predicted by the TLM, MLM and Salustowicz' model increase exponentially near the Earth's surface and goes to infinity.

The TLM predicts the radial stress distribution, which is very close to the pressure distribution reported by Anderson and Hart (1976). While the tangential stress by the TLM is equal to the radial one throughout the core, its distribution varies within the mantle. At the core–mantle interface, there is a discontinuity between tangential stresses of the core and the mantle, and the tangential stress is less than the radial stress in the mantle. However, it becomes greater than the radial stress, and it has a finite value at the Earth's surface while the radial stress is nil there, which implies that the ratio of the horizontal stress to the vertical stress should be infinite at the Earth's surface.

The vertical stress distribution obtained from Terzaghi's proposal is close to that of the radial stress, but it becomes very large in the core, which is quite different from those by other solutions. The lateral stress distribution is also different from the tangential stress distribution obtained from other solutions, and it could not explain that why lateral stresses should be large at the Earth's surface. Amadei and Savage (1985) proposed a model to explain large horizontal stresses near the Earth's surface by introducing the anisotropy of elastic constants of rock mass while using this one-dimensional column model. However, it must be noted that the large horizontal stresses near the Earth's surface can also be obtained by taking into account the sphericity of the Earth even the rock mass remains isotropic.

A better solution for the stress state in the Earth is provided by the multilayered model (Fig. 7.2). In the calculations, the Earth is assumed to be consisted of nine concentric spherical layers. While the radial stress distribution almost coincides with those by the two-layered model and pressure model, the tangential stress distribution is remarkably different from that by the two-layered model. Since the mechanical and physical properties used in the analysis differ in each layer, the tangential stress at each interface becomes discontinuous. Furthermore, the value of the tangential stress is much less than that predicted by the two-layered model or Salustowicz's model.

7.1.3 Numerical approaches

7.1.3.1 The approach assuming that the Earth is in fluid state

The original approach is based on the assumption that the Earth is in fluid state. This approach was proposed by Jeffrey and Bullen (1940), and Anderson and Hart (1976) used a better model for the density distribution of the Earth and utilized the finite different technique and spherical symmetry and obtained the pressure distribution in the Earth.

7.1.3.2 The approach assuming that the Earth is a spherical symmetric thermo-elasto-plastic body

Triaxial tests on rocks were undertaken by many experimenters in the fields of geomechanics and geophysics under high confining pressure and high temperature regimes (i.e. von Karman, 1911; Edmonton and Paterson, 1972; Byerlee, 1978; Shimada, 1993; Hirth and Tullis, 1994; etc.). These experiments showed that the mechanical behavior of rocks changes from a brittle behavior to a ductile behavior as the confining pressure increases (Fig. 7.3(a)). Furthermore, the strength decreases as temperature increases, as shown in Figure 7.3(b). Taking into account these facts, Aydan (1995) proposed a thermo-plastic yield criterion:

$$\sigma_1 - \sigma_3 = [S_\infty - (S_\infty - S_0)e^{-b_1\sigma_3}]e^{-b_2 T} \tag{7.15}$$

where S_0 is the uniaxial compressive strength at room temperature, S_∞ is the ultimate deviatoric compressive strength at room temperature, T is temperature, b_1, b_2 are physical constants, σ_1 is the maximum principal stress, and σ_3 is the minimum principal stress (confining pressure). Figure 7.3 shows a plot of the preceding yield criterion in the space of temperature T and confining pressure σ_3.

Aydan (1995) assumed that the Earth is a spherical symmetric body so that the governing equilibrium equation is given by:

$$\frac{\partial \sigma_r}{\partial_r} + 2\frac{\sigma_r - \sigma_\theta}{r} = \rho g \tag{7.16}$$

where ρ and g are density and gravitational acceleration, which may vary with depth. The constitutive law for elastic behavior is written as:

$$\begin{Bmatrix} \sigma_r \\ \sigma_\theta \end{Bmatrix} = \begin{bmatrix} 2\mu + \lambda & 2\lambda \\ \lambda & 2(\mu + \lambda) \end{bmatrix} \begin{Bmatrix} \varepsilon_r \\ \varepsilon_\theta \end{Bmatrix} \quad \text{or} \quad \{\sigma\} = [D]\{\varepsilon\} \tag{7.17}$$

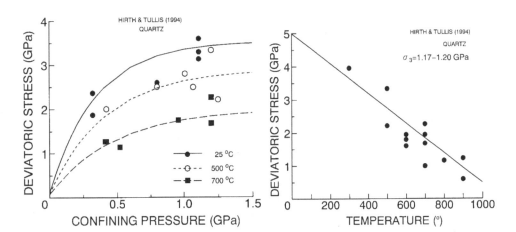

Figure 7.3 Triaxial compression tests on quartz

Source: Data from Hirth and Tullis, 1994

It should be noted that Lamé's constants in the preceding equation can vary with depth. Using the general procedure of finite element discretization and taking variations on δu, with the use of Gauss divergence theorem and the following conditions:

$$\delta u = 0 \quad on \quad \Gamma_u, \quad and \quad \hat{t} = 0 \quad on \quad \Gamma_t$$

the weak form of Equation (7.16) takes the following form:

$$\int_\Omega (\sigma_r \delta\varepsilon_r + 2\sigma_\theta \delta\varepsilon_\theta)d\Omega = \int_\Omega \rho g \delta u d\Omega \tag{7.18}$$

where $d\Omega = 4\pi r^2 dr$.

Let us assume that the displacement in a given element is approximated by the following expression:

$$u = [N]\{U\} \tag{7.19}$$

where $[N]$ is shape function, and $\{U\}$ is nodal displacement vector. Using the preceding approximate form and the constitutive law (7.17), the following finite element form is obtained for a typical finite element:

$$[K]\{U\} = \{F\} \tag{7.20}$$

where

$$[K] = \int_{\Omega_e} [B]^T [D][B]d\Omega \quad \{F\} = \int_{\Omega_e} \rho g [N]^T d\Omega$$

If a linear type shape function is chosen as given here:

$$N^i = \frac{r_j - r}{L}, N^j = \frac{r - r_i}{L}, L = r_j - r \tag{7.21}$$

the stiffness matrix $[K]$ and load vector $\{F\}$ given in Equation (7.20) are specifically obtained as follows:

$$[K] = \frac{4\pi}{L^2} \begin{bmatrix} K_{11} & K_{12} \\ K_{21} & K_{22} \end{bmatrix}, \{F\} = \frac{\rho g \pi}{3} \begin{Bmatrix} r_j^3 + r_j^2 r_i + r_j r_i^2 - 3r_i^3 \\ 3r_j^3 + r_j^2 r_i + r_j r_i^2 - r_i^3 \end{Bmatrix} \tag{7.22}$$

where

$$K_{11} = \frac{D_1 + 2D_2 + D_1^*}{3}(r_j^3 - r_i^3) - (D_2 + D_1^*)r_j(r_j^2 - r_i^2) + D_1^* r_j^2 (r_j - r_i)$$

$$K_{22} = \frac{D_1 + 2D_2 + D_1^*}{3}(r_j^3 - r_i^3) - (D_2 + D_1^*)r_i(r_j^2 - r_i^2) + D_1^* r_i^2 (r_j - r_i)$$

$$K_{12} = K_{21} = -\frac{D_1 + 2D_2 + D_1^*}{3}(r_j^3 - r_i^3) - (D_2 + D_1^*)\frac{r_j + r_i}{2}(r_j^2 - r_i^2) + D_1^* r_i r_j (r_j - r_i)$$

$$D_1 = 2\mu + \lambda, D_2 = 2\lambda \quad D_1^* = 4(\mu + \lambda)$$

In analyses, it was assumed that the stress–strain response of rocks constituting the Earth exhibit an elastic–perfectly plastic behavior. The elastic constants of rocks were taken from a report by Anderson and Hart (1976). The initial stiffness technique was chosen as the iteration technique to deal with the plastic behavior of rocks in finite element analyses (Owen and Hinton, 1980). The temperature distribution of the Earth was input as known by using a distribution reported in a textbook (p. 248) by Fowler (1990) (Fig. 7.4).

A series of case studies, given here, were carried by Aydan (1995) (Fig. 7.5):

CASE 1: The Earth was in liquid state (hydrostatic).
CASE 2: The crust and mantle were elastic solids, and the core was in liquid state (nonhydrostatic).
CASE 3: The crust and mantle were elasto-plastic solids, and the core was in liquid state (isothermic: room temperature).
CASE 4: The crust and mantle were thermo-elasto-plastic solids, and the core was in liquid state (nonisothermic).

Parameters b_1, b_2, S_o, S_∞ of the yield criterion were chosen as 0.2 GPa^{-1}, 0.0014 $^\circ C^{-1}$, 0 MPa, 5 GPa, respectively.

Figure 7.6 shows the distributions of radial and tangential stresses and the ratio of tangential stress to radial stress in the lithosphere for each case. The elasto-plastic analyses showed that the whole crust and mantle became plastic. Large tangential stresses seen in Case 2 were dissipated when the plastic behavior of the crust and mantle was considered. The tangential stresses dissipated in the mantle increased both radial and tangential stresses in the core. The consideration of the thermo-elasto plastic behavior of the mantle and the crust of the Earth further decreased the deviatoric stresses in the mantle and in the crust.

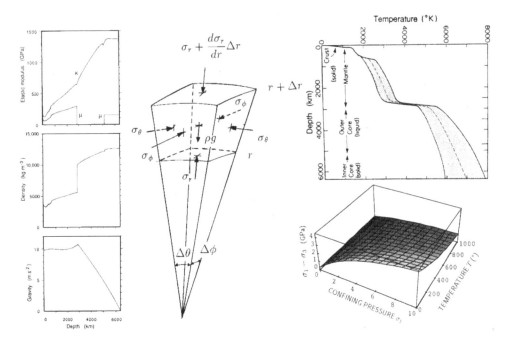

Figure 7.4 Illustration of spherical model of the Earth and material properties

Source: Arranged from Aydan, 1994; Fowler, 1990

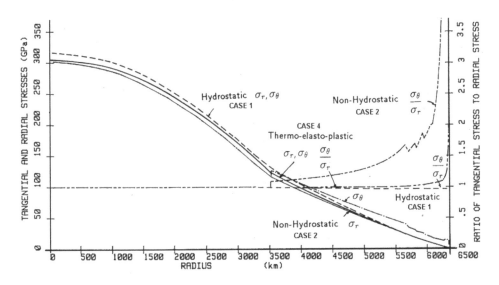

Figure 7.5 The stress state of the Earth

Source: From Aydan, 1995

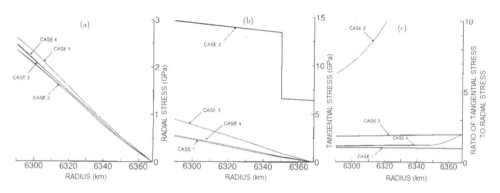

Figure 7.6 Computed radial and tangential stresses in lithosphere

The lateral stress coefficient, which has a value of infinity at the ground surface for CASE 2, also decreased in magnitude when the plastic behavior was considered.

In the next series of parametric studies, the uniaxial compressive strength parameter S_o was varied from 0 to 40 MPa in increments of 20 MPa, while the other parameters were kept the same as those in the previous thermo-elasto-plastic analysis. The computed distributions of radial and tangential stresses and the ratio of tangential stress to radial stress, together with *in-situ* observations (Aydan and Paşamehmetoğlu, 1994) and predictions by Shimada (1993) for a depth of 4 km in the Earth's crust, are plotted in Figure 7.7. The radial stress (vertical stress) almost coincided with each other (Fig. 7.7(a)). They are almost equal to the vertical stress calculated from $\rho g H$ (Shimada, 1993), and it is a good fit to *in-situ* measurements.

The tangential stresses (horizontal stress) for each value of S_o fit fairly well to *in-situ* measurements for a depth of 1 km from the ground surface (Fig. 7.7(b)). However, finite element computations overestimate the tangential stresses at depths greater than 1 km.

The ratio of tangential stress to radial stress for each value of S_o provides upper bounds for *in-situ* measurements (Fig. 7.7(c)). The discrepancy between computed and measured values may be eliminated if a thermo-elasto-visco-plastic constitutive law is employed in finite element analysis.

7.2 *In-situ* stress measurement methods

7.2.1 *Stress relief (overcoring) method*

The overcoring method has been used for many decades. This method is based on the principle of releasing the stress state of instrumented rock core inside a borehole (Fig. 7.8) by overcoring. It is also a widely used technique all over the world. There are two variations of this technique:

- Leeman's method or doorstopper method, which utilizes a flat-ended borehole
- Sugawara-Obara's method, which utilizes a semi-spherical-ended borehole. Another variation to this approach is proposed by Kobayashi and Mizuta, who used a conical-ended borehole (conical-ended borehole overcoring (CCBO).

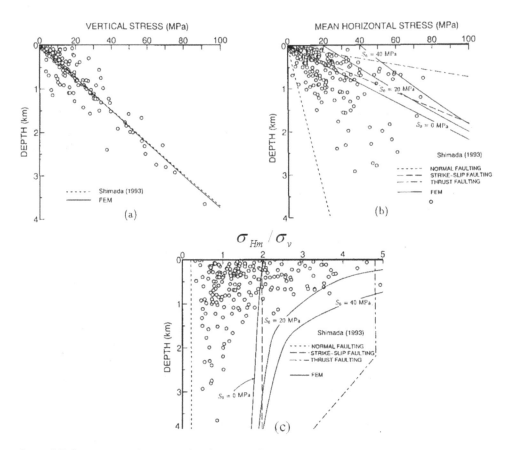

Figure 7.7 Comparison of computed and measured *in-situ* stresses
Source: From Aydan and Kawamoto, 1997

The method involves creating a circular hole, placing some device for measuring strains or displacements into the hole, and then drilling over the top of that hole to relieve the stress and thus cause a deformation change. The measured deformation, together with the rock modulus and Poisson's ratio measured in the laboratory, is used to calculate the magnitude and directions of the stresses existing in the rock by considering the geometry and elasticity of surrounding rock.

7.2.2 Flat jack method

This method is based on the principle of releasing the stress state of the cavity first and then restoring it by pressurizing the flat jack (Fig. 7.9). It is also a widely used technique all over the world.

7.2.3 Hydro-fracturing and sleeve fracturing method

The hydro-fracturing method of stress measurement involves pressurizing a fluid into a section of borehole bounded by two packers sufficient to fracture the rock around the

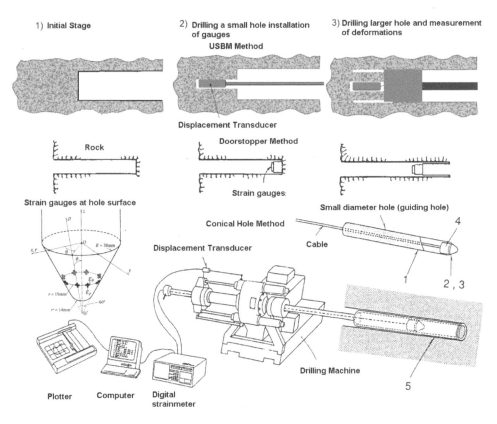

Figure 7.8 Various stress measurement techniques

Source: Arranged from Sugawara and Obara, 1993; Obert *et al.*, 1962; Leeman, 1964

hole. The fracture is intended to occur parallel to the axis of the borehole. The method assumes that one of the principal stresses (generally vertical stress) is parallel to the borehole axis, and then the maximum and minimum stresses in the plane perpendicular to the borehole axis are obtained from pressure readings (Fig. 7.10). In addition, the tensile strength of the surrounding rock is necessary for computing stress components. The sleeve-fracturing method applies the internal pressure in the borehole through mechanical jacks instead of pressurized fluid.

7.2.4   Acoustic emission (AE) Method

The acoustic emission method utilizes the Kaiser effect for inferring the stress state. Since the stress tensor is a symmetric second-order tensor, it has six independent components. As a result, it is necessary to perform uniaxial or triaxial tests in six different directions. According to the Kaiser effect (Kaiser, 1953), it is expected that the acoustic emission response will differ when the (deviatoric) stress level exceeds the one that material was previously subjected to, as illustrated in Figure 7.11. The fundamental complexity in rock

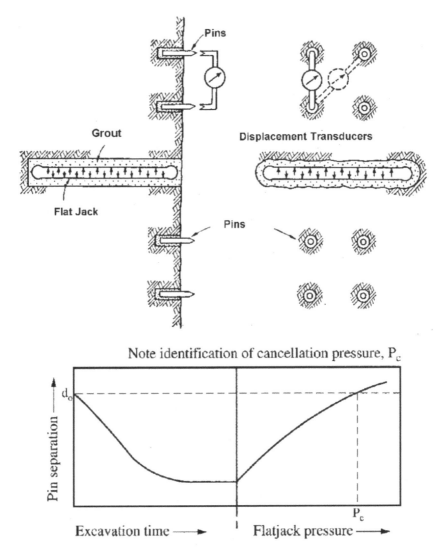

Figure 7.9 Illustration of stress measurement by the flat jack method
Source: Arranged from ISRM, 1986

mechanics is that the Earth's crust has a stress history. In actual experiments, one may find one or several stress levels. The question is how to select or to define the one, which reflects the current stress level that the rock was subjected to before unloading.

The acoustic emission method (AEM) was first suggested by Kanagawa *et al.* (1976, 1981) for inferring the *in-situ* stresses. Since then, many attempts were made to measure the stress state by this method. Nevertheless, the cost of acoustic emission measurement equipment was quite high in the past, and it could not become a widely accepted method. Recently, the cost of equipment becomes less, and the experiments can be easily performed

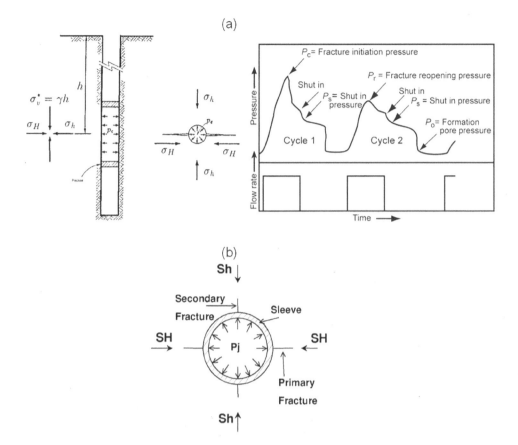

Figure 7.10 Illustration of stress measurement by (a) the hydrofracturing and (b) sleeve fracturing
     methods

Source: Arranged from Haimson, 1987; Stephanson *et al.*, 1986

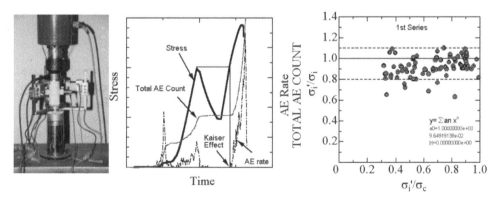

Figure 7.11 Illustration of stress measurement by AE method and Kaiser effect

Source: Arranged from Daido *et al.*, 2003

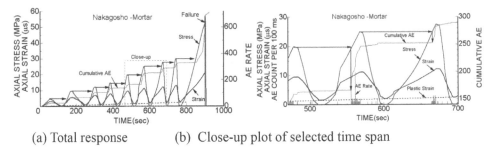

(a) Total response   (b) Close-up plot of selected time span

Figure 7.12 Acoustic emission response of mortar during cyclic loading

under laboratory conditions. As a result, there is a growing interest in stress measurements by this method (Holcomb, 1993; Hughson and Crawhord, 1987; Seto *et al.*, 1999; Wang *et al.*, 2000; Watanabe and Tano, 1999; Watanabe *et al.*, 1994, 1999).

The validity of the AE method is always questioned as other methods of *in-situ* stress measurements were already carried out at a given site. Daido *et al.* (2003) carried out some fundamental experiments, and they found that the Kaiser effect is a sound concept to infer the stress state to which rock has been previously subjected. However, there may be some deviations depending upon the stress level with respect to the strength of the rock (Fig. 7.12). Particularly, if the stress level is greater than the unstable crack threshold value defined by Bieniawski (1967), the inferred stresses may be less than the actual level that rock was subjected to. On the other, the inferred stress levels may be greater than the actual ones if the stress level is lower than the unstable crack propagation threshold stress level. Although many researchers associate AE events in rocks with micro-cracking, it would be better to utilize the concept of permanent straining associated with the plastic behavior of materials, macroscopically as also seen in Figure 7.12. If such a concept is used, the fundamental of the acoustic emission method would be based on the firm principles of mechanics.

7.3 *In-situ* stress inference methods

7.3.1 *Borehole breakout method*

When induced stresses exceed the strength of rock around the borehole, the methods based on the theory of elasticity cannot be used. Borehole breakouts are formed by spalling of fragments of the wellbore in a direction parallel to the minimum stress. Borehole spalling occurs during drilling and progresses with time. The identification and analysis of borehole breakouts as a technique for *in-situ* measurement of stress orientation and magnitude and for identifying orientation of both naturally occurring and induced fractures have received a great deal of attention. The inference of dimensions of borehole breakout is based on the Kirsch solution for circular holes with the use of the Mohr-Coulomb yield criterion. In practice, the dimensions of borehole breakout can be measured by borehole camera, laser scanning or mechanical caliper. This method is based on the principle of determining the stress state by observing the borehole breakouts in boreholes (Fig. 7.13). Zoback

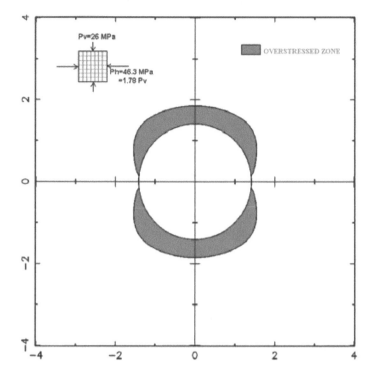

Figure 7.13 Borehole breakout method for stress estimation

applied the Kirsch solution together with Mohr-Coulomb yield criterion to estimate the shape, extent and orientation of breakouts to estimate the stress state. Although this technique is attributed to Zoback, the main principle was already proposed by Kastner (1962) for circular tunnels and was used by many (Talobre, 1957).

7.3.2 Fault type method

Possible stress states in the crust associated with faulting were first discussed by Hubbert (1951) and Anderson (1951). This approach assumes that the crust is in a plastic state and obeys a Mohr-Coulomb–type yield criterion. The stress states associated with various faulting regimes are outlined and discussed in a textbook by Jaeger and Cook (1979). When this technique is used, one of the stress components is always indeterminate. However, the vertical stress σ_z for strike–slip faulting regime is assumed to be equal to $(\sigma_H + \sigma_h)/2$ (Sibson, 1974). Parameters of the Mohr-Coulomb yield criterion may be temperature and/or confining pressure dependent. This fact is also considered in some publications (i.e. Brace and Kohlstedt, 1980; Shimada, 1993).

7.3.3 Fault striation method

Aydan (2000a) proposed a new method to infer the crustal stresses from the striations of the faults or other structural geological features, which may be quite useful in studying the

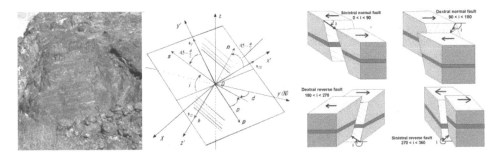

Figure 7.14 View of fault striation at Tanyeri near Erzincan and illustrations for notation for fault
striation method

stress state associated with past and current earthquakes. Figure 7.14 illustrates the notation
used in this method, and the final expression to infer the stress state is:

$$
\begin{bmatrix}
-q & l_1^2 & m_1^2 & 2l_1m_1 & 2l_1n_1 & 2m_1n_1 \\
-\beta & l_2^2 & m_2^2 & 2l_2m_2 & 2l_2n_2 & 2m_2n_2 \\
-1 & l_3^2 & m_3^2 & 2l_3m_3 & 2l_3n_3 & 2m_3n_3 \\
0 & l_1l_2 & m_1m_2 & (l_1m_2+l_2m_1) & (l_1n_2+l_2n_1) & (m_1n_2+m_2n_1) \\
0 & l_1l_3 & m_1m_3 & (l_1m_3+l_3m_1) & (l_1n_3+l_3n_1) & (m_1n_3+m_3n_1) \\
0 & l_2l_3 & m_2m_3 & (l_2m_3+l_3m_2) & (l_2n_3+l_3n_2) & (m_2n_3+m_3n_2)
\end{bmatrix}
\begin{Bmatrix}
N_{III} \\ N_{xx} \\ N_{yy} \\ N_{xy} \\ N_{xz} \\ N_{yz}
\end{Bmatrix}
=
\begin{Bmatrix}
-n_1^2 \\ -n_2^2 \\ -n_3^2 \\ -n_1n_2 \\ -n_1n_3 \\ -n_2n_3
\end{Bmatrix}
\qquad (7.23)
$$

where

$$
\mathbf{s}_I = \begin{Bmatrix} l_1 \\ m_1 \\ n_1 \end{Bmatrix} = \begin{bmatrix} s_x & s_y & s_z \\ n_x & n_y & n_z \\ b_x & b_y & b_z \end{bmatrix}^{-1} \begin{Bmatrix} \cos\left(45-\dfrac{\phi}{2}\right) \\ \cos\left(45+\dfrac{\phi}{2}\right) \\ 0 \end{Bmatrix};
$$

$$
\mathbf{s}_{III} = \begin{Bmatrix} l_3 \\ m_3 \\ n_3 \end{Bmatrix} = \begin{bmatrix} s_x & s_y & s_z \\ n_x & n_y & n_z \\ b_x & b_y & b_z \end{bmatrix}^{-1} \begin{Bmatrix} \cos\left(135-\dfrac{\phi}{2}\right) \\ \cos\left(45-\dfrac{\phi}{2}\right) \\ 0 \end{Bmatrix}
$$

$\mathbf{n} = \{n_x, n_y, n_z\}; s = \{s_x, s_y, s_z\}, n_x = \sin p \sin d; n_y = \sin p \cos d; n_z = \cos p \sin i$

$s_x = -\cos p \sin d \sin i; s_y = -\cos p \cos d \sin i; s_z = \cos p \sin i$

$\mathbf{b} = \mathbf{s} \times \mathbf{n}; \mathbf{b} = \{b_x, b_y, b_z\}; b_x = s_y n_z - s_z n_y; b_y = s_z n_x - s_x n_z; b_z = s_x n_y - s_y n_x$

$\beta = \dfrac{\sigma_{II}}{\sigma_{III}} = \dfrac{1}{2}(b - \sqrt{b^2 - 4c}), N_{III} = \dfrac{\sigma_{III}}{\sigma_z}; N_{xx} = \dfrac{\sigma_{xx}}{\sigma_z}, N_{yy} = \dfrac{\sigma_{yy}}{\sigma_z}, N_{zz} = 1, N_{xy} = \dfrac{\sigma_{yz}}{\sigma_z};$

$N_{yz} = \dfrac{\sigma_{yz}}{\sigma_z}; N_{xz} = \dfrac{\sigma_{xz}}{\sigma_z}, \sigma_z = \gamma h, b = \dfrac{(1+6\alpha^2)(q+1)}{1-3\alpha^2},$

$c = \dfrac{(1-3\alpha^2)(q^2+1) - q(1+6\alpha^2)}{1-3\alpha^2}, \alpha = \dfrac{2\sin\phi}{\sqrt{3}(3+\sin\phi)}, q = \dfrac{1+\sin\phi}{1-\sin\phi}$

and p is plunge, d is dip direction, i is striation angle, h is overburden, γ is unit weight, ϕ is friction angle, and c is cohesion.

7.3.4 Focal plane solution method

Aydan (Aydan and Kim, 2002; Aydan, 2003; Aydan and Tokashiki, 2003) recently advanced this method to infer the stress state of the Earth crust from focal plane solutions. The focal plane solutions used in geoscience are derived by assuming that the pure-shear condition holds. As a result of this assumption, one of the principal stresses is compressive while the other one is tensile in focal plane solutions. This condition may also imply that the friction angle of the fault is nil. Therefore, the principal stresses are inclined at an angle of 45 degrees with respect to the normal of slip direction. This condition is used to determine the *P*-axis and *T*-axis in focal plane solutions. Each focal plane solution involves the fault plane on which the sliding takes place and the auxiliary plane. The normal of the auxiliary plane corresponds to the slip vector, and it is orthogonal to the neutral plane on which the *P*-axis and *T*-axis exist.

7.3.5 Core-disking method

When drilling is done in brittle rock under high stress, it is often reported that core disking occurs. Jaeger and Cook (1963) was first to investigate the stress state resulting in core-disking (Fig. 7.15a). The stress state is very complicated during the drilling operation as the compressive and torsional tractions are imposed on the rock stubs by the drill bit in the close vicinity of the borehole end. Numerical analyses clearly indicate local high-tensile stresses in the closed vicinity of the borehole end. Nevertheless, the ridges on the surface of disking fractures are indicative of maximum *in-situ* stress perpendicular to the borehole axis (Fig. 7.15b).

7.3.6 Blast hole damage method

Aydan (2013) proposed a method to infer the *in-situ* stresses from the damage zones around the blast holes in this chapter, the blast hole-damage method (BDM). The several

(a) Core disking   (b) Surface of fracture plane

Figure 7.15 Core disking at Kaore Underground Powerhouse site

applications of the method to several sites where *in-situ* stress states are obtained by using direct or indirect techniques, and its validity is discussed in view of the measurements from other methods.

The stress state around a circular cavity in an elastic medium under biaxial far-field stresses are first obtained Kirsch (1898). These solutions are modified to incorporate the effect of uniform internal pressures (Jaeger and Cook, 1979). In a polar coordinate system, the radial, tangential and shear stresses around the circular cavity can be written in the following forms:

$$\sigma_r = \frac{\sigma_{10} + \sigma_{30}}{2}\left(1 - \left(\frac{a}{r}\right)^2\right) - \frac{\sigma_{10} - \sigma_{30}}{2}\left(1 - 4\left(\frac{a}{r}\right)^2 + 3\left(\frac{a}{r}\right)^4\right)\cos2(\theta - \beta) + p_i\left(\frac{a}{r}\right)^2 \quad (7.24a)$$

$$\sigma_\theta = \frac{\sigma_{10} + \sigma_{30}}{2}\left(1 + \left(\frac{a}{r}\right)^2\right) + \frac{\sigma_{10} - \sigma_{30}}{2}\left(1 + 3\left(\frac{a}{r}\right)^4\right)\cos2(\theta - \beta) - p_i\left(\frac{a}{r}\right)^2 \quad (7.24b)$$

$$\tau_{r\theta} = \frac{\sigma_{10} - \sigma_{30}}{2}\left(1 - 4\left(\frac{a}{r}\right)^2 + 3\left(\frac{a}{r}\right)^4\right)\sin2(\theta - \beta) \quad (7.24c)$$

where σ_{10}, σ_{30} is far-field principal stresses, a is radius of hole, r is radial distance, β is inclination of σ_{10} far-field stress from horizontal, θ is angle of the point from horizontal, and P_i is internal pressure applied onto the hole perimetry.

The yield criteria available in rock mechanics are:

Mohr-Coulomb

$$\sigma_1 = \sigma_c + q\sigma_3 \quad (7.25a)$$

Drucker-Prager

$$\alpha I_1 + \sqrt{J_2} = k \quad (7.25b)$$

Hoek and Brown (1980)

$$\sigma_1 = \sigma_3 + \sqrt{m\sigma_c\sigma_3 + s\sigma_c^2} \quad (7.25c)$$

Aydan (1995)

$$\sigma_1 = \sigma_3 + [S_\infty - (S_\infty - \sigma_c)e^{-b_1\sigma_3}]e^{-b_2 T} \quad (7.25d)$$

where

$$I_1 = \sigma_I + \sigma_{II} + \sigma_{III}; J_2 = \frac{1}{6}\left((\sigma_I - \sigma_{II})^2 + (\sigma_{II} - \sigma_{III})^2 + (\sigma_{III} - \sigma_I)^2\right)$$

$$\alpha = \frac{2\sin\phi}{\sqrt{3}(3 + \sin\phi)}, k = \frac{6c\cos\phi}{\sqrt{3}(3 + \sin\phi)}, c \text{ is cohesion}, \phi \text{ is friction angle}, q = \frac{1 + \sin\phi}{1 - \sin\phi},$$

σ_∞ is ultimate deviatoric strength, T is temperature, and m, s, b_1, b_2 are empirical constants.

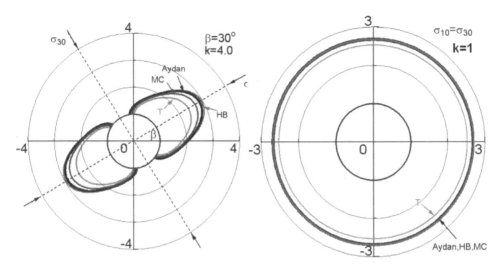

Figure 7.16 Estimated yield zones around the blast hole

Mohr-Coulomb and Drucker-Prager yield criterion are a linear function of confining or mean stress, while the criteria of Hoek-Brown and Aydan are of the nonlinear type (Fig. 7.16). Furthermore, Aydan's criterion also accounts for the effect of temperature. However, the effect of temperature is omitted in Figure 7.16 for the sake of comparison. If the yield criterion is chosen to be a function of minimum and maximum principal stresses, they can be given in the following form in terms of stress components given by Equation (7.24):

$$\sigma_1 = \frac{\sigma_\theta + \sigma_r}{2} + \sqrt{\left(\frac{\sigma_\theta - \sigma_r}{2}\right)^2 + \tau_{r\theta}^2} \tag{7.26a}$$

$$\sigma_3 = \frac{\sigma_\theta + \sigma_r}{2} - \sqrt{\left(\frac{\sigma_\theta - \sigma_r}{2}\right)^2 + \tau_{r\theta}^2} \tag{7.26b}$$

The damage zone around the blast hole under high internal pressure can be estimated using one of the yield criteria just listed. It should be noted that the yielding is induced by the high internal pressure in the blast hole, which is essentially different from *in-situ* stress-induced borehole-breakout. In other words, there will always be a damage zone around the blast hole perimeter when the blasting technique is employed. The blast hole pressure depends on the characteristics of the surrounding medium, the amount, layout and type of explosive, blasting velocity and the geometry of the blast hole. The blast hole pressure ranges from 100 to 10 GPa (i.e. Jaeger and Cook, 1979; Brady and Brown, 1985).

First we assume that the properties of surrounding rock have the values as given in Table 7.1, and the blast hole (internal) pressure has a value of 400 MPa. The rock chosen roughly corresponds to an igneous rock such as granitic rock. The maximum far-field stress is inclined at an angle of 30 degrees from horizontal, and the lateral stress

Table 7.1 Values of *in-situ* stress parameters and properties of yield functions

σ_{10} (MPa)	k	σ_c (MPa)	σ_t (MPa)	ϕ (o)	m	σ_∞ (MPa)	b
50.0	4.0	100.0	5.0	60	20	400	31.39

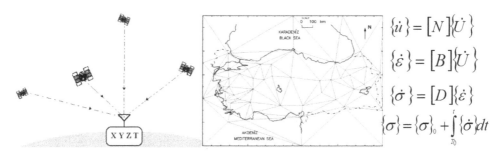

$$\{\ddot{u}\} = [N]\{\ddot{U}\}$$

$$\{\dot{\varepsilon}\} = [B]\{\dot{U}\}$$

$$\{\dot{\sigma}\} = [D]\{\dot{\varepsilon}\}$$

$$\{\sigma\} = \{\sigma\}_0 + \int_{t_0}^{t}\{\dot{\sigma}\}dt$$

Figure 7.17 Illustration of computation of strain and stress increments using the crustal deformations
Source: Arranged from Aydan, 2000b

coefficient has a value of 4. Figure 7.16 shows the example of computation for the given conditions. The largest yield zone is obtained for Hoek-Brown (HB) criterion while the tension cutoff criterion (T) results in smaller yield zone. The criterion of the Aydan estimates a slightly larger yield zone than the Mohr-Coulomb (MC) criterion. It is very interesting to note that the yielding propagates in the direction of maximum far-field stress. In other words, the elongation direction of the yield zone would be the best indicator of the maximum far-field stress in the plane of the blast hole.

7.3.7 Global positioning system method

Although it is difficult to obtain total stresses from GPS measurements, it may be possible to determine principal stress variations and their orientations in the tangential plane to the Earth surface from the processing of GPS measurements as proposed by Aydan (2000b, 2004) and Aydan *et al.* (2011, 2013c). However, if the coaxiality condition between stress rate and total stress in the crust holds, it may be also possible to infer the orientation of maximum and minimum horizontal stresses in the crust. It should be noted that it is geometrically possible to compute strain rate components in a plane tangential to the Earth's surface from the variation of positions of stations at a given time interval, although it is unlikely to obtain the components of strain rate components on other planes. For this purpose, a mesh and interpolation technique similar to those in the finite element method are utilized, and the nodes are selected to be corresponding to the location of GPS antennas (Fig. 7.17). The element type can be selected as desired with the consideration of the GPS network. Displacement increments of each node for a given time interval are used to compute the strain increments first, and then stress increments were computed with the incorporation of constitutive models of the Earth's crust. Aydan (2000b) used Hooke's law as a preliminary approach to compute the strain and stress increments.

7.4   Comparisons

In this section, the applications of the method are described briefly, together with direct stress measurements.

7.4.1   Location-based comparison

7.4.1.1   Yucca mountains (USA)

The first application was done to the site, the Yucca mountains Nuclear Waste Disposal Site. It should be noted that none of the stress relief methods can be applied since the rock around the hole becomes plastic at such depths. The same methods were also used at the Yucca mountain site for the same reasons (Stock *et al.*, 1992). Faults in this site are generally normal faults, and their average dip directions and dip are 285 degrees and 76 degrees, respectively. The striation direction was chosen as 90 degrees. An earthquake occurred near Yucca mountain on June 14, 2002. Computed results for the fault striation (FSM) and focal mechanism parameters (FMS) of that earthquake are given Table 7.2, together with *in-situ* stress measurements (Meas.). From the comparison of computed and measured results, it can be said that the computations are quite close to the measurements.

7.4.1.2   Underground powerhouse of Okumino pumped storage scheme (Japan)

The next example of application was done to the site called Okumino in Gifu Prefecture of Japan. The Neodani fault exists at a distance of 5 km in the southwest of Okumino Powerhouse. In 1891, an earthquake with a magnitude 8 occurred along this fault. The maximum horizontal displacement of the fault was 8 m, and the northeast side of the fault was downthrown except one location called Midori, where the northeast side was thrown 6 m upward and 4 m laterally. Matsuda (1974) suggested that such a movement occurs as a result of compressive forces at the fault bend as seen in the Midori site. Therefore, this fault is regarded to be a left-lateral strike-slip fault, and the striation direction was taken as 0 degrees. Stress measurements were carried out by Chubu Electric Power Company during the construction of a powerhouse (Tsuchiyama *et al.*, 1993). Computed results, together with *in-situ* stress measurements, are given in Table 7.3. From the comparison of computed and measured results, it can be said that the computations are once again quite close to the measurements.

Table 7.2 Comparison of inferred and measured *in-situ* stresses at Yucca Mountain Nuclear Waste Site

Method	σ_1			σ_2			σ_3			$\frac{\sigma_h}{\sigma_v}$	$\frac{\sigma_H}{\sigma_v}$	d_{σ_H}
	$\frac{\sigma_1}{\sigma_v}$	d_1	P_1	$\frac{\sigma_2}{\sigma_v}$	d_2	P_2	$\frac{\sigma_3}{\sigma_v}$	d_3	P_3			
FSM	1.05	285	74	0.73	195	0	0.35	105	16	0.41	0.73	15
Meas.	–	–	–	–	–	–	–	–	–	0.51	0.81	20–30
FMS	1.14	164	58	0.79	27	25	0.38	288	19	0.45	0.86	12

Table 7.3 Comparison of computed results with measurements for Okumino

Method	σ_1			σ_2			σ_3			$\frac{\sigma_h}{\sigma_v}$	$\frac{\sigma_H}{\sigma_v}$	d_{σ_H}
	$\frac{\sigma_1}{\sigma_v}$	d_1	P_1	$\frac{\sigma_2}{\sigma_v}$	d_2	P_2	$\frac{\sigma_3}{\sigma_v}$	d_3	P_3			
FSM	1.47	110	5	1.01	230	80	0.49	20	9	0.50	1.46	290
Meas.	1.50	104	4	1.07	204	73	0.29	13	17	0.11	1.50	283

Table 7.4 Comparison of *in-situ* stress measurements by various methods

Method	σ_1			σ_2			σ_3			$\frac{\sigma_h}{\sigma_v}$	$\frac{\sigma_H}{\sigma_v}$	d_{σ_H}
	$\frac{\sigma_1}{\sigma_v}$	d_1	P_1	$\frac{\sigma_2}{\sigma_v}$	d_2	P_2	$\frac{\sigma_3}{\sigma_v}$	d_3	P_3			
FSM	2.30	90	16	1.59	187	23	0.76	329	62	1.44	2.21	84
BDM	3.02	105	16	1.69	197	6	0.83	307	73	1.67	2.86	104
CBT	2.49	89	25	1.01	353	13	0.66	239	62	0.99	2.18	91
HFM	2.70	89	1	1.59	179	28	0.84	358	63	1.43	2.70	88
AEM										0.94	1.96	102

7.4.1.3   Planned underground powerhouse of Kaore pumped storage scheme (Japan)

An extensive *in-situ* stress measurement program was carried out at this site since the preliminary *in-situ* stress measurement program by the overcoring method yielded unusually high *in-situ* stresses that were not observed previously at other powerhouse construction sites in Japan (Ishiguro *et al.*, 1997). *In-situ* stress measurements involve using overcoring methods, such as the borehole deformation method (BDM), conically ended borehole method (CBT), hydrofracturing method (HFM) and acoustic emission method (AEM). In addition to these methods, a stress inference method called fault striation method (FSM) (Aydan, 2000a) based on fault striations was also used. The dominant fault on the site was denoted as F-1. Its dip direction and dip were 297 and 51, respectively. The rake angle (striation angle) was 24 degrees. The friction angle of the fault was set to 30 degrees for the stress inference computations. Table 7.4 compares the inferred *in-situ* stress with the measurements obtained from various methods.

7.4.1.4   Antique underground city of Derinkuyu (Turkey)

The fourth example of application was associated with Derinkuyu Underground City. The author and his research group carried out stress measurements for Derinkuyu using the AE method, GPS method and FSM (Aydan *et al.*, 1999; Watanabe *et al.*, 1999; Aydan and Ulusay, 2013). A normal fault with a 10–20 cm thick brecciate zone at the seventh floor of Derinkuyu Underground City was observed. The faults in this site are generally normal faults, and their average dip directions and dip are 290 degrees and 83 degrees. The striation direction was chosen as 85 degrees. Computed results, together with *in-situ* stress measurements, are given in Table 7.5. From the comparison of computed and measured results, it can be once again said that the computations are quite close to the measurements. These results are confirmed by those obtained from the GPS method (Aydan *et al.*, 1999; Aydan, 2000b). Figure 7.18 shows computer output for this site, together with a view

Table 7.5 Comparison of computed results with measurements for Derinkuyu

Method	σ_1			σ_2			σ_3			$\frac{\sigma_h}{\sigma_v}$	$\frac{\sigma_H}{\sigma_v}$	d_{σ_H}
	$\frac{\sigma_1}{\sigma_v}$	d_1	P_1	$\frac{\sigma_2}{\sigma_v}$	d_2	P_2	$\frac{\sigma_3}{\sigma_v}$	d_3	P_3			
FMS	1.04	84	77	0.71	181	1	0.35	271	13			
FSM	1.12	301	67	0.77	199	5	0.37	107	23	0.48	0.77	15
AEM	1.12	305	62	0.68	183	17	0.33	85.0	23	0.38	1.04	18

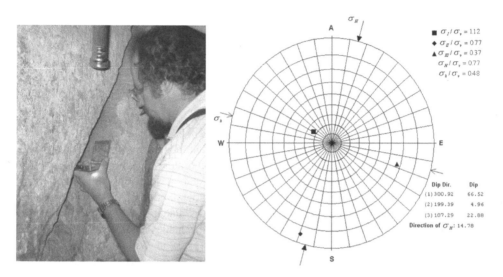

Figure 7.18 View of the fault in the antique Derinkuyu Underground City and inferred stress state

of the fault. Furthermore, this example of application indicates that this technique could be useful in archeological underground structures. Similar studies were also done on the stability assessment of underground tombs of pharaohs in the Luxor region of Egypt by the author and his group (Aydan *et al.*, 2008; Aydan and Geniş, 2004).

7.4.2 Application to regional crustal stress estimations

7.4.2.1 Turkey

The crustal stresses were measured at only three locations in Turkey until 1996, and two of these measurements were done by the Middle East Technical University of Turkey. Three additional measurements by the acoustic emission (AE) method were performed through an international joint research study on the living environment of the Derinkuyu Underground City of Turkey, which was supported by the Monbusho grant-in-aid (09044154). There is an urgent necessity to clarify the present stress state of Turkey and to monitor its variation in time and in space thereafter.

Aydan and his colleagues from Nihon University, Hacettepe University, and Istanbul Technical University have been continuing stress measurements by using the acoustic

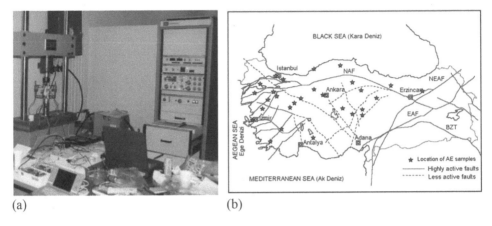

Figure 7.19 (a) View of stress measurement by the AE method at Hacettepe University, (b) sampling locations in Turkey for the AE method

emission technique since1998 (Aydan *et al*., 1999). The acoustic emission method was chosen because it is quite cheap to perform under laboratory conditions. At the initial stage of the research, the samples gathered from several locations in Turkey were brought to Japan, and stress measurements were carried out at Nihon University (Watanabe *et al*., 1999). Later, Watanabe and Tano (1999) Tano and Watanabe (1998) developed a portable system, and the tests were carried out at the rock mechanics laboratories of Hacettepe University (HU) and Istanbul Technical University (ITU) in Turkey (Fig. 7.19). The number of locations is presently 28, and the locations are shown in Figure 7.19. Table 7.6 summarizes stress measurements at several locations by the AE method. Figure 7.20 compares the stress measurements by the AE method, together with available *in-situ* stress measurements worldwide (Aydan, 1995) and some empirical relations for horizontal stress ratio over the overburden pressure developed by Hoek and Brown (1980) and Aydan and Paşamehmetoğlu (1994), as well as those from the thermo-elasto-plastic finite element analyses by Aydan (1995). It is interesting to note that the stress measurements by the AE method for Turkey are quite consistent with measurements using other methods and empirical formulas (Aydan and Kawamoto, 1997).

Since the stress inference from the AE method is often questioned, the stress state inferred from focal plane solutions of large earthquakes, fault striations and GPS measurements are compared to check the validity of AE stress measurements (Aydan *et al*., 2011). Figure 7.21 compares the maximum horizontal stress directions inferred from different methods. So far, the *in-situ* stress inference results from the AE method reasonably agree with those inferred from other methods. It seems that the AE may be a quite useful tool for both engineers and geoscientists to infer crustal stresses. Furthermore, the study carried out by the author and his colleagues may be quite unique in which there is no applications of the AE method of this scale in any country. It would be challenging and interesting to compare the results reported in this chapter with the measurements by other direct *in-situ* stress measurements to be performed in the future.

The North Anatolian Fault (NAF) and North-East Anatolian Fault (NEAF) run almost parallel to the shoreline of the Black Sea. The NAF is known to be 1500 km long and

Table 7.6 Principal stresses, maximum and minimum horizontal stresses and their directions for the different sampling locations in Turkey by the AE method

Location	σ_1 (MPa)	P_1 (°)	d_1 (°)	σ_2 (MPa)	P_2 (°)	d_2 (°)	σ_3 (MPa)	P_3 (°)	d_3 (°)	H (m)
Eynez	4.82	93	69	2.79	3	8	1.58	270	20	177
İzmir	3.72	146	50	2.03	45	9	0.50	308	39	80
Çayırhan	5.81	297	36	4.11	32	6	3.40	130	54	223
Küre	4.80	58	67	4.00	177	14	3.26	278	18	81
Dodurga	5.00	10	68	4.48	254	10	4.18	161	10	252
Zonguldak	12.7	141	12	12.0	265	68	9.47	047	24	505
İstanbul	1.89	342	52	1.28	75	1	0.70	165	38	60
Ankara	0.43	297	59	0.33	55	16	0.16	153	26	10
Bayburt	1.25	15	6	0.88	130	75	0.64	284	14	15
Denizli	0.92	188	4	0.79	283	33	0.68	089	56	23
Kırşehir	1.24	91	4	0.96	184	40	0.80	357	50	28
Sivas	1.41	283	16	1.33	176	42	0.92	029	43	25
Bigadiç	2.31	249	36	1.66	147	16	1.00	038	50	70
Kestelek	1.83	260	11	1.68	357	35	0.72	155	53	50
Eskişehir	0.55	185	39	0.34	279	5	0.18	015	51	10
Seydişehir (Doğankuzu)	5.14	70	14	3.20	180	53	2.11	331	33	100
Seydişehir (Mortaş)	3.47	341	26	2.40	241	21	2.31	117	56	90
Ordu	9.96	357	29	5.06	104	27	3.46	229	48	200
Emet	1.66	34	38	1.42	141	20	0.54	252	45	45
Kayseri	0.62	336	17	0.61	075	25	0.58	216	59	20
Orhaneli	3.91	352	54	2.99	101	18	1.44	210	29	100
Demirbilek	2.93	133	50	2.62	349	36	2.10	246	20	110
Marmara	1.70	92	49	1.53	327	21	0.46	232	34	45
Gebze	1.21	95	78	0.90	359	3	0.47	269	12	50
Avanos	0.37	256	79	0.25	50	4	0.10	321	4	18
Derinkuyu	0.45	305	62	0.27	183	17	0.13	85	23	20

Source: From Watanabe et al., 1999, 2003; Aydan et al., 1999

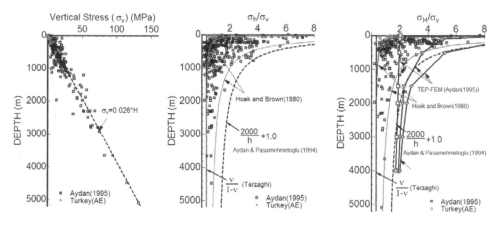

Figure 7.20 Comparison of stress measurements by the AE method with those from other methods

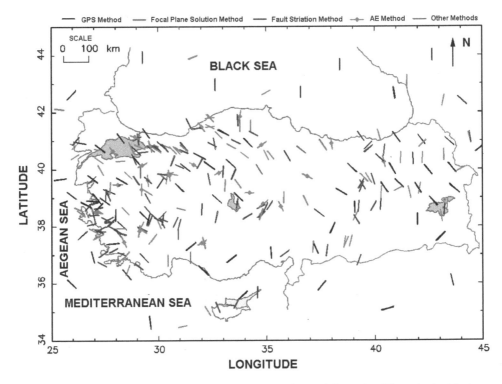

Figure 7.21 Comparison of maximum stress directions obtained from the AE method with those from other direct measurement methods and indirect stress techniques for Turkey

splays into several branches in western Turkey. It produces very large earthquakes and presents a great seismic danger to Turkey. The NEAF is less active and produces lesser earthquakes, which are smaller in magnitude as compared with those of the NAF. The Trakya fault (TF), which is thought to be an extension of the Tuzgölü and Eskişehir fault zone in Trakya (Thrace), is also a less active fault, and it produced the Edirne earthquake (Ms 5.2) in 1953.

Despite the great importance in the seismicity of the NAF, there are very few stress measurement studies in the vicinity of these three faults. The AE method as a direct stress measurement technique and focal mechanism solutions method (FMSM), fault striation method (FSM) and GPS method were used as indirect stress inference technique to infer the stress state of the northern part of Turkey, which includes the major active faults NAF, NEAF and TF and is delimited by latitudes 39–43 and longitudes 25–45.

Figure 7.22 shows the computed and measured results of maximum crustal horizontal stress direction for the northern part of Turkey. Figure 7.23 shows the maximum and minimum horizontal stress ratios normalized by the vertical stress for the same region. Although the data for the AE method and fault striations are still limited for the region, the results are promising to infer the stress state in a regional scale.

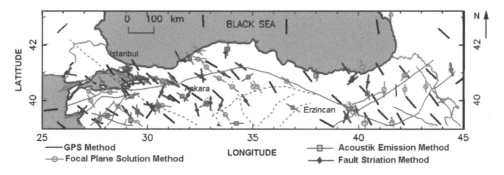

Figure 7.22 Directions of the maximum horizontal stress in Northern Anatolia

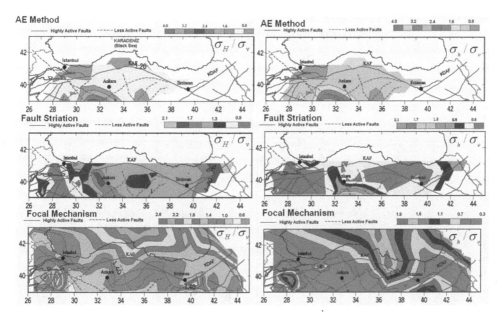

Figure 7.23 Comparison of measured and inferred maximum and minimum horizontal stress ratios normalized by the vertical stress

7.4.2.2 *Japan*

The approach used for Northern Turkey has been also applied to sites in Japan, where *in-situ* stress measurements are available, and their validity has been checked (Aydan, 2013b). The details of the data used in computations can be found in Aydan (2013b). The friction angle of the faults is assumed to be 30 degrees by considering the experimental data of Byerlee (1978). Figure 7.24 shows the stress states associated with the 1891 Nobi-Beya earthquake, which is the greatest intraplate earthquake, and the 2011 Great East Japan earthquake, which is the greatest subduction earthquake, respectively.

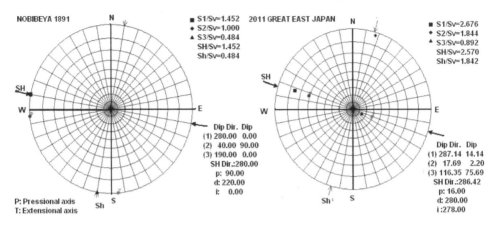

Figure 7.24 Stress states associated with the greatest intraplate and interplate earthquakes in Japan: (a) 1891 Nobi-Beya earthquake, (b) 2011 Great East Japan earthquake

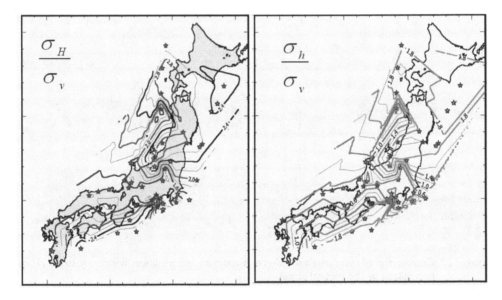

Figure 7.25 Computed maximum and minimum horizontal stress ratios normalized by vertical stress using the focal mechanism solutions

Source: From Aydan, 2013b

The maximum lateral stress ratio ranged between 1.3 and 3, while the minimum lateral stress coefficients range between 0.48 and 2.05. It was interesting to note that these values are in accordance with values assumed in the design of large caverns in Japan (Hibino, 2007). Figure 7.25 shows the contours of maximum and minimum lateral stress coefficients, respectively. It seems that lateral stress coefficients are higher in the close vicinity of the subduction zones of major plates in the close vicinity of Japan. From the figure, one

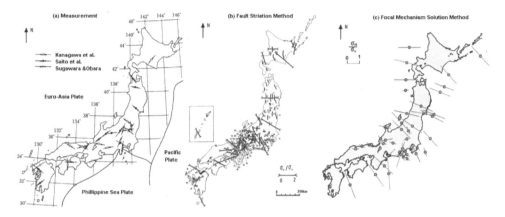

Figure 7.26 Active faults of Japan and the inference of directions and magnitudes of inferred maximum horizontal stresses from the proposed method

Source: Arranged from Aydan (2003, 2013b)

can infer that the lateral stresses are quite high in Northern Japan. These results are also in accordance with inferences from reported previously active faults of Japan (Aydan, 2003).

Figure 7.26 compares the directions of maximum horizontal stresses obtained from this study with those reported by Saito (1993) and Sugawara and Obara (1993). It is very interesting to note that the computed results are quite similar to those of the *in-situ* stress measurements. Furthermore, these results confirm the huge difference between the stress states of Euro-Asian and North American plates.

7.4.2.3   Ryukyu Archipelago

There are almost no *in-situ* stress measurements in the Ryukyu Islands except the location of the Yanbaru pumped storage scheme, although the Ryukyu Islands and their close vicinity are subjected to very large earthquakes from time to time. The authors have been recently investigating fault outcrops in the Ryukyu Islands and its close vicinity (Fig. 7.27a). In this subsection, the crustal stresses in the Ryukyu Islands from the striations of faults, and focal plane solutions are inferred and compared with reported stress measurements. The epicenters of earthquakes obtained by HARVARD (Seismological Division) along the Ryukyu arc are shown in Figure 7.27(c), and the parameters of the earthquakes nearby the Ryukyu Islands were chosen such that they satisfy the conditions that they have a focal depth less than of 30 km and moment magnitude greater Mw 6.0. The normalized stress tensor components by the vertical stress from the fault striations were obtained by using the method described previously.

The inferred maximum horizontal stress and its direction from the fault striations and focal mechanism solutions of earthquakes are shown in Figure 7.27, together with the *in-situ* stress measurement. Figure 7.28 shows the ratio of the maximum horizontal stress ratio as a function of orientation from north. As seen from the figures, the directions and magnitudes of the maximum horizontal stress indicate almost the same tendency. Particularly, the direct stress measurements are very close to those inferred from the focal

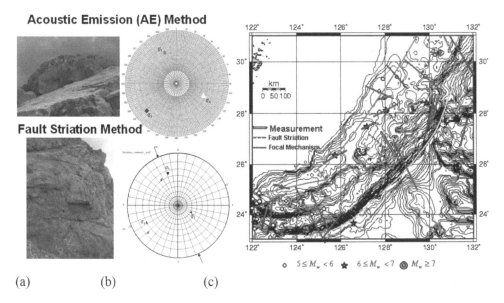

Figure 7.27 (a) Views of sampling and fault striation, (b) measured and inferred stresses for Motobu, (c) inferred directions of the maximum horizontal stress and its normalized value

Source: Arranged from Aydan and Tokashiki, 2003

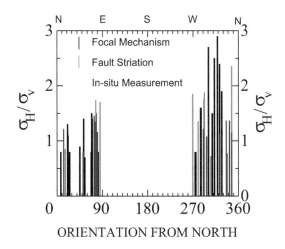

Figure 7.28 Comparison of the variation of measured and inferred maximum horizontal stress ratio with orientation

plane solutions and fault striations in the same vicinity. This conclusion is quite similar to those in other sites by the first author (Aydan, 2000a; Aydan and Kim, 2002; Aydan *et al.*, 2002; Watanabe *et al.*, 1999, 2003). The compressive horizontal stress is high in the vicinity of the Ryukyu trench, while its magnitude decreases along Okinawa trough.

7.5   Integration of various direct measurement and indirect techniques for *in-situ* stress estimation

As mentioned in previous sections, there are several techniques to measure or to infer the stress state in the Earth's crust. Each method or technique has its own merits and drawbacks. Furthermore, it is very difficult to obtain the true values of *in-situ* stresses, which are generally scattered. Therefore, the evaluation of *in-situ* stresses must be cross-checked using several methods or techniques. In any site, it is very likely to encounter several geological features. If the fault striation measurements are possible, the striations should be documented and ordered from new events to old events, and stress inferences should be carried out as a first step.

The second step is to search the focal plane solutions of earthquakes with a hypocenter depth less than 33 km deep. The most difficult aspect is how to select the plane, which is associated with the earthquake. If the earthquake caused some surface disturbance, those surface disturbances would yield information to determine the causative fault. Furthermore, the aftershock activity (if it is available) could provide very useful information for selecting the causative fault and determining its geometry (i.e. Aydan *et al.*, 1998). If this information is not available, it must be selected using some available documents on the geology, which may be slightly biased.

The third step is the utilization of direct stress measurement method. If the *in-situ* stress measurement is critical for the structure, then one or several of the direct stress measurement methods should be selected on the basis of economic, environmental and technological considerations. The depth of structure and accessibility to rock block sampling would be quite important factors. As mentioned previously, the acoustic emission (AE) method could be used for this purpose.

The utilization of borehole breakouts, fracturing in the vicinity of boreholes, damage around blast holes (Zoback *et al.*, 1985; Aydan, 2012, 2013a; Aydan *et al.*, 2013) could be used as supplementary information for obtaining or checking the results of direct and indirect *in-situ* stress evaluations. Furthermore, the stress variations utilizing the crustal deformation obtained from GPS measurements with the assumption that increments are coaxial with principal stresses could be one of the important tools to check the *in-situ* stress evaluations.

The final step would be designating the *in-situ* stress state of the site on the basis of information from direct and indirect techniques. If one or several direct stress measurement method are used, their results should be selected in view of the information from indirect stress inference techniques. If direct stress measurements are not available, the information from the active or recently moved faults and or focal mechanism solutions should be used to evaluate the *in-situ* stresses.

It should be born in mind that *in-situ* stress state would not remain the same in time (Aydan *et al.*, 2012; Aydan, 2015, 2016). This may be very important issue for long and large engineering structures such as tunnels, underground powerhouses, suspension bridges and arch and gravity dams. The variations may be estimated from the computation of stress rates using the GPS measurements and/or displacements from the processing of acceleration records by the EPS method. In addition, the inferences from the use of focal plane solutions may be used for the variations of the regional stress state with time.

7.6 Crustal stress changes

The GPS method may be used to monitor the deformation of the Earth's crust continuously with time. From these measurements, one may compute the strain rates and probably the stress rates. The stress rates derived from the GPS displacement rates can be effectively used to locate the areas with high seismic risk as proposed by Aydan (2000a, 2006, 2013a). Thus, daily variations of derived strain–stress rates from dense, continuously operating GPS networks in Japan and the United States may provide high-quality data to understand the behavior of the Earth's crust preceding earthquakes.

This approach is applied to the 2011 Great East Japan earthquake using the area shown in Figure 7.29 and an element consisting of Oshika, Wakuya and Rifu GPS stations of the GEONET (Aydan, 2013a). Figure 7.30a shows the computed principal stresses, maximum shear stress or disturbing stress and the orientation of the principal stress change. As noted from the figure, there is a release of the principal stress up to 5 MPa. The orientation of maximum principal stress change component was 269.2 degrees, and the orientation was 270.8 degrees after the earthquake while the coseismic change was less than 2 degrees. These changes should be added to those of the original stress field, which is not usually known unless *in-situ* stress measurements are carried out at the given area.

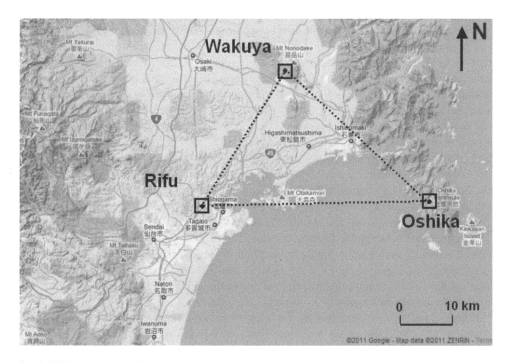

Figure 7.29 Locations of GPS stations used in computations

Sakaguchi *et al.* (2013) carried out *in-situ* stress measurements at a depth of 290 m at Kamaishi mine 170 km away to the northeast of the epicenter. The measured *in-situ* stress were almost twice the ones measured before the earthquake. This implies that the changes of crustal stresses were tremendous. The stress measurement attempt by Lin *et al.* (2013) above the epicentral area of the earthquake using the borehole breakout method (BBM) did not yield decisive results. Nevertheless, the maximum horizontal stress direction inferred from the BBM was 319 ± 23 degrees.

Aydan (2013b) recently computed the stress state of four major earthquakes in the vicinity of the epicentral area, including that the 2011 Great East Japan earthquake from their focal plane solutions using the method of Aydan (Aydan, 2000b; Aydan and Kim, 2002; Aydan *et al.*, 2002). The results of computation are given in Table 7.7, and the associated stereo projection of principal stresses is shown in Figure 7.30(b). The maximum horizontal stress orientations inferred from the focal mechanism solutions are also close to the results reported by Lin *et al.* (2013). The deviations of measured orientation reported by Lin *et al.* (2013) from those computed by Aydan (2013b) may be related to the heterogeneity of rock units encountered during the drilling.

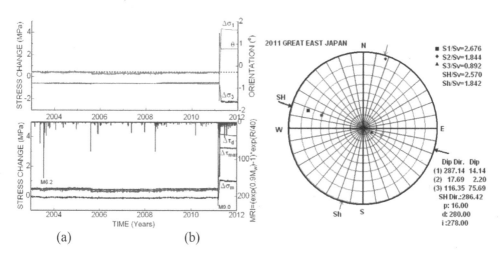

Figure 7.30 (a) Variations of stress changes computed from GPS measurements, (b) stress state inferred from the focal plane solution

Source: From Aydan, 2013

Table 7.7 Inferred crustal stresses in Japan

Method	$\frac{\sigma_1}{\sigma_v}$	d_1	P_1	$\frac{\sigma_2}{\sigma_v}$	d_2	P_2	$\frac{\sigma_3}{\sigma_v}$	d_3	P_3	$\frac{\sigma_h}{\sigma_v}$	$\frac{\sigma_1}{\sigma_v}$	θ
1978 Miyagi-oki	2.79	292.3	10.5	1.92	23.2	4.7	0.93	137.1	78.4	1.92	2.7	291.2
2003 Miyagi-hokubu	2.89	277.4	8	1.99	7.6	1.8	0.96	110.6	81.8	1.99	2.8	277.1
2008 Iwate-Miyagi	2.83	109.4	8.4	1.95	200.4	6.9	0.94	329.3	79.1	1.94	2.8	288.2
2011 Great East Japan	2.68	287.1	14.1	1.84	17.7	2.2	0.89	116.3	75.7	1.84	2.6	286.4

References

Amadei, B. & Stephansson, O. (1997) *Rock Stress And Its Measurement*. Chapman & Hall, London.

Anderson, E.M. (1951) *Dynamics of Faulting*. Oliver & Boyd, Edinburgh.

Anderson, L.D. & Hart, R.S. (1976) An earth model based on free oscillations and body waves. *Journal. Geophysical Research.*, 81, 1461–1475.

Angellier, J. (1984) Tectonic analysis of fault slip data sets. *Journal Geophysical Research.*, 89(B7), 5834–5848.

Aydan, Ö. (1993) A consideration on the stress state of the earth due to the gravitational pull. *The 7th Annual Symposium on Computational Mechanics, Tokyo*. pp. 243–252.

Aydan, Ö. (1995) The stress state of the earth and the earth's crust due to the gravitational pull. *35th US Rock Mechanics Symposium, Lake Tahoe*. pp. 237–243.

Aydan, Ö. (2000a) A stress inference method based on structural geological features for the full-stress components in the earth's crust. *Yerbilimleri*, (22), 223–236.

Aydan, Ö. (2000b) A stress inference method based on GPS measurements for the directions and rate of stresses in the earth' crust and their variation with time. *Yerbilimleri*, 22, 21–32.

Aydan, Ö. (2003) The Inference of crustal stresses in Japan with a particular emphasis on Tokai region. *Int. Symp. on Rock Stress, Kumamoto*. pp. 343–348.

Aydan, Ö. (2004) Implications of GPS-derived displacement, strain and stress rates on the 2003 Miyagi-Hokubu earthquakes. *Yerbilimleri*, (30), 91–102.

Aydan, Ö. (2012) Inferring in-situ stresses from damage around blasted holes. *Asian Rock Mechanics Symposium, Seoul*. pp. 157–167.

Aydan, Ö. (2013a) In-situ stress inference from damage around blasted holes. *Journal of Geo-system Engineering, Taylor and Francis*, 16(1), 83–91.

Aydan, Ö. (2013b) Inference of contemporary crustal stresses from recent large earthquakes and its comparison with other direct and indirect methods. *6th International Symposium on in Situ Rock Stress (RS2013), Sendai, Paper No. 1051*. pp. 420–427.

Aydan, Ö. (2013c) The applicability of crustal deformation monitoring by global positioning system to real-time earthquake prediction. *Bull. Inst. Oceanic Research And Development, Tokai University, No. 34*. pp. 1–16.

Aydan, Ö. (2015) Crustal stress changes and characteristics of damage to geo-engineering structures induced by the Great East Japan Earthquake of 2011. *Bull Eng Geol Environ*, 74(3), 1057–1070.

Aydan, Ö. (2016) The state of art on large cavern design for underground powerhouses and some long-term issues. In: *Lehr/Wiley Encyclopedia Energy: Science, Technology and Applications*. John Wiley and Sons, New York.

Aydan, Ö. & Geniş, M. (2004) Surrounding rock properties and openings stability of rock tomb of Amenhotep III (Egypt). *ISRM Regional Rock Mechanics Symposium, Sivas*. pp. 191–202.

Aydan, Ö. & Kawamoto, T. (1994) The stress state of the interior and the crust of the earth (in Japanese). *The 9th National Symposium in Rock Mechanics of Japan, Tokyo*. pp. 635–640.

Aydan, Ö. & Kawamoto, T. (1997) The general characteristics of the stress state in the various parts of the earth's crust. *Int. Symp. Rock Stress, Kumamoto*. pp. 369–373.

Aydan, Ö. and Kim, Y. (2002) The inference of crustal stresses and possible earthquake faulting mechanism in Shizuoka Prefecture from the striations of faults. *Journal of the School of Marine Science and Technology*, Tokai University, (54), 21–35.

Aydan, Ö. & Paşamehmetoğlu, G. (1994) In-situ measurements and lateral stress coefficients in various parts of the earth. *Kaya Mekaniği Bülteni*, 10, 1–17.

Aydan, Ö. & Tokashiki, N. (2003) The inference of crustal stresses in Ryukyu Islands. *3rd International Symposium on Rock Stress, Kumamoto*. pp. 349–354.

Aydan, Ö. & Ulusay, R. (2013) Geomechanical evaluation of Derinkuyu Antique Underground City and its implications in geoengineering. *Rock Mechanics and Rock Engineering*, Springer, 46(4), 731–754.

Aydan, Ö., Ulusay, R., Kumsar, H., Sonmez, H. & Tuncay, E. (1998) *A Site Investigation of Adana-Ceyhan Earthquake of June 27, 1998.* Turkish Earthquake Foundation, Istanbul. Report No. TDV/ DR 006-30, 131p.

Aydan, Ö., Kawamoto, T., Yüzer, E., Ulusay, R., Erdoğan, M., Akagi, T., Ito, T., Tokashiki, N. & Tano, H. (1999) *A Research on the Living Environment of Derinkuyu Underground City, Central Turkey.* Report of MONBUSHO Research Project No.: 09044154, Japan (in Japanese).

Aydan, Ö., Kumsar, H. & Ulusay, R. (2002) How to infer the possible mechanism and characteristics of earthquakes from the striations and ground surface traces of existing faults. *JSCE, Earthquake and Structural Engineering Division,* 19(2), 199–208.

Aydan, Ö., Daido, M., Tano, H., Nakama, S. & Matsui, H. (2006) The failure mechanism of around horizontal boreholes excavated in sedimentary rock. 50th US Rock mechanics Symposium, Paper No. 06-130 (on CD).

Aydan, Ö., Tano, H., Geniş, M., Sakamoto, I. & Hamada, M. (2008) Environmental and rock mechanics investigations for the restoration of the tomb of Amenophis III. *Japan: Egypt Joint Symposium New Horizons in Geotechnical and Geoenvironmental Engineering, Tanta, Egypt.* pp. 151–162.

Aydan, Ö., Ohta, Y., Daido, M., Kumsar, H., Genis, M., Tokashiki, N., Ito, T. & Amini, M. (2011) Chapter 15: Earthquakes as a rock dynamic problem and their effects on rock engineering structures. In: Zhou, Y. & Zhao, J. (eds.) *Advances in Rock Dynamics and applications.* CRC Press, Taylor and Francis Group. London, pp. 341–422.

Aydan, Ö., Uehara, F. & Kawamoto, T. (2012) Numerical study of the long-term performance of an underground powerhouse subjected to varying initial stress states, cyclic water heads, and temperature variations. *International Journal of Geomechanics, ASCE,* 12(1), 14–26.

Aydan, Ö., Sato, T., Hikima, R. & Tanno, T. (2013) Inference of in-situ stress by Blasthole Damage Method (BDM) at Mizunami URL and its comparison with other direct and indirect methods. *6th International Symposium on In Situ Rock Stress (RS2013), Sendai, Paper No. 1050.* pp. 359–368.

Bieniawski, Z.T. (1967) Mechanism of brittle fracture of rocks. *Int. J. Rock Mech. Min. Sci.,* 4, 365–430.

Brace, W.F. & Kohlstedt, D.L. (1980) Limits on lithospheric stress imposed by laboratory experiments. *Journal Geophysical Research.,* 85, 6248–6252.

Brady, B.H.G. & Brown, E.T. (1985) *Rock Mechanics for Underground Mining.* George Allen & Unwin, 527p.

Brown, E.T. and Hoek, E. (1978) Trends in relationships between measured in-situ stresses and depth. *Int. J. Rock Mech. Min. Sci.,* 15, 211–215.

Byerlee, J. (1978) Friction of rocks. *Pure Applied Geophysics,* 116, 615–626.

Daido, M., Aydan, Ö., Kuwae, H. & Sakoda, S. (2003) An experimental study on the validity of Kaiser effect in stress measurements by acoustic emissions method (AEM): Rock stress. In: Sugawara, Obara & Saito (eds.). pp. 271–274.

Fowler, C.M.R. (1990) *The Solid Earth: An Introduction to Global Geophysics.* Cambridge University Press, Cambridge.

Handin, J. (1966) Strength and ductility. *Geol. Soc. Am. Mem.,* 97, 223–289.

HARVARD (2013) *Focal Plane Solutions.* www.seismology.harvard.edu/.

Hast, N. (1969) The state of stress in the upper part of the Earth's crust. *Tectonophysics,* 8, 169–211.

Heim, A. (1878) *Mechanismus der Gebirgsbildung.* Bale.

Herget, G. (1986) Changes of ground stresses with depth in the Canadian Shield. *Int. Symp. on Rock Stress and Rock Stress Measurements, Stockholm.* pp. 61–68.

Hibino, S. (2007) *Necessary Knowledge of Rock Mass for Engineers (Gijutsusha ni hitsuyo na ganban no chishiki).* Kajima Pub. Co., Tokyo (in Japanese).

Hirth, G. & Tullis, J. (1994) The brittle-plastic transition in experimentally deformed quartz aggregates. *J. Geophysical Research,* 99, 11, 731–11, 747.

Hoek, E. & Brown, E.T. (1980) *Underground Excavations in Rock.* Inst. of Mining Metal., London. 527p.

Holcomb, D.J. (1993) General theory of the Kaiser effect. *International Journal of Rock Mechanics Mining Sciences and Geomechanics Abstracts*, 30(7), 929–935.

Hughson, D.R. & Crawhord, A.M. (1987) Kaiser effect gauging: The influence of confining stress on its response. In: Herget, G. & Vongpaisal, S. (eds.) *Proc. of the 6th ISRM International Congress on Rock Mechanics*, Balkema, Montreal. pp. 981–985.

Ishiguro, Y., Nishimura, H., Nishino, K. & Sugawara, K. (1997) Rock stress measurement for design of underground powerhouse and considerations In: Sugawara, E. & Obara, Y. (eds.) *Rock Stress*. Balkema, Rotterdam. pp. 491–498.

ISRM (1986) Suggested method for deformability determination using a large flat jack technique. *International Journal of Rock Mechanics Mining Sciences and Geomechanics Abstracts*, 23, 131–140.

Jaeger, J.C., and N.G.W. Cook (1963), Pinching-off and disking of rocks, J. Geophys. Res. 68, 6, 1759–1765,

Jaeger, J.C. & Cook, N.G.W. (1979) *Fundamentals of Rock Mechanics*, 1st edition. Methuen, London. p. 593.

Jeffreys, H. & Bullen, K.E. (1940) *Seismological Tables*. British Association for the Advancement of Science, London.

Kaiser, J. (1953) Erkentnisse und Folgerungen aus der Messung Von Gerauschen bei Zugbeanspruchung von Metallischen Werkstoffen. *Archiv Fur das Eisenhuttenwesen*, 24, 43–45.

Kanagawa, T., Hayashi, M. & Nakasa, H. (1976) Estimation of spatial geo-stresses in rock samples using the Kaiser effect of Acoustic emission. *Proceedings of Third Acoustic Emission Symposium, Tokyo, Japan*. pp. 229–248.

Kanagawa, T., Hayashi, M. & Kitahara, Y. (1981) Acoustic emission and overcoring methods. *Proceedings of the International Symposium on Weak Rock, Tokyo*. pp. 1205–1210.

Kastner, H. (1962) *Statik des Tunnel-und Stollenbaues*. Springer, Berlin Heidelberg, New York.

Kirsch, G. (1898) Die Theorie der Elastizitat und die Bedurfnisse der Festigkeitslehre. *VDI*, 2(42), 707.

Leeman, E.R. (1964) The measurement of stress in rock, part I, the principles of rock stress measurements; part II, borehole rock stress measuring instruments. *Journal of the South African Institute of Mining and Metallurgy*, 65(2), 45–114.

Lin, W. *et al.* (2013) Stress state in the largest displacement area of the 2011 Tohoku-Oki earthquake. Science 339, 687–690

Matsuda, T. (1974) Surface faults associated with Nobi (Mino-Owari) earthquake of 1891, Japan. *Bulletin of Earthquake Research Institute of Tokyo University*, 13, 85–126.

McGarr, A. (1980) Some constraints on levels of shear stress in the crust from observations and theory. *Journal Geophysical Research*, 85(B11), 6231–6238.

Nadai, A.L. (1950) *Theory of Flow and Fracture of Solids*, Vol. 2. pp. 623–624.

Obert, L., Merrill, R.H. & Morgan, T.A. (1962) *Borehole Deformation Gauge for Determining the Stress in Mine Rock*. US Bureau of Mines Report of Investigation, RI 5978.

Owen, D.R.J. & Hinton, E. (1980) *Finite Element in Plasticity: Theory and Practice*, Pineridge Press Ltd, Swansea.

Saito, T. (1993) Chapter 8: In-situ stresses. In: Kobayashi, S. (ed.) *Rock Mechanics*. Maruzen, Tokyo.

Sakaguchi, K., Koyano, T. & Yokoyama, T. (2013) Change in stress field in the Kamaishi Mine, associated with the 2011 Tohoku-oki earthquake. Proc. the 13th Japan Symposium on Rock Mechanics, Okinawa, 513–517 (in Japanese).

Salustowicz, A. (1968) Einige Bemerkungen über den Spannungszustand im Inneren der Erdkugel. *Felsmekanik*, Suppl. 7, 93–98.

Seto, M., Nag, D.K. & Vutukuri, V.S. (1999) In-situ rock stress measurement from rock cores using the acoustic emission method and deformation rate analysis. *Geotechnical and Geological Engineering*, 17, 241–266.

Shimada, M. (1993) Two types of brittle fracture of silicate rocks and scale effect on rock strength: Their implications in the earth crust. *Proc. Scale Effects in Rock Masses*, Lisbon, 55–62.

Sibson, R.H. (1974) Frictional constraints on thrust, wrench and normal faults. *Nature*, 249, 542–544.

Stephanson, O., Sarkka, P. & Myrvang, A. (1986) State of stress in Fennoscandia. *Int. Symp. on Rock Stress and Rock Stress Measurements*, Stockholm, 21–32.

Stock, J.M., Healy, J.H., Hickman, S.H. & Zoback, M.D. (1992) Hydraulic fracturing stress measurements at Yucca Mountain, Nevada, and relationship to regional stress field. *Journal. Geophysical Research*, 90, 8691–8706.

Sugawara, K. & Obara, Y. (1993) Measuring rock stress. In *Comprehensive Rock Engineering*, Chapter 21, Volume 3, Pergamon Press. pp. 533–552.

Talobre, J. (1957) *The Mechanics of Rocks, Dunod* (in French), Paris.

Terzaghi, K. & Richart, F.E. (1952) Stresses in rock about cavities. *Geotechnique*, 3, 57–90.

Tsuchiyama, S., Aydan, Ö. & Ichikawa, Y. (1993) Deformational behaviour of a large underground opening and its back analysis. *Int. Symp. Assessment and Prevention of Failure Phenomena in Rock Engineering, Istanbul*. Pp. 865–870.

Watanabe, H. & Tano, H. (1999) In-situ stress estimation of Cappadocia Region using the increment of AE event count rate. *Journal of College of Engineering*, Nihon University, 41(1), 35–42 (in Japanese).

Watanabe, H., Tano, H. & Akatsu, T. (1994) Fundamental study on pre-stress measurement of triaxial compressed rock. *Journal of College of Engineering*, Nihon University, 35(A), 11–19 (in Japanese).

Watanabe, H., Tano, H., Ulusay, R., Yüzer, E., Erdoğan, E. & Aydan, Ö. (1999) The initial stress state in Cappadocia. In: K. Matsui & H. Shimada (eds) *Proc of the '99 Japan-Korea Joint Symposium on Rock Engineering, Fukuoka, Japan*. Pp. 249–260.

Watanabe, H., Tano, H., Aydan, Ö., Ulusay, R. & Bilgin, A.H., Seiki, T. (2003) The measurement of the in-situ stress state by Acoustic Emission (AE) method in weak rocks. *RS-Kumamoto, Int. Symp. On Rock Stress, Kumamoto*.

Zoback, M.D., *et al.* (1985) Well bore breakouts and in-situ stress. *Journal Geophysical Research*, 90, 5523–5530.

Zoback, M.D. & Healy, J.H. (1992) In-situ stress measurements to 3.5 km depth in the Cajon Pass scientific research borehole: Implications for the mechanics of crustal faulting. *Journal Geophysical Research*, 97, 5039–5057.

Zoback, M.L. & Zoback, M.D. (1980) State of stress in the conterminous United States. 85, 6113–6156, 1980.

Chapter 8

Analytical methods

The solution of governing equations of coupled or uncoupled motion, mass transportation and energy transport phenomena requires certain methods, which may be analytical and numerical. When the resulting equations, including initial and/or boundary conditions, are simple to solve, the analytical methods are preferred. As the resulting equations and initial and/or boundary conditions are generally complex in many rock engineering problems, the use of numerical methods such as finite difference, finite element or boundary element methods becomes necessary.

8.1 Basic approaches

The fundamental governing equations presented in Chapter 4 are in the form of ordinary differential or partial differential equations for mass, momentum, energy conservation laws with chosen constitutive laws. The analytical methods basically attempt to solve the equations in their original form, and they can be categorized as intuitive or separation of variable techniques.

8.1.1 Intuitive function methods

This method is fundamentally based on choosing a function intuitively, which satisfy boundary/initial conditions. For example, the well-known Kirsch's solutions for circular opening under biaxial far-field stresses are based on this approach, and it utilizes Airy's stress function. The readers are advised to several books such as Jaeger and Cook (1979) on this aspect.

8.1.2 Separation variable method

This method assumes that the solution consists of the convolution of several functions of the independent variables (e.g. Kreyszig, 1983). The insertion of these functions to partial differential equations results in separated ordinary differential equations. Then the integral coefficients are determined so that the boundary and/or initial conditions are satisfied. Particularly, the determination of integral coefficients may be quite cumbersome. This method is also known as Fourier's method.

8.1.3 Complex variable solution

This method is a very powerful tool for the solution of many problems in elasticity. The method was originally devised by Kolosov (1909), and it is further expanded by several SSCB mathematicians. For example, Muskhelishvili (1962) provided a comprehensive textbook on this solution method. Similar textbooks by Milne-Thomson (1960), Green and Zerna (1968) and England (1971) can be found in literature. The well-known textbook by Kreyszig (1993) describes this method. The method is quite powerful in solving the partial differential equations subjected to very complex far-field boundary stress conditions for anisotropic elastic materials. Several applications of this method are described in the textbook by Jaeger and Cook (1979). Gerçek (1996) provided the application of the method for stress concentration around cavities having different shapes under biaxial far-field stresses, in which the integral constants are obtained numerically. The textbook by Verruijt (1970) describes the utilization of this method in seepage problems.

8.2 Analytical solutions for solids

In this section, several specific applications of the analytical solutions are described.

8.2.1 Visco-elastic rock sample subjected to uniaxial loading

Aydan (1997) proposed a method to model the dynamic response of rock samples during loading. In this subsection, this method and several examples of its application to some typical situations are presented.

(a) Theoretical formulation

Let us consider a sample under uniaxial loading as shown in Figure 8.1(a). The force equilibrium of such a sample can be written in the following form (Fig. 8.1(b)):

$$\sigma = D_r \varepsilon + C_r \dot{\varepsilon} + \rho H \ddot{u} \tag{8.1}$$

where D_r, C_r, ρ, H are elastic modulus, viscosity coefficient, density and sample height. If acceleration $\ddot{u}$ is uniform over the sample and its strain ε is defined as:

$$\varepsilon = \frac{u}{H} \tag{8.2}$$

Equation (8.1) becomes:

$$\sigma = D_r \varepsilon + D_r \dot{\varepsilon} + \rho H^2 \ddot{\varepsilon} \tag{8.3}$$

Let us assume that stress is applied to the sample in the following form (Fig. 8.2):

for $0 \leq t \leq T_0$

$$\sigma = \frac{\sigma_0}{T_0} t \tag{8.4}$$

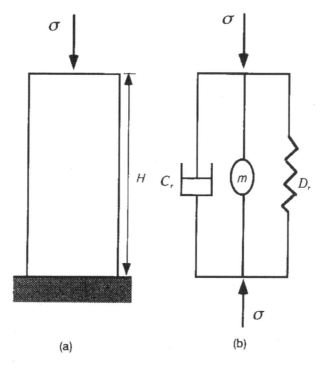

Figure 8.1 (a) Uniaxial compression test, (b) its mechanical model

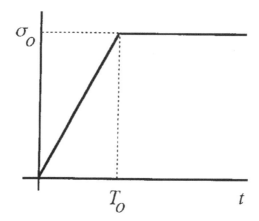

Figure 8.2 Time-history of uniaxial loading

for $t \geq T_0$

$$\sigma = \sigma_0 \tag{8.5}$$

The solutions of this ordinary differential equation are:

Case 1: Roots are real.

$$\varepsilon = C_1 e^{\lambda_1 t} + C_2 e^{\lambda_2 t} + \varepsilon_p \tag{8.6}$$

where

$$\lambda_1 = \frac{1}{2\rho H^2}\left(-C_r + \sqrt{C_r^2 - 4D_r \rho H^2}\right), \lambda_2 = \frac{1}{2\rho H^2}\left(-C_r - \sqrt{C_r^2 - 4D_r \rho H^2}\right)$$

for $0 \leq t \leq T_0$

$$\varepsilon_p = \frac{\sigma_0}{\rho H^2 T_0} \frac{1}{\lambda_1^2 \lambda_2^2}[\lambda_1 \lambda_2 t + (\lambda_1 + \lambda_2)]$$

for $0 \leq t \leq T_0$

$$\varepsilon_p = \frac{\sigma_0}{\rho H^2} \frac{1}{\lambda_1 \lambda_2}$$

Case 2: Roots are same.

$$\varepsilon = [C_1 + C_2 t]e^{\lambda t} + \varepsilon_p \tag{8.7}$$

where

$$\lambda = -\frac{C_r}{2\rho H^2}$$

for $0 \leq t \leq T_0$

$$\varepsilon_p = \frac{1}{\lambda^3} \frac{\sigma_0}{\rho H^2 T_0}$$

for $0t \geq T_0$

$$\varepsilon = \frac{1}{\lambda^2} \frac{\sigma_0}{\rho H^2}$$

Case 3: Roots are complex.

$$\varepsilon = e^{pt}[A cos qt + B \sin qt] + \varepsilon_p \tag{8.8}$$

where

$$p = -\frac{C_r}{2\rho H^2}, \quad q = \sqrt{4D_r\rho H^2 - C_r^2}$$

for $0 \leq t \leq T_0$

$$\varepsilon_p = \frac{1}{(p^2 + q^2)^2} \frac{\sigma_0}{\rho H^2 T_0} [(p^2 + q^2)t + 2p]$$

for $0t \geq T_0$

$$\varepsilon_p = \frac{1}{p^2 + q^2} \frac{\sigma_0}{\rho H^2}$$

Integration constants C_1 and C_2 can be determined from the following initial conditions:

for $0 \leq t < T_0$

$$\varepsilon = 0 \quad \text{at} \quad t = 0$$
$$\dot{\varepsilon} = 0 \quad \text{at} \quad t = 0 \tag{8.9}$$

for $t \geq T_0$

$$\varepsilon = \varepsilon_0 \quad \text{at} \quad t = T_0$$
$$\dot{\varepsilon} = \dot{\varepsilon}_0 \quad \text{at} \quad t = T_0 \tag{8.10}$$

Integration constants can be easily obtained for these conditions for each case. However, their specific forms are not presented as they are too lengthy.

(b) Applications

Several applications of the theoretical relations derived in the previous section are given herein to investigate the effects of viscosity coefficient, elasticity coefficient, loading rate, and sample height.

1   *The effect of viscosity coefficient*: Figure 8.3 shows the effect of the viscosity coefficient on the deformation responses of a sample. It is of great interest that when rock is elastic, an oscillating behavior must be observed. Furthermore, stress–strain relation is not linear, and it also oscillates as the applied stress is linearly increased. However, this oscillating behavior is suppressed as the viscosity coefficient increases.

2   *The effect of elasticity coefficient*: Figure 8.4 shows the effect of elasticity coefficient on the deformation responses of a sample with a viscosity coefficient of $0\ GPa \cdot s$. The amplitudes of the oscillating part and stationary part of strain decrease as the value of elasticity coefficient increases. Nevertheless, the oscillating behavior is apparent for each case.

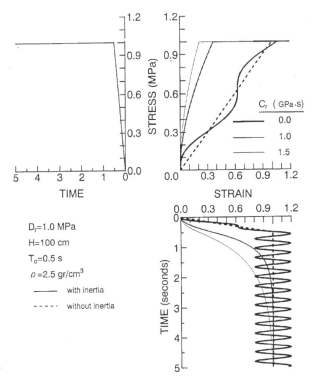

Figure 8.3 Effect of viscosity coefficient on dynamics response of a sample subjected to uniaxial compression

3   *The effect of loading rate:* Figure 8.5 shows the effect of loading rate on the deformation responses of a sample with a viscosity coefficient of 0 $GPa \cdot s$. While the amplitude of the stationary part of strain remain the same, the amplitude of the oscillating part of strain decreases as the loading rate decreases. Although the oscillating behavior could not be suppressed, the effect of oscillation tends to become smaller.

4   *The effect of sample height:* Figure 8.6 shows the effect of sample height on the deformation responses of a sample with a viscosity coefficient of 0 $GPa \cdot s$. The amplitudes of the oscillating part and stationary part of strain remain the same while the period of oscillations becomes larger as the value of sample height increases.

8.2.2   *Visco-elastic layer on an incline*

A semi-infinite slab on an incline is considered and is assumed to be subjected to instantaneous gravitational loading (Fig. 8.7(a)). The original formulation was developed by Aydan (1994), and it is adopted herein for assessing the dynamic response of a semi-infinite layer on an incline, which is a very close situation to the slope stability assessment of Terzaghi (1960).

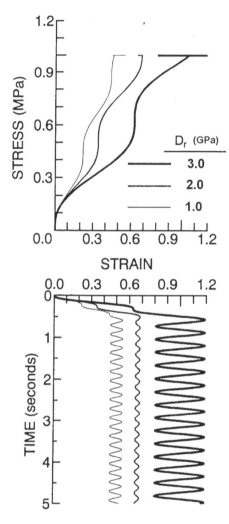

Figure 8.4 Effect of elastic coefficient on dynamics response of a sample subjected to uniaxial compression

Let us assume that the deformation is purely due to shearing under gravitational loading and that the slab behaves in a visco-elastic manner of Kelvin-Voigt type given by (Fig. 8.7(b)):

$$\tau = G\gamma + \eta\dot{\gamma} \tag{8.11}$$

where G is the elastic shear modulus and η is the viscos shear modulus. This model is known as the Voigt-Kelvin model (Eringen, 1980). When $G = 0$, then it simply corresponds to a Newtonian fluid. On the other hand, when $\eta = 0$, it corresponds to a Hookean solid.

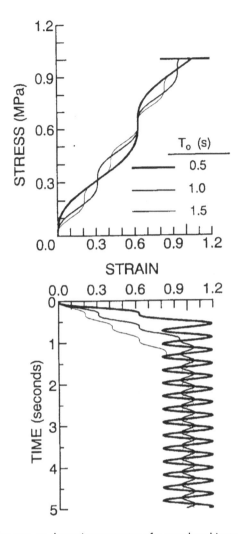

Figure 8.5 Effect of loading rate on dynamics response of a sample subjected to uniaxial compression

Let us consider an infinitesimal element within an inclined infinitely long layer as illustrated in Figure 8.8. The governing equation takes the following form by considering the equilibrium of the element by applying Newton's second law (Eringen, 1980):

$$\frac{\partial \tau}{\partial y} - \frac{\partial p}{\partial x} + \rho g \sin \alpha = \rho \ddot{u} \tag{8.12}$$

If the thickness of the liquified layer does not vary with x and the medium consists of the same material, then $\partial p / \partial x = 0$, and the preceding equation becomes:

$$\frac{\partial \tau}{\partial y} + \rho g \sin \alpha = \rho \ddot{u} \tag{8.13}$$

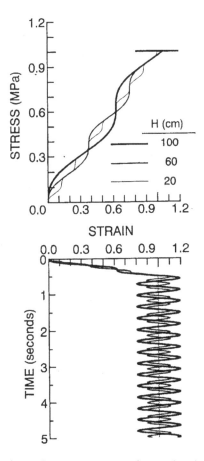

Figure 8.6 Effect of sample height on dynamics response of a sample subjected to uniaxial compression

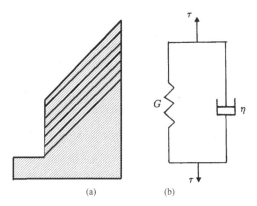

Figure 8.7 (a) Illustration of semi-infinite slope, (b) constitutive model
Source: (a) modified from Terzaghi, 1960)

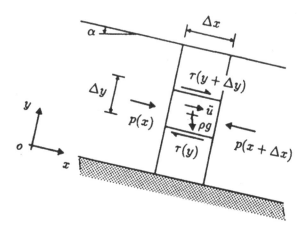

Figure 8.8 Mechanical model for shearing of semi-infinite slope

(a)   Closed-form solutions

Assuming that shear strain and shear strain rate can be defined as:

$$\gamma = \frac{\partial u}{\partial y}, \quad \dot{\gamma} = \frac{\partial \dot{u}}{\partial y} \tag{8.14}$$

and introducing the constitutive law given by Equation (8.11) into Equation (8.13) yields the following partial differential equation:

$$\rho \frac{\partial^2 u}{\partial t^2} + \eta \frac{\partial^2}{\partial y^2} \left(\frac{\partial u}{\partial t} \right) + G \frac{\partial^2 u}{\partial y^2} = \rho g \sin \alpha \tag{8.15}$$

Let us assume that the solution of this partial differential equation using the separation of variable technique is given as (i.e. Kreyszig, 1983; Zachmanoglou and Thoe, 1986):

$$u(y,t) = Y(y) \cdot T(t) \tag{8.16}$$

As a particular case based on intuitive approach, $Y(y)$ is assumed to be of the following form by considering an earlier solution of the equilibrium equation without the inertial term for semi-infinite slab with free-surface boundary conditions (Fig. 8.9(a)):

$$Y(y) = y\left(H - \frac{y}{2}\right) \tag{8.17}$$

Inserting this relation into Equation (8.15), we have:

$$\rho \frac{\partial^2 T}{\partial t^2} y\left(H - \frac{y}{2}\right) + \eta \frac{\partial T}{\partial t} + GT = \rho g \sin \alpha \tag{8.18}$$

Integrating the preceding equation with respect to y for bounds $y = 0$ and $y = H$ results in the following second-order nonhomogeneous ordinary differential equation:

$$\frac{\partial^2 T}{\partial t^2} - \frac{3\eta}{\rho H^2} \frac{\partial T}{\partial t} + \frac{3G}{\rho H^2} T = \frac{3g \sin \alpha}{H^2} \tag{8.19}$$

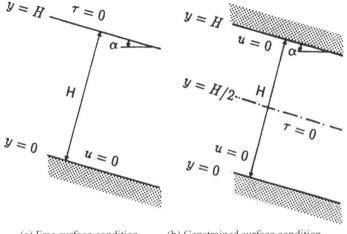

(a) Free surface condition   (b) Constrained surface condition

Figure 8.9 Boundary conditions

The solutions of this differential equation are:

Case 1: Roots are real.

$$T = C_1 e^{-\lambda_1 t} + C_2 e^{-\lambda_2 t} + \frac{1}{\lambda_1 \lambda_2} \frac{3g \sin \alpha}{H^2} \tag{8.20}$$

where

$$\lambda_1 = \frac{1}{2\rho H^2}(-3\eta + \sqrt{9\eta^2 - 12G\rho H^2}), \ \lambda_2 = \frac{1}{2\rho H^2}(-3\eta - \sqrt{9\eta^2 - 12G\rho H^2})$$

Case 2: Roots are same.

$$T = [C_1 + C_2 t]e^{-\lambda t} + \frac{1}{\lambda^2} \frac{3g \sin \alpha}{H^2} \tag{8.21}$$

where

$$\lambda = -\frac{3\eta}{2\rho H^2}$$

Case 3: Roots are complex.

$$T = e^{p} {}^{1mmt}[A\cos qt + B\sin qt] + \frac{1}{p^2 + q^2} \frac{3g \sin \alpha}{H^2} \tag{8.22}$$

where

$$p = -\frac{3\eta}{2\rho H^2}, \quad q = \sqrt{12G\rho H^2 - 9\eta^2}$$

Integration constants C_1 and C_2 can be determined from the following initial conditions:

$$u(y,t) = 0 \quad \text{at} \quad t = 0$$

$$\dot{u}(y,t) = 0 \quad \text{at} \quad t = 0 \tag{8.23}$$

For the preceding initial conditions, the integration constants for each case are:

Case 1: Roots are real.

$$C_1 = -\frac{1}{\lambda_1(\lambda_2 - \lambda_1)}\frac{3g\sin\alpha}{H^2}, \quad C_2 = \frac{1}{\lambda_2(\lambda_2 - \lambda_1)}\frac{3g\sin\alpha}{H^2} \tag{8.24}$$

Case 2: Roots are same.

$$C_1 = -\frac{1}{\lambda^2}\frac{3g\sin\alpha}{H^2}, \quad C_2 = \frac{1}{\lambda^2}\frac{3g\sin\alpha}{H^2} \tag{8.25}$$

Case 3: Roots are complex.

$$C_1 = -\frac{1}{p^2 + q^2}\frac{3g\sin\alpha}{H^2}, \quad C_2 = \frac{p}{q}\cdot\frac{1}{p^2 + q^2}\frac{3g\sin\alpha}{H^2} \tag{8.26}$$

Integration constants for the constrained boundary conditions can be obtained in a similar manner. This situation may be quite relevant to the response of soft layers sandwiched between two relatively rigid layers in underground excavations and trapdoor experiment used particularly for underground openings in soil (Terzaghi, 1946).

Figure 8.10 compares the solutions obtained from the closed-form solution and FEM for a 4 m thick semi-infinite slab. Using the solutions presented in this section, one may compare the expected responses under different circumstances. Such a comparison has been already done by (Aydan, 1994–1998). The negligence of the inertia component in Equation (8.15) results in a parabolic partial differential equation. Neglecting the viscous effect in the resulting equation would result in a differential equation of elliptical form. Figure 8.10 compares the responses obtained for three situations of the differential equation. As noted from the figure, all solutions converge to the solution obtained from the elliptical form (static case). The inertia component implies that displacement as well as resulting stresses and strains responses would be greater than those of the elliptical form.

8.2.2 One-dimensional bar

The equation of motion for the axial responses of rock bolts and rock anchors, together with the consideration of inertia component including mass proportional damping, can be written in the following form (Fig. 8.11):

$$\rho\frac{\partial^2 u_b}{\partial t^2} + h_a^*\frac{\partial u_b}{\partial t} = \frac{\partial\sigma}{\partial x} + \frac{2}{r_b}\tau_b \tag{8.27}$$

The analytical solutions for Equation (8.27) are extremely difficult for the given constitutive laws, boundary and initial conditions. However, it is possible to obtain solutions for simple cases, which may be useful for the interpretation of results of site investigations.

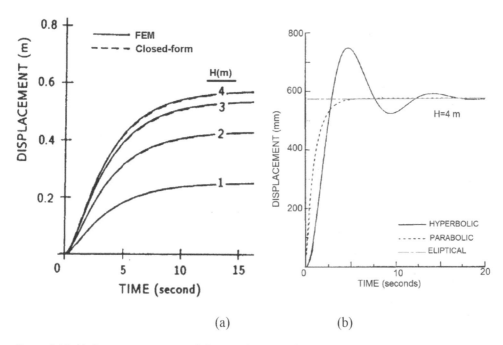

Figure 8.10 (a) Dynamic response of 4 m thick semi-infinite slab under instantaneously applied gravitation load, (b) comparison of responses obtained for hyperbolic, parabolic and elliptical forms of the differential equation for a 4 m thick slab under instantaneous gravitational loading

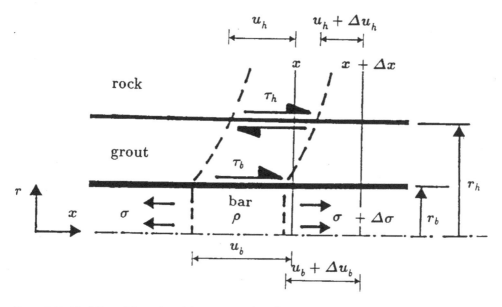

Figure 8.11 Modeling of dynamic axial response of tendons

Equation (8.27) may be reduced to the following form by omitting the effect of damping and interaction with surrounding rock as:

$$\frac{\partial^2 u_b}{\partial t^2} = V_p^2 \frac{\partial^2 u_b}{\partial x^2} \tag{8.28}$$

where

$$V_p = \sqrt{\frac{E_b}{\rho}}$$

The general solution of partial differential Equation (8.28), which is also known as the D'Alambert solution, may be given as:

$$u_b = h(x - V_p t) + H(x + V_p t) \tag{8.29}$$

For a very simple situation, the solution may be given as follows:

$$u_b = A\sin\frac{2\pi}{L}(x \pm V_p t) \tag{8.30}$$

Where L is tendon length. Thus, the eigenvalues of tendon may be obtained as follows:

$$f_p = n\frac{1}{2L}V_p, \quad n = 1, 2, 3 \tag{8.31}$$

Similarly, the eigenvalues of traverse vibration of the tendon under a given prestress may be obtained as follows:

$$f_T = n\frac{1}{2L}V_T, \quad n = 1, 2, 3 \tag{8.32}$$

where

$$V_T = \sqrt{\frac{\sigma_o}{\rho}}$$

8.2.3 Circular cavity in elastic rock under far-field hydrostatic stress

An analytical solution is herein presented in the case of a circular underground opening excavated in a hydrostatic state of stress. To start the derivations, the following equations are set:

Then, the governing equations for bolted and unbolted sections are (Fig. 8.12):

$$\frac{d\sigma_r}{dr} + \frac{\sigma_r - \sigma_\theta}{r} = 0 \tag{8.33}$$

where r is the distance from opening center, σ_r is the radial stress, and σ_θ is the tangential stress.

The constitutive law between stresses and strains of rock is of the following form:

$$\left\{\begin{array}{c} \varepsilon_r \\ \varepsilon_\theta \end{array}\right\} = \frac{1 - v_r^2}{E_r} \begin{bmatrix} 1 & -\dfrac{v_r}{1 - v_r} \\ -\dfrac{v_r}{1 - v_r} & 1 \end{bmatrix} \left\{\begin{array}{c} \sigma_r \\ \sigma_\theta \end{array}\right\}$$

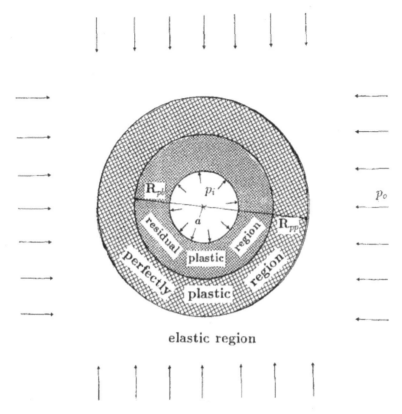

Figure 8.12 Notations

Source: From Aydan et al., 1993

or

$$\left\{ \begin{array}{c} \sigma_r \\ \sigma_\theta \end{array} \right\} = \frac{E_r(1 - v_r)}{(1 + v_r)(1 - 2v_r)} \left[\begin{array}{cc} 1 & -\dfrac{v_r}{1 - v_r} \\ -\dfrac{v_r}{1 - v_r} & 1 \end{array} \right] \left\{ \begin{array}{c} \varepsilon_r \\ \varepsilon_\theta \end{array} \right\} \tag{8.34}$$

where

E_r is the elastic modulus of rock, ε_r is the radial strain, ε_θ is the tangential strain in u:

$$\varepsilon_r = \frac{du}{dr}, \quad \varepsilon_\theta = \frac{u}{r} \tag{8.35}$$

The stresses and displacements in the bolted and unbolted sections can be obtained by solving the governing equations (Equation (8.33)) together with the constitutive law (Equation (8.34)) and the boundary conditions. Substituting the constitutive law (Equation (8.34)) in governing equations (Equation (8.33)), together with the relations (Equation (8.35)), results in the following differential equations:

$$r^2 \frac{d^2u}{dr^2} + r\frac{du}{dr} - u = 0 \tag{8.36}$$

The general solutions of the preceding differential equations are of the following forms:

$$u = A_1 r + A_2 \frac{1}{r} \tag{8.37}$$

By introducing the following boundary conditions for each section:

* Unbolted section

$$\sigma_r = P_b \quad \text{at} \quad r = b$$

$$\sigma_r = \sigma_0 \quad \text{at} \quad r = \infty$$

integration constants A_1 and A_2 are obtained as:

$$A_1 = \frac{(1 + v_r)(1 - 2v_r)}{E_r}\sigma_0, \quad A_2 = \frac{(1 + v_r)}{E_r}b^2(\sigma_0 - P_b).$$

Inserting the preceding constants in Equations (8.37) result in the following expressions after some manipulations:

$$u = \frac{1 + v_r}{E_r}\left[(1 - v_r)\sigma_0 r + (\sigma_0 - P_b)\frac{b^2}{r^2}\right] \tag{8.38}$$

The preceding displacement fields are for a body that is initially unstressed. As the initial displacement field has already taken place in the rock mass before the excavation of openings, this initial displacement field has to be subtracted from the preceding expressions. The initial displacement field is the state when P_b was equal to the far-field stress σ_0. Inserting these identities in the preceding expressions yields:

$$u_0 = \frac{1 + v_r}{E_r}(1 - v_r)\sigma_0 r \tag{8.39}$$

Finally, one obtains the following expressions for displacement fields due to the excavation of the opening:

$$u_e = \frac{1 + v_r}{E_r}(\sigma_0 - P_b)\frac{b^2}{r^2} \tag{8.40}$$

where subscript e denotes excavation. Variations in displacement fields will, in turn, bring about variations in radial and tangential stresses. These variations are obtained from the preceding relations together with Equation (8.35) and the constitutive law (Equation (8.34)) as:

$$\sigma_{re} = -(\sigma_0 - P_b)\frac{b^2}{r^2} \tag{8.41}$$

$$\sigma_{\theta e} = (\sigma_0 - P_b)\frac{b^2}{r^2} \tag{8.42}$$

The stress state in the surrounding medium is therefore defined as the sum of variation in the postexcavation state and the preexcavation stress state given as:

$$\sigma_r = \sigma_0 + \sigma_{re} = \sigma_0 - (\sigma_0 - P_b)\frac{b^2}{r^2} \tag{8.43}$$

$$\sigma_\theta = \sigma_0 + \sigma^*_{\theta e} = \sigma_0 + (\sigma_0 - P_b)\frac{b^2}{r^2} \tag{8.44}$$

Note: The preceding relations for stresses can also be directly obtained from the governing equations (Equation (8.33)) with the use of following identities:

$$\sigma_r + \sigma_\theta = 2\sigma_0 \tag{8.45}$$

and the same boundary conditions.

8.2.4 Unified analytical solutions for circular/spherical cavity in elasto-plastic rock

(a) General solution

CONSTITUTIVE LAWS

Generalized form of the constitutive law between stresses and strains of rock in the elastic region for radially symmetric problem (cylindrical and spherical openings) can be given as:

$$\left\{ \begin{array}{c} \sigma_r \\ \sigma_\theta \end{array} \right\} = \left[\begin{array}{cc} \lambda + 2\mu & n\lambda \\ \lambda & n\lambda + 2\mu \end{array} \right] \left\{ \begin{array}{c} \varepsilon_r \\ \varepsilon_\theta \end{array} \right\} \tag{8.46}$$

where n is the shape coefficient and has a value of 1 for a cylindrical opening and 2 for a spherical opening; σ_r is radial stress; σ_θ is tangential stress; ε_r is radial strain; ε_θ is tangential strain. λ and μ are Lamé constants and are given as:

$$\lambda = \frac{Ev}{(1+v)(1-2v)}, \quad \mu = \frac{E}{2(1+v)} \tag{8.47}$$

Where E is the elastic modulus of rock, and v is Poisson's ratio of rock.

EQUILIBRIUM EQUATION

When the problem is radially symmetric, the momentum law for static case takes the following form:

$$\frac{d\sigma_r}{dr} + n\frac{\sigma_r - \sigma_\theta}{r} = 0 \tag{8.48}$$

where r is distance from opening center.

COMPATIBILITY CONDITION

The compatibility condition between strain components for radially symmetric openings is given as:

$$\frac{d\varepsilon_\theta}{dr} + \frac{\varepsilon_\theta - \varepsilon_r}{r} = 0 \tag{8.49}$$

Relations between strain components and radial displacement (u) are:

$$\varepsilon_r = \frac{du}{dr}, \quad \varepsilon_\theta = \frac{u}{r} \tag{8.50}$$

BEHAVIOR OF ROCK MATERIAL

An elastic-perfect-residual plastic model as shown in Figure 8.13 approximates the behavior of rock. Although it is possible to consider the strain-softening behavior, it is extremely difficult to obtain closed-form solutions, and numerical techniques would be necessary. Rock was assumed to obey the Mohr-Coulomb yield criterion. Although it is possible to derive solutions for the Hoek-Brown criterion, it is not intentionally done as the generalized Hoek-Brown criterion violates the Euler theorem used in the classical theory of plasticity for constitutive modeling of rocks. Failure zones about radially symmetric openings excavated in rock mass for elastic-perfect-residual plastic behavior and yield functions for each region are illustrated in Figure 8.14 and given by:

$$\sigma_1 = q\sigma_3 + \sigma_c, \quad q = \frac{1 + \sin\phi}{1 - \sin\phi} \quad \text{perfectly plastic region} \tag{8.51a}$$

$$\sigma_1 = q^*\sigma_3 + \sigma_c^*, \quad q^* = \frac{1 + \sin\phi^*}{1 - \sin\phi^*} \quad \text{residual plastic region} \tag{8.51b}$$

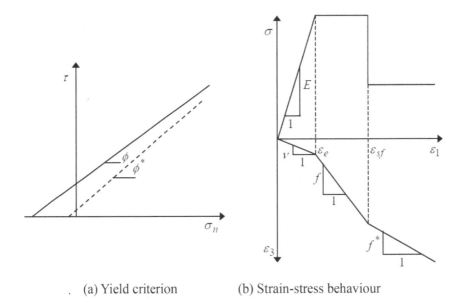

(a) Yield criterion   (b) Strain-stress behaviour

Figure 8.13 Mechanical models for rock mass

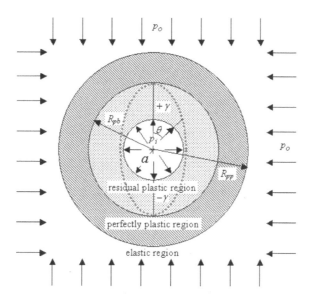

Figure 8.14 States about an opening and notations (Gravity is considered in dotted zone.)

where σ_1 is the maximum principal stress, σ_3 is the minimum principal stress, σ_c is the uniaxial compressive strength of intact rock, σ_c^* is the uniaxial compressive strength of broken rock, ϕ is the internal friction angle of intact rock and ϕ^* is the internal friction angle of broken rock.

Relations between total radial and tangential strains in plastic regimes are assumed to be of the following form:

$$\varepsilon_r = -f\varepsilon_\theta \quad \text{for perfectly plastic region} \tag{8.52a}$$

$$\varepsilon_r = -f^*\varepsilon_\theta \quad \text{for residual plastic region} \tag{8.52b}$$

where f and f^* are physical constants obtained from the tests. These constants may be interpreted as plastic Poisson's ratios (Aydan *et al.*, 1993).

(I)   STRESS AND STRAIN FIELD AROUND OPENING

(1)   RESIDUAL PLASTIC ZONE $(a \leq r \leq R_{pb})$

Inserting the yield criterion Equation (8.52b) into the governing Equation (8.48) with $\sigma_3 = \sigma_r$ and $\sigma_1 = \sigma_\theta$ yields:

$$\frac{d\sigma_r}{dr} + n(1 - q^*)\frac{\sigma_r}{r} = n\frac{\sigma_c^*}{r} \tag{8.53}$$

The solution of the preceding differential equation is:

$$\sigma_r = Cr^{n(q^*-1)} - \frac{\sigma_c^*}{q^* - 1} \tag{8.54}$$

The integration constant C is obtained from the boundary condition $\sigma_r = p_i$ at $r = a$ as:

$$C = \left(p_i + \frac{\sigma_c^*}{q^* - 1}\right) \frac{1}{a^{n(q*-1)}} \tag{8.55}$$

where p_i is internal or support pressure. Thus, the stresses now take the following forms:

$$\sigma_r = \left(p_i + \frac{\sigma_c^*}{q^* - 1}\right) \left(\frac{r}{a}\right)^{n(q^*-1)} - \frac{\sigma_c^*}{q^* - 1} \tag{8.56}$$

$$\sigma_\theta = q^* \left(p_i + \frac{\sigma_c^*}{q^* - 1}\right) \left(\frac{r}{a}\right)^{n(q^*-1)} - \frac{\sigma_c^*}{q^* - 1} \tag{8.57}$$

Solving the differential equation obtained by inserting the relation given by Equation (8.52b) in Equation (8.50) yields:

$$\varepsilon_\theta = \frac{A}{r^{f^*+1}} \tag{8.58}$$

The integration constant A is determined from the continuity of the tangential strain at perfect-residual plastic boundary $r = R_{pb}$ as:

$$A = \varepsilon_\theta^{pb} R_{pb}^{f^*+1} \tag{8.59}$$

ε_θ^{pb} in Equation (8.59) is the tangential strain at perfect-residual plastic boundary ($r = R_{pb}$) and it is specifically given by:

$$\varepsilon_\theta^{pb} = \eta_{sf} \varepsilon_\theta^{ep} \quad \eta_{sf} = \frac{\varepsilon_{sf}}{\varepsilon_e} \tag{8.60}$$

where η_{sf} is the tangential strain level at the perfect-residual plastic boundary (Fig. 8.13); ε_θ^{ep} is the tangential strain at the elastic-perfect plastic boundary as:

$$\varepsilon_\theta^{ep} = \frac{1 + v}{nE} (p_0 - \sigma_{rp}) \tag{8.61}$$

σ_{rp} in Equation (8.61) is the radial stress at elastic-perfect plastic boundary. As a result, the tangential strain in the surrounding rock becomes:

$$\varepsilon_\theta = \frac{1 + v}{nE} (p_0 - \sigma_{rp}) \eta_{sf} \left(\frac{R_{pb}}{r}\right)^{f^*+1} \tag{8.62}$$

(2) PERFECTLY PLASTIC ZONE ($R_{pb} \leq r \leq R_{pp}$)

Inserting the yield criterion Equation (8.51a) into the governing Equation (8.48) with $\sigma_3 = \sigma_r$ and $\sigma_1 = \sigma_\theta$ gives:

$$\frac{d\sigma_r}{dr} + n(1 - q)\frac{\sigma_r}{r} = n\frac{\sigma_c}{r} \tag{8.63}$$

The solution of the preceding differential equation is:

$$\sigma_r = Cr^{n(q-1)} - \frac{\sigma_c}{q - 1} \tag{8.64}$$

The integration constant C is obtained from the boundary condition $\sigma_r = \sigma_{rp}$ at $r = R_{pp}$ as:

$$C = \left(\sigma_{rp} + \frac{\sigma_c}{q-1}\right)\frac{1}{R_{pp}^{n(q-1)}} \tag{8.65}$$

Thus, the stresses now take the following forms:

$$\sigma_r = \left(\sigma_{rp} + \frac{\sigma_c}{q-1}\right)\left(\frac{r}{R_{pp}}\right)^{n(q-1)} - \frac{\sigma_c}{q-1} \tag{8.66}$$

$$\sigma_\theta = q\left(\sigma_{rp} + \frac{\sigma_c}{q-1}\right)\left(\frac{r}{R_{pp}}\right)^{n(q-1)} - \frac{\sigma_c}{q-1} \tag{8.67}$$

Since the derivation of the tangential strain is similar to the previous case, the final expression takes the following form:

$$\varepsilon_\theta = \frac{1+v}{nE}(p_0 - \sigma_{rp})\left(\frac{R_{pp}}{r}\right)^{f+1} \tag{8.68}$$

The relation between the plastic zone radii is also found from the requirement of the continuity of tangential strain at $r = R_{pb}$ and relation Equation (8.62) as:

$$\frac{R_{pp}}{R_{pb}} = n_{sf}^{\frac{1}{f+1}} \tag{8.69}$$

(3) ELASTIC ZONE ($R_{pp} \leq r$)

The derivation of stresses and displacement expressions for a cylindrical opening was previously given in detail with the consideration of initially stressed elastic medium by a far-field hydrostatic *in-situ* stress (p_0). The final forms of the expressions for radially symmetric openings are of the following forms:

$$\sigma_r = p_0 - (p_0 - \sigma_{rp})\left(\frac{R_{pp}}{r}\right)^{n+1} \tag{8.70}$$

$$\sigma_\theta = p_0 + \frac{1}{n}(p_0 - \sigma_{rp})\left(\frac{R_{pp}}{r}\right)^{n+1} \tag{8.71}$$

$$\varepsilon_\theta = \frac{1+v}{nE}(p_0 - \sigma_{rp})\left(\frac{R_{pp}}{r}\right)^{n+1} \tag{8.72}$$

The specific form for σ_{rp} is obtained from the continuity condition of tangential stresses at $r = R_{pp}$ by equality Equation (8.68) and Equation (8.72) as:

$$\sigma_{rp} = \frac{p_0 + n(p_0 - \sigma_c)}{1 + nq} \tag{8.73}$$

(II)   PLASTIC ZONES RADIUS AROUND OPENING

(1)   PERFECTLY PLASTIC-RESIDUAL PLASTIC ZONE BOUNDARY RADIUS (R_{pb})

The perfectly plastic-residual plastic zone boundary radius is found from the requirement of the continuity of radial stresses, i.e. by the equality of Equations (8.73) and (8.67), at $r = R_{pb}$ as:

$$\frac{R_{pb}}{a} = \left\{ \frac{\frac{(1+n)[(q-1)+\alpha]}{(1+nq)(q-1)} \left(\eta_{sf} \right)^{\frac{n(1-q)}{f+1}} - \frac{\alpha}{q-1} + \frac{\alpha^*}{q^*-1}}{\beta + \frac{\alpha^*}{q^*-1}} \right\}^{\frac{1}{n(q^*-1)}} \tag{8.74}$$

where β is the support pressure normalized by overburden pressure and given as:

$$\beta = \frac{p_i}{p_0} \tag{8.75}$$

and α is also the competency factor as:

$$\alpha = \frac{\sigma_c}{p_0} \tag{8.76}$$

(2)   PERFECTLY PLASTIC AND ELASTIC ZONE BOUNDARY RADIUS (R_{pp})

The perfectly plastic and elastic zone boundary radius is also found by inserting σ_{rp} given by Equation (8.73) in the radial stress Equation (8.68) with $\sigma_r = p_i$ at $r = a$ as:

$$\frac{R_{pp}}{a} = \left\{ \frac{(1+n)[(q-1)+\alpha]}{(1+nq)[(q-1)\beta+\alpha]} \right\}^{\frac{1}{n(q-1)}} \tag{8.77}$$

(III)   NORMALIZED OPENING WALL STRAINS

(1)   ELASTIC STATE

The tangential strain at the opening wall can be obtained as:

$$\varepsilon_\theta^a = \frac{1+v}{nE} (p_0 - p_i) \tag{8.78}$$

$$\sigma_\theta^a = \frac{n+1}{n} p_0 - \frac{1}{n} p_i \tag{8.79}$$

If the opening is strained to its elastic limit, then $\sigma_\theta^a = \sigma_c$ for $p_i = 0$. Thus, we have the elastic strain limit as:

$$\varepsilon_\theta^e = \frac{1+v}{E} \cdot \frac{\sigma_c}{n+1} \tag{8.80}$$

Using the preceding relation in Equation (8.78), one obtains the normalized opening wall strain (ξ) as:

$$\xi = \frac{\varepsilon_\theta^a}{\varepsilon_\theta^e} = \frac{n+1}{n}\left(\frac{1-\beta}{\alpha}\right) \leq 1 \tag{8.81}$$

(2) PERFECTLY-PLASTIC STATE

The tangential strain at the opening wall can be obtained as:

$$\varepsilon_\theta^a = \frac{1+v}{nE}(p_0 - \sigma_{rp})\left(\frac{R_{pp}}{a}\right)^{f+1} \tag{8.82}$$

The elastic limit is given as:

$$\varepsilon_\theta^e = \frac{1+v}{nE}(p_0 - \sigma_{rp}) \tag{8.83}$$

Using the preceding relation in Equation (8.82), one obtains the normalized opening wall strain as:

$$\xi = \frac{\varepsilon_\theta^a}{\varepsilon_\theta^e} = \left\{\frac{(1+n)[(q-1)+\alpha]}{(1+nq)[(q-1)\beta+\alpha]}\right\}^{\frac{f+1}{n(q-1)}} \tag{8.84}$$

(3) RESIDUAL PLASTIC STATE

The tangential strain at the opening wall can be obtained as:

$$\varepsilon_\theta^a = \frac{1+v}{nE}(p_0 - \sigma_{rp})\eta_{sf}\left(\frac{R_{pb}}{a}\right)^{f^*+1} \tag{8.85}$$

Using Equations (8.81) and (8.83), one obtains the normalized opening wall strain as:

$$\xi = \frac{\varepsilon_\theta^a}{\varepsilon_\theta^e} = \eta_{sf}\left\{\frac{\frac{(1+n)[(q-1)+\alpha]}{(1+nq)(q-1)}\left(\eta_{sf}\right)^{\frac{n(1-q)}{f+1}} - \frac{\alpha}{q-1} + \frac{\alpha^*}{q^*-1}}{\beta + \frac{\alpha^*}{q^*-1}}\right\}^{\frac{f^*+1}{n(q^*-1)}} \tag{8.86}$$

where

$$\alpha^* = \frac{\sigma_c^*}{p_0} \tag{8.87}$$

(B) CONSIDERATION OF SUPPORT SYSTEM

Although shotcrete, rock bolts and steel ribs are the principal support members and are widely used, it is very rare to find any fundamental study on the proper design method for the reinforcement effect of support systems consisting of shotcrete, rock bolts and steel ribs. As a certain displacement of ground takes place before the installation of supports, the resulting internal pressure provided by the support members must be evaluated

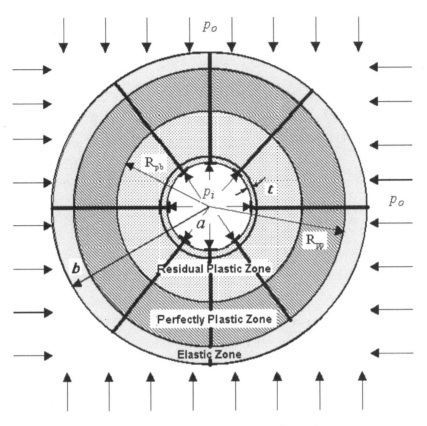

Figure 8.15 Support system for a radially symmetric opening and notations

in terms of relative displacement. Depending on the state of rock before the installation of supports, there may be several combinations. In this section, it is assumed that rock is in elastic state at the time of installation of rock bolts, shotcrete and steel ribs (Fig. 8.15) and that the relative displacement at each region is evaluated for this special case as discussed here (refer to Aydan (1989) for other combinations).

At the time of installation of support members, if rock behaves elastically, the radial deformation of the tunnel is given by

$$u_{in} = \frac{1+v}{nE} r(p_0 - \sigma_{rp}) \left(\frac{a}{r}\right)^{n+1} \tag{8.88}$$

Subtracting this deformation from equations 7.62, 7.68 and 7.72, we obtain the following relations:

1 Flow region ($a \leq r \leq R_{pb}$)

$$\Delta u = \frac{1+v}{nE} r \left[(p_0 - \sigma_{rp}) \eta_{sf} \left(\frac{R_{pb}}{r}\right)^{f*+1} - (p_o - p_{in}) \left(\frac{a}{r}\right)^{n+1} \right] \tag{8.89}$$

2 Perfectly plastic region ($R_{pb} \leq r \leq R_{pp}$)

$$\Delta u = \frac{1+v}{nE} r \left[(p_0 - \sigma_{rp}) \left(\frac{R_{pp}}{r} \right)^{f+1} - (p_o - p_{in}) \left(\frac{a}{r} \right)^{n+1} \right] \tag{8.90}$$

3 Elastic region ($R_{pp} \leq r$)

$$\Delta u = \frac{1+v}{nE} r \left[(p_0 - \sigma_{rp}) \left(\frac{R_{pp}}{r} \right)^{n+1} - (p_o - p_{in}) \left(\frac{a}{r} \right)^{n+1} \right] \tag{8.91}$$

(l) MODELING ROCK BOLTS

For a prescribed displacement (Δu_h) at the grout-rock interface, the governing equation of rock bolts takes the following form (Aydan, 1989):

$$\frac{d^2 u_{ax}}{d\xi^2} - \alpha^2 (u_{ax} - \Delta u_h) = 0, \alpha^2 = \frac{2}{r_b^2 ln(r_h/r_b)} \frac{G_g}{E_b} \tag{8.92}$$

where $\xi = r$, r_h is the borehole radius, r_b is the bar radius, G_g is the shear modulus of grout, and E_b is the elastic modulus of rock bolt. If the relative displacements of rock are introduced in the preceding nonhomogeneous second-order differential equation, the solution of the preceding equation by elementary methods of integration becomes impossible, and the use of a numerical integration becomes necessary. For the sake of simplification, we introduce the following form for differential displacement:

$$f(\xi) = C_o^* e^{-D\xi} \tag{8.93}$$

Constants C_o^* and D are determined from the values of Δu_h at each respective interface of regions. Specifically, they are as follows:

1 Flow region ($a \leq r \leq R_{pb}$)

$$D_{pb} = \frac{ln(\Delta u_a / \Delta u_{pb})}{(R_{pb} - a)}, C_o^{*pb} = \Delta u_{pb} e^{D_{pb} R_{pb}} \tag{8.94a}$$

$$\Delta u_a = \frac{1+v}{nE} a \left[(p_0 - \sigma_{rp}) \eta_{sf} \left(\frac{R_{pb}}{a} \right)^{f^*+1} - (p_o - p_{in}^b) \right] \tag{8.94b}$$

$$\Delta u_{pb} = \frac{1+v}{nE} R_{pb} \left[(p_0 - \sigma_{rp}) \eta_{sf} - (p_o - p_{in}^b) \left(\frac{a}{R_{pb}} \right)^{n+1} \right] \tag{8.94c}$$

2 Perfectly plastic region ($R_{pb} \leq r \leq R_{pp}$)

$$D_{pp} = \frac{ln(\Delta u_{pb} / \Delta u_{pp})}{(R_{pp} - R_{pb})}, C_o^{*pp} = \Delta u_{pp} e^{D_{pp} R_{pp}}, \tag{8.95a}$$

$$\Delta u_{pp} = \frac{1+v}{nE} \left[(p_0 - \sigma_{rp}) R_{bp} \eta_{sf} \left(\frac{R_{pp}}{R_{pb}} \right)^{f*+1} - (p_o - p_{in}^b) \frac{a^{n+1}}{R_{pp}^n} \right] \tag{8.95b}$$

3   Elastic region $(R_{pp} \leq r)$

$$D_{pe} = \frac{ln(\Delta u_b / \Delta u_{pe})}{(b - R_{pp})}, C_o^{*pe} = \Delta u_b e^{D_{pe}b} \tag{8.96a}$$

$$\Delta u_{pe} = \frac{1+v}{nE} R_{pe} \left[(p_0 - \sigma_{rp}) \eta_{sf} - (p_o - p_{in}^b) \left(\frac{a}{R_{pe}} \right)^{n+1} \right] \tag{8.96b}$$

$$\Delta u_b = \frac{1+v}{nE} b \left[(p_0 - \sigma_{rp}) \left(\frac{R_{pe}}{b} \right)^{n+1} \frac{R_{pb}}{b} \eta_{sf} - (p_o - p_{in}^b) \left(\frac{a}{b} \right)^{n+1} \right] \tag{8.96c}$$

The axial displacement and axial stress of a rock bolts take the following form:

$$u_{ax} = A_1 e^{-\alpha\xi} + A_2 e^{\alpha\xi} + C_o e^{-D\xi} \tag{8.97a}$$

$$\sigma_b = E_b \alpha \left[A_1 e^{-\alpha\xi} - A_2 e^{\alpha\xi} + \frac{D}{\alpha} C_o e^{-D\xi} \right] \tag{8.97b}$$

The integration constants (A_1 and A_2) at each region are determined from the continuity and boundary conditions. Accordingly, the internal pressure provided by the rock bolts can be obtained from the well-known formula:

$$\Delta p_i^b = \sigma_b(r=a) \frac{A_b}{e_t e_l} \tag{8.98}$$

where A_b, e_t and e_l are the cross-section area and spacing of a typical rock bolt.

(II)   MODELING SHOTCRETE

The thin wall cylinder or thick wall cylinder approach is commonly used to assess the internal pressure effect of shotcrete in tunneling. For radially symmetric situations, a similar approach can be adopted. One can easily derive the relation for the displacement and outer pressure p_{io}^s acting on the shotcrete with internal pressure $p_{ii}^s = 0$:

$$u = \frac{1+v_s}{nE_s} p_{io}^s \frac{a_o^{n+2}}{a_o^{n+1} - a_i^{n+1}} \left[\frac{1-2v_s}{1-v_s+nv_s} + \frac{1}{n} \left(\frac{a_i}{a_o} \right)^{n+1} \right] \tag{8.99}$$

The incremental form of the preceding equation is:

$$\Delta u = \frac{1+v_s}{nE_s} \Delta p_{io}^s \frac{a_o^{n+2}}{a_o^{n+1} - a_i^{n+1}} \left[\frac{1-2v_s}{1-v_s+nv_s} + \frac{1}{n} \left(\frac{a_i}{a_o} \right)^{n+1} \right] \tag{8.100}$$

and its inverse is:

$$\Delta p_{io}^s = K_s \Delta u \tag{8.101}$$

where

$$K_s = \frac{nE_s}{1+v_s} \frac{a_o^{n+1} - a_i^{n+1}}{a_o^{n+2}} \left[\frac{(1-v_s+nv_s)a_o^{n+1}}{n(1-2v_s)a_o^{n+1} + (1-v_s+nv_s)a_i^{n+1}} \right] \tag{8.102}$$

If the thickness of shotcrete is negligible as compared with the radius of opening, then we have the following:

$$K_s = \frac{nE_s}{1+v_s} \frac{t}{a_o^2} \frac{(1-v_s+nv_s)}{(1-v_s)} \tag{8.103}$$

(III) MODELING STEEL RIBS

If the steel rib is modeled as a one-dimensional rib, its radial deformation is given by:

$$u = \frac{1}{nE_{rb}} p_i^{rb} \frac{a_o^2 e_l}{A_{rb}} \tag{8.104}$$

Its incremental form becomes:

$$\Delta u = \frac{1}{nE_{rb}} \Delta p_i^{rb} \frac{a_o^2 e_l}{A_{rb}} \tag{8.105}$$

Or inversely, we have:

$$\Delta p_i^{rb} = K_{rb} \Delta u \quad \text{and} \quad K_{rb} = nE_{rb} \frac{A_{rb}}{a_o^2 e_l} \tag{8.106}$$

The total internal pressure of the support system may be given as:

$$p_i^{ss} = \Delta p_i^b + \Delta p_{io}^s + \Delta p_i^{rb} \tag{8.107}$$

If support members yield during the deformation of surrounding rock, their behaviors are assumed to be elastic-perfectly plastic. This particularly requires further formulation of the axial stress evaluation of rock bolts.

(C) CONSIDERATION OF BODY FORCES IN PLASTIC ZONE

In general sense, the consideration of body forces violates the radial symmetry of the governing equation. However, there are some proposals in literature for this purpose (i.e. Fenner, 1938; Hoek and Brown, 1980; Aydan, 1989; Sezaki et al., 1994). Some slight modifications to the developed solutions are described in order to consider the effect of body forces in residual-plastic zones. As the perfectly plastic region sustains its original strength, the effect of body forces should be negligible in this region. However, the effect of body forces may be important in the residual plastic (flow) region. Using a similar approach proposed by Aydan (1989), the maximum body force on the support system in the residual plastic region may be approximately obtained as follows (Fig. 8.15):

$$p_{ib} = \left(1 - \left(\frac{a}{R_{pb}}\right)^{n(q^*-1)}\right) \frac{\gamma a \cos\theta}{n(q^*-1)-1} \tag{8.108}$$

where γ is the unit weight of residual plastic zone. The values of the angle (θ) for the crown, sidewall and invert are 0, $\pi/2$ and π, respectively. It should be noted that this assumption is only valid provided that the radial symmetry of the problem is not violated.

(D) CONSIDERATION OF CREEP FAILURE

It is well-known that the deformation and strength characteristics of rocks depend upon the stress rate or strain rate used in tests (i.e. Bieniawski, 1970; Lama and Vutukuri, 1978; Aydan *et al.*, 1994, 2010). Creep tests and relaxation tests are commonly used to determine the time-dependent characteristics of rocks. A common conception that the creep of rocks does not occur unless the applied stress is greater than a threshold stress value is called the creep threshold (Aydan *et al.*, 1995a, 1995b; Aydan and Nawrocki, 1998). This threshold stress level is generally related to the stress level at which fractures are initiated. The so-called transient creep is likely to be a result of the actual visco-elastic behavior of rock. The secondary creep, on the other hand, is due to the stable crack propagation, and the tertiary creep is due to the unstable crack propagation. Therefore, the secondary and tertiary creep is a visco-plastic phenomenon rather than a visco-elastic phenomenon as it involves energy dissipation by fracturing. Figure 8.16 shows the normalized uniaxial creep stress by the short-term uniaxial strength for various rocks as a function of the failure time (time elapsed until the failure). The shrinkage of the uniaxial strength of rock may be represented by the following functional form:

$$\frac{\sigma_c}{\sigma_c^o} = F(t, t_s) \tag{8.109}$$

where t and t_s are time and short-term test duration. σ_c^o is the short-term uniaxial strength of rock mass. Aydan *et al.* (1995a, 1995b) explored several specific forms of function (*F*) to

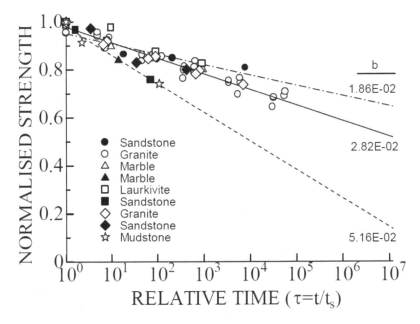

Figure 8.16 Creep strength of various rocks

Source: From Aydan and Nawrocki, 1998

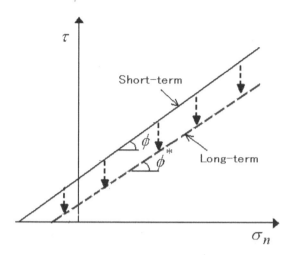

Figure 8.17 Illustration of triaxial long-term strength
Source: Modified from Aydan et al., 1995b

evaluate the experimental results. Figure 8.16 compares the experimental results with the following functional form:

$$\frac{\sigma_c}{\sigma_c^o} = 1 - b \ln\left(\frac{t}{t_s}\right) \tag{8.110}$$

This simple concept may be also extended to multiaxial stress state in which the yield surface of rocks under multiaxial loading conditions is modeled as the shrinkage of the yielding surface. An illustration of this concept on the Mohr-Coulomb yield criterion is shown in Figure 8.17.

If the time-dependent characteristics of the uniaxial compressive strength of rocks is known, it may be possible to determine the time-dependent variation of various mechanical properties of rocks from the relationships obtained by Aydan et al. (1993, 1995a, 1995b). Using the approach originally proposed by Ladanyi (1974), together with the time-dependent variation of mechanical properties involved in the equations given in the previous subsection, it is possible to determine the time-dependent deformation of tunnels. Under multiaxial initial stress conditions and complex tunnel geometry, excavation scheme and boundary conditions, the use of numerical techniques is necessary. For numerical analyses, the time-dependent behavior of rocks may be modeled by an approach proposed by Aydan et al. (1995b, 1996).

(F) APPLICATIONS

Elasto-plastic responses of cylindrical and spherical openings were obtained by using the analytical solution, which is described in this chapter. It is assumed that a cylindrical and spherical opening with a diameter of 6 m is excavated at a depth of below 1000 m from ground surface in rock mass having $\sigma_c = 20$ MPa and $\gamma = 25$ kN/m^3. Empirical equations given by Aydan et al. (1993, 1996) were used for determining properties of rock mass.

Table 8.1 Properties of strength and deformation of rock material and rock mass used in the parametric study

σ_c MPa	σ_c^* MPa	E GPa	υ	γ kN/m³	ϕ (°)	ϕ^* (°)	η_{sf}	f	f^*	p_0 MPa
20	0.05	5.5	0.25	25	42.3	54.8	1.311	2.60	3.96	25

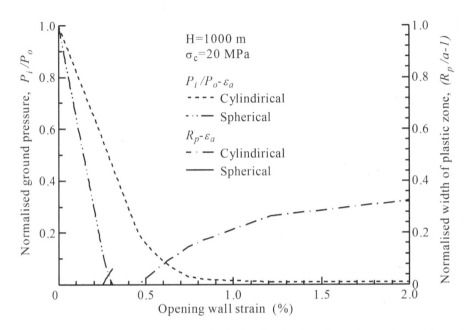

Figure 8.18 Strain, plastic zone radius around cylindrical and spherical openings and ground reaction curves

In these equations, the properties of rock mass were related to the uniaxial compressive strength of rock material. The selected parameters used in the analyses are given in Table 8.1.

Firstly, the ground reaction curves were obtained for the cylindrical and spherical openings. The opening wall strain was related to the normalized ground pressure and normalized plastic radius about the opening (Fig. 8.18). At the same internal support pressure, the plastic zone radius and wall strain around the cylindrical openings are greater than those for the spherical opening (Fig. 8.18 and 8.19). In the present example, the elastic-perfect plastic zone radius and residual-perfect plastic zone radius about cylindrical openings are 5.84 m and 5.07 m, respectively. On the other hand, no residual plastic zone was observed, but an elastic-perfect plastic zone radius with 3.19 m was found about the spherical opening.

Then the distribution of tangential and radial stresses is obtained around the cylindrical and spherical openings for the condition; that is, the internal pressure is zero. It is clear that tangential stresses around the cylindrical openings are greater than the spherical openings. In addition, the radial and tangential stresses around the spherical opening approach the *in-situ* stress faster as a function of distance from the ground surface. As a result, the

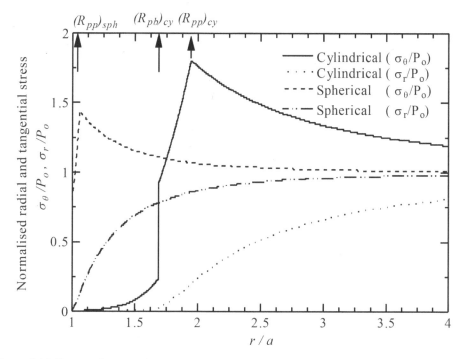

$(R_{pp})_{sph}$ $(R_{pb})_{cy}$ $(R_{pp})_{cy}$

Legend:
—— Cylindrical (σ_θ/P_o)
⋯⋯ Cylindrical (σ_r/P_o)
‐‐‐ Spherical (σ_θ/P_o)
‐··‐ Spherical (σ_r/P_o)

Y-axis: Normalised radial and tangential stress σ_θ/P_o, σ_r/P_o

X-axis: r/a

Figure 8.19 Tangential and radial stresses distribution around cylindrical and spherical openings ($P_i = 0$)

radius of the plastic zone about the spherical opening is smaller than that for the cylindrical one (Fig. 8.19).

The next comparison was concerned with the comparison of elastic-perfectly plastic-residual plastic model with a strain-softening model. The tunnel was assumed to be 200 m below the ground surface in a rock mass having a uniaxial strength of 2.5 MPa. Empirical equations given by Aydan *et al.* (1993, 1996) were used for determining other properties of rock mass for analyses. Figures 8.20 and 8.21 show the computed ground reaction curve and stress distributions for two constitutive models. As expected, there is almost no difference regarding ground response curves if the consideration of the softening part of the constitutive model is taken into account by the proposed scheme. The only difference is associated with the distribution of tangential stress in rock mass as seen in Figure 8.21.

8.2.5 Foundations: bearing capacity

The stress and strain field induced in the impression experiments is close to the compression of the rock under a rigid indenter (Fig. 8.22(b)). Timoshenko and Goodier (1951) developed the following relation for a circular rigid indentation of elastic half-space problem:

$$\frac{\delta}{D} = \frac{\pi}{4}\frac{1-v^2}{E}p \quad \text{with} \quad p = \frac{4F}{\pi D^2} \tag{8.111}$$

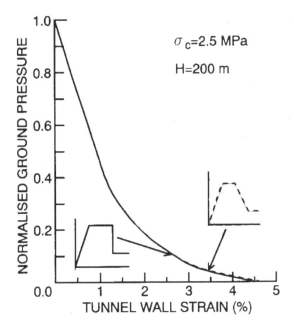

Figure 8.20 Comparison of ground response curves for different constitutive relations

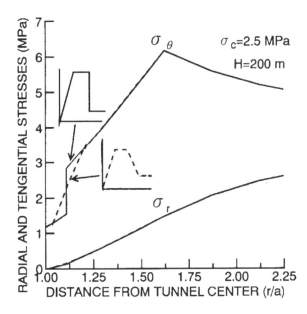

Figure 8.21 Comparison of stress distributions for different constitutive relations

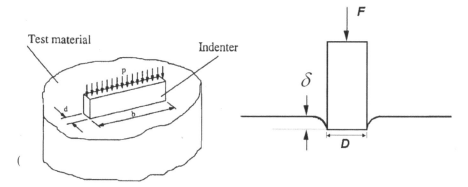

Figure 8.22 Geometrical illustration of various models

where F, υ and E are applied load, Poisson ratio and elastic modulus of rock. While the displacement distribution is uniform beneath the indenter, the contact pressure induced by the indenter would not be uniform. Jaeger and Cook (1979) discussed the initiation of yielding due to compression, and they suggested that the yielding stress level under compression should correspond to one to two times the uniaxial compressive strength of rock, and the yielding would occur at a depth of the order of the radius of the indenter.

Aydan *et al.* (2008) showed that, with the use of spherical cavity approach, the following relation should exist between the uniaxial compression stress–strain rate ($\sigma - \varepsilon$) and applied pressure (p) and the nominal strain of the indenter with a cylindrical flat end ($\varepsilon_i = \delta/D$):

$$\frac{\sigma}{\varepsilon} = \frac{1+v}{4}\frac{p}{\varepsilon_i} \tag{8.112}$$

where v is Poisson's ratio of rock and D is the diameter of the indenter. It should be noted that radial stress and strain of impression experiments are analogous to stress–strain of uniaxial stress state. As discussed by Aydan *et al.* (2008), there may be at least two stress levels for initiating the yielding of rock beneath the indenter. The first yield stress level would correspond to twice the tensile strength level of rock, and the other one would correspond to the uniaxial compressive strength level. However, the effect of tensile yielding is generally difficult to differentiate as the deformation moduli before and after yielding in tension remain fairly the same. The overall deformation modulus may change after yielding in compression. The experiments also indicate that the ultimate strength value (p_u) cannot be greater than a stress level given by:

$$p_u = \frac{2}{1 - \sin\phi}\sigma_c \tag{8.113}$$

The preceding equation implies that the ultimate strength for a frictionless cohesive medium would be twice its uniaxial strength or four times its cohesion. However, it should be noted that this type equation implies that considerable yielding should take

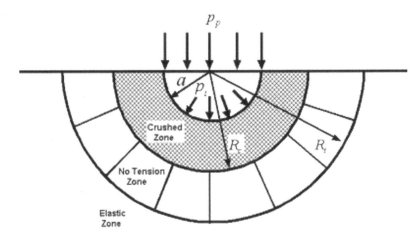

Figure 8.23 Illustration of zones formed beneath loading plate and notation

place beneath the indenters. Aydan *et al.* (2008) developed the following formulas for three different situations of rock beneath the indenter (Fig. 8.23):

Elastic behavior ($p_i \leq 2\sigma_t$)

$$\frac{u_a}{a} = \frac{1+v}{2E} p_i \tag{8.114}$$

Radially ruptured (no tension) plastic behavior ($2\sigma_t < p_i \leq \sigma_c$)

$$\frac{u_a}{a} = \frac{1+v}{2E} p_t \frac{R_t}{a} + \frac{p_i}{2E}\left(1 - \frac{a}{R_t}\right); \frac{R_t}{a} = \left(\frac{p_i}{p_t}\right)^{1/2}; p_t \leq 2\sigma_t \tag{8.115}$$

Crushed plastic behavior ($p_i > \sigma_c$)

$$\frac{u_a}{a} = \left[\frac{1+v}{2E} p_t \frac{R_t}{a} + \frac{p_c}{2E}\frac{R_c}{a}\left(1 - \frac{R_c}{R_t}\right)\right]\left(\frac{p_i}{p_c}\right)^{\frac{q}{(q-1)}}; \frac{R_c}{a} = \left(\frac{p_i}{p_c}\right)^{\frac{q}{2(q-1)}}; \frac{R_c}{R_t} = \left(\frac{p_t}{p_c}\right)^{1/2} \tag{8.116}$$

Furthermore, the applied pressure is equal to radial pressure on the walls of a spherical body in view of the equivalence of work done by the pressure of the indenter to that induced by the wall of the spherical body on the surrounding medium as:

$$p_p = p_i \tag{8.117}$$

Assuming that the volume of the hemispherical body beneath the indenter remains for a given impression displacement (δ), the outward displacement (u_a) of the hemispherical cavity wall can be easily related to the impression displacement (δ) as follows:

$$\delta = 2u_a \tag{8.118}$$

8.2.6 Two-dimensional analytical methods

It is generally difficult to derive closed-form solutions for surface and underground excavations with complex geometry and complex material behavior. Most solutions would be limited to one-dimension in space, and time may be incorporated in certain solutions. For deformation-stress analyses, it is rare to find closed-form solutions for surface structures while there are some closed-form solutions for underground openings in two-dimensional elastic space. The most famous solution is probably that of Kirsch (1898) for a circular hole under biaxial stress state. Solutions for arbitrary shape openings are also developed, and the reader is advised to consult the textbooks by Timoshenko and Goodier (1951 Savin (1965), Muskhelishvili (1962), Jaeger and Cook (1979). It becomes more difficult to obtain analytical solutions when the surrounding media start to behave in an elasto-plastic manner. The simple yet often used closed solutions are for openings with a circular geometry excavated in elasto-plastic media. Several solutions are developed using different yield criteria and postyielding models (e.g. Talobre, 1957; Terzaghi, 1946). There are some solutions for underground openings in elasto-plastic rock supported by rock bolts, shotcrete and steel ribs (Hoek and Brown, 1980; Aydan et al., 1993). Galin (1946) was first to obtain closed-form solutions around circular openings enclosed completely by a plastic zone under bi-axial stress state in Tresca-type perfectly plastic materials. Detournay (1986) attempted to obtain solutions for the same situation with the Mohr-Coulomb yield criterion and discussed several cases using the same solution technique.

There are also limited number of analytical solutions for the problems of seepage, heat flow and diffusion since most solutions would be valid for the given boundary and initial conditions.

The analytical solutions for displacement, strain and stress field around cavities exhibiting nonlinear behavior under nonhydrostatic conditions are generally difficult to obtain. However, some analytical solutions were obtained by Kirsch (1889) for circular cavities, by Ingliss (1913) for elliptical cavities and by Mindlin for circular cavities in gravitating media when the surrounding medium behaves elastically. Muskhelishvili Muskhelishvili (1962) devised a general method based on complex variable functions for arbitrary shape cavities.

The stress state around a circular cavity in an elastic medium under a biaxial far-field stresses were first obtained by Kirsch (1898) using Airy's stress function. These solutions are modified to incorporate the effect of uniform internal pressures (Jaeger and Cook, 1979). In a polar coordinate system, the radial, tangential and shear stresses around the circular cavity can be written in the following forms (Fig. 8.24):

$$\sigma_r = \frac{\sigma_{10} + \sigma_{30}}{2}\left(1 - \left(\frac{a}{r}\right)^2\right) - \frac{\sigma_{10} - \sigma_{30}}{2}\left(1 - 4\left(\frac{a}{r}\right)^2 + 3\left(\frac{a}{r}\right)^4\right)\cos 2(\theta - \beta) + p_i\left(\frac{a}{r}\right)^2$$

$$\sigma_\theta = \frac{\sigma_{10} + \sigma_{30}}{2}\left(1 + \left(\frac{a}{r}\right)^2\right) + \frac{\sigma_{10} - \sigma_{30}}{2}\left(1 + 3\left(\frac{a}{r}\right)^4\right)\cos 2(\theta - \beta) - p_i\left(\frac{a}{r}\right)^2 \qquad (8.119)$$

$$\tau_{r\theta} = \frac{\sigma_{10} - \sigma_{30}}{2}\left(1 - 4\left(\frac{a}{r}\right)^2 + 3\left(\frac{a}{r}\right)^4\right)\sin 2(\theta - \beta)$$

where σ_{10}, σ_{30} are the far-field principal stresses, a is the radius of hole, r is radial distance, β is the inclination of σ_{10} far-field stress from horizontal, θ is the angle of the point from horizontal, and p_i is the internal pressure applied onto the hole perimetry.

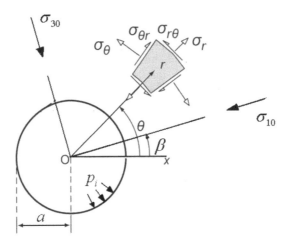

Figure 8.24 Stress tensor components and far-field stresses around a circular hole

The yield criteria available in rock mechanics are:

Mohr-Coulomb

$$\sigma_1 = \sigma_c + q\sigma_3 \tag{8.120a}$$

Drucker-Prager

$$\alpha I_1 + \sqrt{J_2} = k \tag{8.120b}$$

Hoek and Brown (1980)

$$\sigma_1 = \sigma_3 + \sqrt{m\sigma_c\sigma_3 + s\sigma_c^2} \tag{8.120c}$$

Aydan (1995)

$$\sigma_1 = \sigma_3 + [S_\infty - (S_\infty - \sigma_c)e^{-b_1\sigma_3}]e^{-b_2 T} \tag{8.120d}$$

where

$$I_1 = \sigma_I + \sigma_{II} + \sigma_{III}, \; J_2 = \frac{1}{6}((\sigma_I - \sigma_{II})^2 + (\sigma_{II} - \sigma_{III})^2 + (\sigma_{III} - \sigma_I)^2$$

$$\alpha = \frac{2\sin\phi}{\sqrt{3}(3 + \sin\phi)}, \; k = \frac{6c\cos\phi}{\sqrt{3}(3 + \sin\phi)}, \; c \text{ is cohesion, } \phi \text{ friction angle, } q = \frac{1 + \sin\phi}{1 - \sin\phi}$$

where σ_∞ is the ultimate deviatoric strength, T is the temperature, and m, s, b_1, b_2 are empirical constants. Mohr-Coulomb and Drucker-Prager yield criteria are a linear function

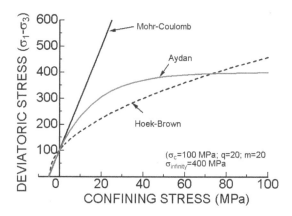

Figure 8.25 Comparison of yield criteria

of confining or mean stress, while the criteria of Hoek-Brown and Aydan are of the non-linear type. Furthermore, Aydan's criterion also accounts for the effect of temperature. Figure 8.25 compares several yield criteria for different rocks. When Aydan's criterion is used, the effect of temperature is omitted in Figure 8.25 for the sake of comparison. It should be also noted that the criterion of Hoek and Brown often fails to represent triaxial strength data if it is required to represent tensile and compressive strength contrary to common belief, that is, the best yield criterion for rocks (Aydan *et al.*, 2012).

If the yield criterion is chosen to be a function of minimum and maximum principal stresses, they can be given in the following form in terms of stress components given by Equation (8.121):

$$\sigma_1 = \frac{\sigma_\theta + \sigma_r}{2} + \sqrt{\left(\frac{\sigma_\theta - \sigma_r}{2}\right)^2 + \tau_{r\theta}^2} \qquad (8.121a)$$

$$\sigma_3 = \frac{\sigma_\theta + \sigma_r}{2} - \sqrt{\left(\frac{\sigma_\theta - \sigma_r}{2}\right)^2 + \tau_{r\theta}^2} \qquad (8.121b)$$

The damage zone around the blast hole under high internal pressure can be estimated using one of these yield criteria. It should be noted that the yielding is induced by the high internal pressure in the blast hole, which is essentially different from *in-situ* stress-induced borehole breakout. In other words, there will always be a damage zone around the blast hole perimeter when the blasting technique is employed. The blast hole pressure depends on the characteristics of the surrounding medium; the amount, layout and type of explosive; the blasting velocity; and the geometry of the blast hole. The blast hole pressure ranges from 100 MPa to 10 GPa (i.e. Jaeger and Cook, 1979; Brady and Brown, 1985).

First we assume that the properties of surrounding rock have the values as given in Figure 8.26 and that the blast hole (internal) pressure has values of 400 and 450 MPa

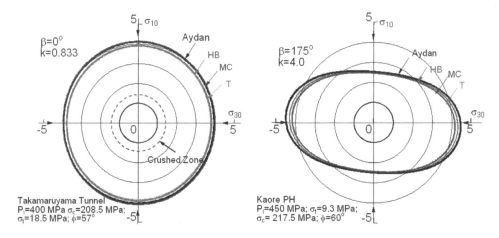

Figure 8.26 Estimated yield zones around the blast hole

for the Takamaruyama tunnel and Kaore Powerhouse, respectively. The rock chosen roughly corresponds to an igneous rock such as granitic rocks. The maximum far-field stress is slightly inclined at an angle of 10 degrees from horizontal and lateral stress coefficient has a value of 4 for Kaore Powerhouse. Figure 8.26 shows the example of computation for the given conditions. The largest yield zone is obtained for Hoek-Brown (HB) criterion while the tension cutoff criterion (T) results in smaller yield zone. The criterion of Aydan estimates a slightly larger yield zone than the Mohr-Coulomb (MC) criterion. It is very interesting to note that the yielding propagates in the direction of maximum far-field stress. In other words, the elongation direction of the yield zone would be the best indicator of the maximum far-field stress in the plane of the blast hole.

In the next example, the far-field stress state is assumed to be isotropic while keeping the values of parameters the same. Except the tension cutoff criterion, all yield criteria estimate almost the same-size yield zones. It should be, however, noted that all yield criteria satisfy the same values of tensile and compressive strength. Therefore, the similarity of the size of yield zones should not be surprising.

If the strength of surrounding rock is anisotropic, the yield functions considering the effect anisotropy should be used. If the elastic constants of the rock are anisotropic, closed-form solutions capable of representing anisotropy should be used instead of those given by Equation (8.121). It should be also noted that the plastic zone developed using the actual elasto-plastic analyses would be larger than those estimated from the elastic solutions (Equation (8.27)). If such discrepancies are expected to be larger, it would be better to use the elasto-plastic finite element method. Nevertheless, the basic conceptual model would be the same.

Gerçek (1996 1997) proposed a seminumerical technique to obtain the integration constant stress functions based on Muskhelishvil's method. Galin (1946, see Savin, 1961 for English version) was first to develop analytical solutions for circular holes in Tresca-type material under nonhydrostatic initial stress state. His solution was extended to Mohr-Coulomb materials by Detournay (1986). He further discussed problems of the

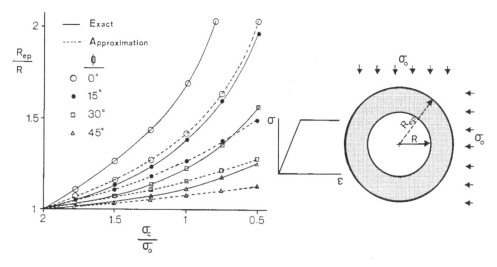

Figure 8.27 Comparison of approximately estimated and exact yield zone
Source: From Aydan, 2018 1987

nonenveloping yield zone around the circular hole. Kastner (1962) proposed a method for estimating the approximate yield zone around circular cavities under nonhydrostatic stress condition using Kirsch's solutions. This method is also employed by Zoback *et al.* (1980) to estimate the shape of borehole breakouts, which was used to infer the *in-situ* stress state. Gerçek (1993) also used the same concept for arbitrarily shaped cavities to estimate the extent of possible yield zone. Although this method estimates the extent of yield zone smaller than the actual one as shown by the first author (Aydan, 1987) for circular cavities under hydrostatic stress state, as shown in Figure 8.27, it yielded the estimated shape of yield zone similar to that by exact solutions. Furthermore, it may also provide some rough guidelines for the anticipated zone for reinforcement by rockbolts. Aydan and Geniş (2010) extended the same concept to estimate overstressed zones about cavities of arbitrary shape based on the stress state computation method proposed by Gerçek (1993) and using strain energy, distortion energy, extension strain and no-tension criteria in addition to Mohr-Coulomb yield criteria. Figure 8.28 shows the overstressed zones around a tunnel subjected to the hydrostatic initial stress state at different stages of excavations and the contours of maximum principal stress. The most critical stress state is during the excavation of the top heading, and the stress state becomes more uniform as the excavation approaches a circular shape. Furthermore, the extent of the tensile stress zone gradually decreases in size as the excavation progresses.

An interesting yield zone developed around a circular opening excavated in a grano-dioritic hard rock at a level of 420 m in Underground Research Laboratory (URL) in Winnipeg, Canada. Figure 8.29 shows the prediction of an overstressed zone around the circular opening at Underground Research Laboratory (URL). Except for the shearing and tension-yielding model, all methods estimated the most likely location of yield zone. In addition to the estimations by the approximate approach, FEM analyses incorporating the yield criteria

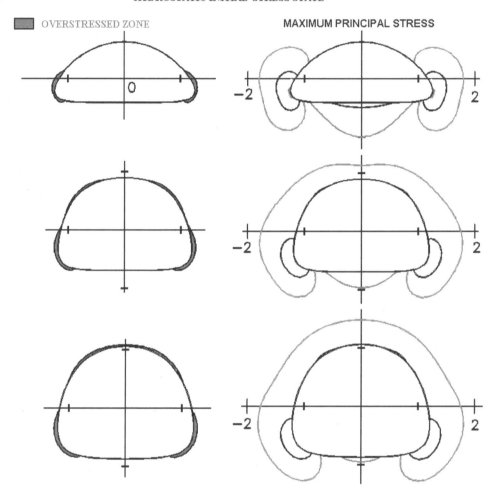

Figure 8.28 Overstressed zones around a tunnel subjected to a hydrostatic initial stress state at different stages of excavations and the contours of maximum principal stress

adopted in the approximate method were performed (Fig. 8.29). As noted from the figure, it seems that the distortion energy concept yields results close to the observations as shown in Figure 8.29.

8.2.7 Three-dimensional analytical solutions

The solutions for three-dimensional situations are quite rare except for a very few solutions. Boussinesq (1885) derived solutions for the distribution of stresses in a half-space resulting from surface loads is largely used in various applications. Kelvin considered a half-space

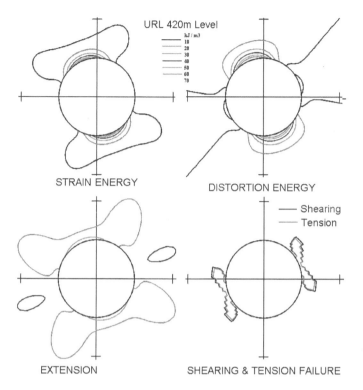

Figure 8.29 View of the opening and estimated yield zones by different methods

problem with a point load of P (Fig. 8.30). The final expressions are given in the following form:

Displacement components

$$u_r = \frac{Pzr}{4\pi G(1-v)R^3}, u_\theta = 0, u_z = \frac{P}{4\pi G(1-v)}\left[\frac{2(1-2v)}{R}+\frac{1}{R}+\frac{z^2}{R^3}\right] \tag{8.122}$$

Stress components

$$\sigma_r = \frac{P}{2\pi(1-v)}\left[\frac{2(1-2v)z}{R^3}-\frac{3r^2z}{R^5}\right], \sigma_\theta = \frac{P(1-2v)z}{2\pi(1-v)R^3}$$

$$\sigma_z = \frac{P}{2\pi(1-v)}\left[\frac{(1-2v)z}{R^3}-\frac{3z^3}{R^5}\right]$$

$$\tau_{r\theta} = \frac{P}{2\pi(1-v)}\left[\frac{(1-2v)z}{R^3}-\frac{3rz^2}{R^5}\right], \tau_{r\theta} = \tau_{z\theta} = 0 \tag{8.123}$$

where

$$R = \sqrt{z^2 + r^2}$$

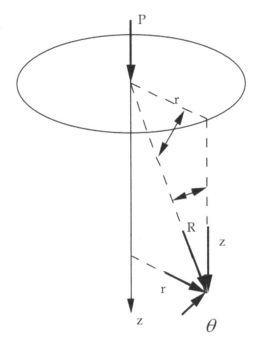

Figure 8.30 Notation for half space under a point load *P*

8.3   Analytical solutions for fluid flow through porous rocks

8.3.1   Some considerations on Darcy's law for rocks and discontinuities

Darcy's law (Equation 5.3) is generally used as a constitutive model for the fluid flow through porous rock and rock discontinuities together with the assumption of laminar flow. A brief description of Darcy's law is presented in this subsection.

Darcy performed a series of experiments on a porous column in 1856. From these experiments, he found that the volume discharge rate Q is directly proportional to the head drop $h_2 - h_1$ and to the cross-sectional area A but that it is inversely proportional to the length difference $l_2 - l_1$. Calling the proportionality constant K as the hydraulic conductivity, Darcy's law is written as:

$$Q = -KA\frac{h_2 - h_1}{l_2 - l_1} \tag{8.124}$$

The negative sign signifies the groundwater flows in the direction of head loss.

Darcy's law is now widely accepted and used in modeling fluid flow in porous or fractured media. It is elaborated and written in a differential form, which is given here for the one-dimensional case as:

$$v = -K\frac{\partial h}{\partial x} \tag{8.125}$$

This law is analogous to Fourier's law in heat flow presented in Chapter 5. Darcy's law is theoretically derived for tube-like pores and slit-like discontinuities in this subsection (Aydan et al., 1997; Üçpırtı and Aydan, 1997).

(a) Darcy's law for rock with cylindrical pores

Equilibrium equation for x-direction is given as:

$$\sum F_x = p\pi[(r+\Delta r)^2 - r^2] - (p+\Delta p)\pi[(r+\Delta r)^2 - r^2] + (\tau+\Delta\tau)2\pi(r+\Delta r)\Delta x - \tau 2\pi r\Delta x = 0$$

$$(8.126)$$

Rearranging the resulting expression and taking the limit and omitting the second-order components yields:

$$\frac{dp}{dx} - \frac{d\tau}{dr} - \frac{\tau}{r} = 0 \tag{8.127}$$

Assuming that the flow is laminar and a linear relationship holds between shear stress and strain rate $\dot{\gamma}$ as:

$$\tau = \eta\dot{\gamma} \quad \dot{\gamma} = \frac{d\dot{u}}{dr} = \frac{dv}{dr} \quad v = \dot{u} = \frac{du}{dt} \tag{8.128}$$

Now, let us insert the preceding relation into Equation (8.129). We have the following partial differential equation:

$$\frac{dp}{dx} - \eta\frac{d^2v}{dr^2} - \frac{\eta}{r}\frac{dv}{dr} = 0 \tag{8.129}$$

Integrating the preceding partial differential equation for r-direction yields the following:

$$v = \frac{1}{\eta}\frac{dp}{dx}\frac{r^2}{4} + C_1\ln r + C_2 \tag{8.130}$$

Introducing the following boundary conditions as:

$$v = v_0 \quad \text{as} \quad r = \frac{D}{2}$$

$$\tau = 0 \quad \text{as} \quad r = 0$$

yields the integration constants C_1 and C_2 as:

$$C_1 = 0, \quad C_2 = v_0 - \frac{1}{\eta}\frac{dp}{dx}\frac{D^2}{16}$$

where D is the diameter of the pore. If velocity v_0 is given in the following form:

$$v_0 = -\alpha\frac{1}{\eta}\frac{dp}{dx}\frac{D^2}{16} \tag{8.131}$$

the integration coefficient C_2 can be obtained as follows:

$$C_2 = -(1+\alpha)\frac{1}{\eta}\frac{dp}{dx}\frac{D^2}{16}$$

The flow rate q passing through the discontinuity for a unit time is:

$$q = \int_0^{2\pi} \int_{r=0}^{y=\frac{D}{2}} vr\,dr\,d\theta \tag{8.132}$$

The explicit form of q is obtained as:

$$q = -\frac{\pi}{\eta} \frac{D^4}{128} \frac{dp}{dx} \tag{8.133}$$

If the flow rate q is redefined in terms of an average velocity $\bar{v}$ over the pore area as

$$q = -\bar{v}\pi \frac{D^2}{4} \tag{8.134}$$

we have the following expression:

$$\bar{v} = -(1+\alpha) \frac{1}{\eta} \frac{D^2}{32} \frac{dp}{dx} \tag{8.135}$$

This relation is known as Hagen-Poiseuille for $\alpha = 0$. As an analogy to Darcy's law, we can rewrite the preceding expression as:

$$\bar{v} = -\frac{k}{\eta} \frac{dp}{dx} \tag{8.136}$$

where

$$k = (1+\alpha)\frac{D^2}{32} \quad or \quad k = (1+\alpha)\frac{a^2}{8}; \quad a = \frac{D}{2}$$

This is known as the actual permeability of the pores. Let us assume that the ratio (porosity) n of the area of pores over the total area is given by (Fig. 8.31a):

$$n = \frac{1}{A_t} \sum_{i=1}^N \pi \frac{D_i^2}{4}, \quad or \quad n = \frac{N\pi\bar{D}^2}{4A_t} \tag{8.137}$$

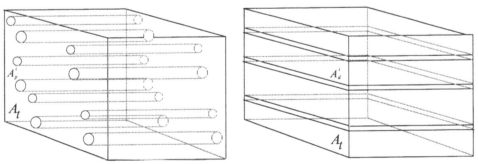

(a) Rock with cylindrical pores   (b) Rock with slit-like discontinuities

Figure 8.31 Geometrical models for Darcy's law

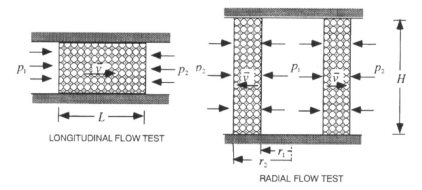

Figure 8.32 Illustration of logitudinal and radial flows

Then, the apparent permeability k_a is related to the actual permeability as:

$$k_a = nk \tag{8.138}$$

(b) Darcy's law for slit-like discontinuities

For x-direction, force equilibrium equation for fluid can be given as follows (Fig. 8.31b):

$$\sum F_x = p_{(x)}\Delta y - p_{(x+\Delta x)}\Delta y + \tau_{(y+\Delta y)}\Delta x - \tau_{(y)}\Delta x = 0 \tag{8.139}$$

where p is pressure and τ is shear stress. Equation (8.140) takes the following partial differential form by taking Taylor expansions of p and τ as:

$$\frac{dp}{dx} - \frac{d\tau}{dy} = 0 \tag{8.140}$$

Assuming that flow is laminar and that the relation between shear stress τ and shear strain rate $\dot{\gamma}$ is linear:

$$\tau = \eta\dot{\gamma}, \quad \dot{\gamma} = \frac{d\dot{u}}{dy} = \frac{dv}{dy}, \quad v = \dot{u} = \frac{du}{dt} \tag{8.141}$$

where η is viscosity and $\dot{u}$ is deformation rate. Substituting the preceding relations into Equation (8.141) yields the following partial differential equation:

$$\frac{dp}{dx} - \eta\frac{d^2v}{dy^2} = 0 \tag{8.142}$$

Integrating the preceding equation for the y-direction yields the following expression for flow velocity v:

$$v = \frac{1}{\eta}\frac{dp}{dx}\frac{y^2}{2} + C_1 y + C_2 \tag{8.143}$$

Introducing the following boundary conditions in Equation (8.144):

$$v = v_o \quad \text{as} \quad y = \frac{h}{2}, \quad \tau = 0 \quad \text{as} \quad y = 0$$

yields the integration constants C_1 and C_2 as:

$$C_1 = 0, \quad C_2 = v_o - \frac{1}{\eta}\frac{dp}{dx}\frac{h^2}{8} \tag{8.144}$$

where h is the aperture of discontinuity. If it is assumed that the following relation exists for v_o:

$$v_o = -\alpha\frac{1}{\eta}\frac{dp}{dx}\frac{h^2}{8} \tag{8.145}$$

then, the integration constant C_2 can be written as:

$$C_2 = -(1+\alpha)\frac{1}{\eta}\frac{dp}{dx}\frac{h^2}{8} \tag{8.146}$$

Total flow rate v_t through the discontinuity at a given time is:

$$v_t = 2\int_{y=0}^{y=\frac{h}{2}} v\,dy \tag{8.147}$$

The explicit form of v_t is obtained as:

$$v_t = -(1+\alpha)\frac{1}{\eta}\frac{h^3}{12}\frac{dp}{dx} \tag{8.148}$$

For $\alpha = 0$, the preceding equation is well-known as a cubic law equation in groundwater hydrology (Snow, 1965), and it is introduced to the field of geomechanics by Polubarinova-Kochina in 1962. Let us redefine the flow rate v_t in terms of an average velocity $\bar{v}$ and the discontinuity aperture h as:

$$v_t = \bar{v}h \tag{8.149}$$

Inserting this equation into Equation (8.148) yields the following:

$$\bar{v} = -\frac{1}{\eta}\frac{h^2}{12}\frac{dp}{dx} \tag{8.150}$$

As an analogy to Darcy's law, the preceding equation may be rewritten as:

$$\bar{v} = -\frac{k_d}{\eta}\frac{dp}{dx} \tag{8.151}$$

where

$$k_d = (1+\alpha)\frac{h^2}{12}$$

where k_d is called the permeability of discontinuity. If discontinuity porosity n_d is defined as the ratio of total area $\sum_{i=1}^{N} A_d^i$ of discontinuities to total area A_t (Fig. 8.31b):

$$n_d = \frac{1}{A_t}\sum_{i=1}^{N} A_d^i \tag{8.152}$$

the following relation between apparent permeability k_{d_a} and actual permeability k_d is obtained as:

$$k_{d_a} = n_d k_d \tag{8.153}$$

8.3.2 Permeability tests based on steady-state flow

8.3.2.1 Pressure difference and flow velocity method

During an experiment, pressures, which are applied at the ends of a test specimen, and flow velocity are measured. If the change of density of fluid with respect to pressure is negligible, permeability can be obtained from the following equation:

$$k = \frac{v_t}{2\pi r_1 H}\frac{\ln(r_2/r_1)\eta}{p_1 - p_2} \tag{8.154}$$

where v_t is flow rate and H is specimen length.

If a gas is used as a permeation fluid, the preceding equation will have the following form:

$$k = \frac{v_t}{\pi H}\frac{\ln(r_2/r_1)}{p_1^2 - p_2^2}p_1\eta \tag{8.155}$$

8.3.2.2 Pressure difference method[1]

For this kind of test, gas is used as a permeation fluid. If the change of pressure with respect to time is linear, the velocity of gas will also change linearly. The relationship between the mass compressibility coefficient and the volumetric compressibility coefficient is given by:

$$c = \rho c^* \tag{8.156}$$

For a compressed gas in a reservoir with a constant volume, the variation of the gas mass for a unit time can be written as:

$$q = \rho c^* V \frac{dp_1}{dt} \tag{8.157}$$

For a given time, the mass passing through a specimen with a cross-section area A may be given by:

$$q = \rho A \bar{v} \tag{8.158}$$

---

1 This test is valid if pressure rate remains constant with time.

Equating Equations (8.158) and (8.159), using Equation (4.80) and rearranging the resulting expression yields the final equation of permeability (Aydan et al., 1997; Aydan, 2016):[2]

$$k = \frac{V\eta}{\pi L} \frac{ln(r_2/r_1)}{(p_1^2 - p_2^2)} \frac{dp_1}{dt} \tag{8.159}$$

where V is the applicable volume (the sum of the volume of the supplemental reservoir, the volume of the pressure injection tubing, and the sample injection hole), η is the gas viscosity, r_2 is the radial distance from the center of the gas injection hole to the periphery of the specimen, r_1 is the radius of the gas injection hole, dp_1/dt is the time rate change of the injection pressure, L is the length of the gas injection hole, and p_1 and p_2 are pressure at reservoirs 1 and 2.

8.3.3 Permeability tests based on non-steady-state flow (transient flow tests)

8.3.3.1 Falling head test method (based on dead weight of liquid)

Experimental setup used for this kind test is shown in Figure 8.33 (Aydan et al., 1997). As seen from the figure, a pipe is placed on the top of the cylindirical hole drilled in the middle of test specimen. The cross-section area of this pipe is denoted by A_h. During the test, the change of pressure and velocity of flow can be measured. A height of water at the outside surface of the test specimen (h_2) is assumed to be constant. When the experiment starts, flow rate inside the pipe can be given as:

$$q = -\rho g A_h \frac{\partial h_1}{\partial t} \tag{8.160}$$

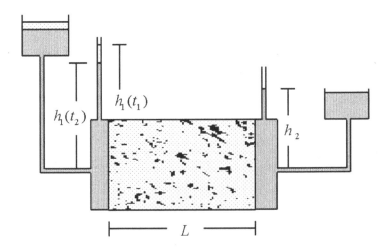

Figure 8.33 Longitudinal transient falling head test

2 This relation was also derived by Zeigler (1976). However, how he derived Eq. (8.160) is not known to the author.

where h_1 is the height of water inside the pipe. At a given time, flow rate through a cross-section area of hole (A_p) inside test specimen is given by:

$$v_t = \bar{v}A_p \tag{8.161}$$

It is assumed that flow rate through the hole perimetry should be equal to the flow rate of the pipe. Then the pressure gradient in the specimen can be given in the following form:

$$\frac{\partial p}{\partial r} \approx -\frac{\partial}{\partial r}(\rho g(h_1 - h_2)) = -\rho g\frac{\partial(h_1 - h_2)}{\partial r} = -\rho g\frac{(h_1 - h_2)}{r\ln(r_o/r_i)} \tag{8.162}$$

Substituting Equation (8.162) together with Equation (5.10) into Equation (8.161) and equalizing the resulting equation to Equation (8.160) yields the following differential equation for the change of water height h_1:

$$\frac{\partial h_1}{h_1 - h_2} = -\frac{k}{\eta}\frac{A_p}{A_h}\frac{1}{r_i\ln(r_o/r_i)}\partial t \tag{8.163}$$

The solution of the preceding differential equation is:

$$h_1 = h_2 + Ce^{-\alpha t} \tag{8.164}$$

where

$$\alpha = \frac{k}{r_i\ln(r_o/r_i)}\frac{A_p}{A_h}\frac{1}{\eta}$$

Introducing the following initial conditions:

$$t = 0 \quad \text{at} \quad h_1 = h_{10}$$

yields the integration constant C as:

$$C = h_{10} - h_2 \tag{8.165}$$

If the integration constant is inserted into Equation (8.165), the following equation is obtained:

$$-\alpha t = \ln(\frac{h_1 - h_2}{h_{10} - h_2}) \tag{8.166}$$

Now, if α is substituted into the preceding equation, the following expression for permeability is obtained:

$$k = \eta r_i\ln(r_o/r_i)\frac{A_h}{A_p}\ln(\frac{h_{10} - h_2}{h_1 - h_2})\frac{1}{t} \tag{8.167}$$

8.3.3.2 Transient pulse method for radial flow

Brace *et al.* (1968) proposed a transient pulse method for longitudinal flow tests. Aydan *et al.* (1997) and Aydan (2016) proposed a permeability test for radial flow. Their method is explained in detail herein. This method is fundamentally very similar to Brace's method (Fig. 8.34). The volumetric strain of fluid inside reservoirs V_1 ve V_2 can be written as follows:

$$\varepsilon_V^1 \approx \frac{\Delta V_1}{V_1}, \quad \varepsilon_V^2 \approx \frac{\Delta V_2}{V_2} \tag{8.168}$$

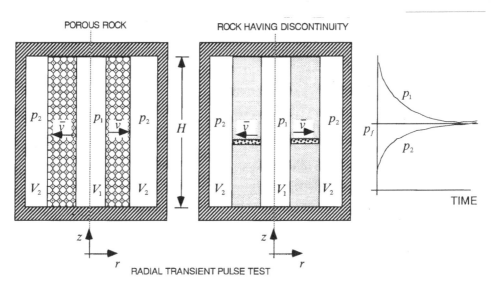

Figure 8.34 Transient pulse radial flow setup for intact rock and discontinuities

Similarly, for the volumetric strain rate of fluid, the following relations can also be written as:

$$\varepsilon_V^1 \approx \frac{\Delta V_1}{V_1}, \quad \varepsilon_V^2 \approx \frac{\Delta V_2}{V_2} \tag{8.169}$$

or

$$\Delta \dot{V}_1 = \dot{\varepsilon}_V^1 V_1, \quad \Delta \dot{V}_2 = \dot{\varepsilon}_V^2 V_2 \tag{8.170}$$

If the following relation exists between the volumetric strain of fluid and pressure:

$$\varepsilon_V^1 = -c_f P_1, \quad \varepsilon_V^2 = -c_f P_2 \tag{8.171}$$

and the compressibility coefficient (c_f) is constant, for volumetric strain rate, the following relation can be also written:

$$\dot{\varepsilon}_V^1 = -c_f \dot{P}_1, \quad \dot{\varepsilon}_V^2 = -c_f \dot{P}_2 \tag{8.172}$$

Flow rate may be given as:

$$v_{t_1} = \Delta \dot{V}_1, \quad v_{t_2} = \Delta \dot{V}_2 \tag{8.173}$$

Using Equations (8.171), (8.172) and (8.173), flow rate can be rewritten in the following form:

$$v_{t_1} = -c_f V_1 \frac{\partial P_1}{\partial t}, \quad v_{t_2} = -c_f V_2 \frac{\partial P_2}{\partial t} \tag{8.174}$$

Introducing the following boundary conditions:

$$r = r_1 \quad \text{as} \quad P = P_1, \qquad r = r_2 \quad \text{as} \quad P = P_2$$

and using Equation (8.174), then the following relation can be obtained for flow rate

$$v_{t_1} = -\frac{kA_{p1}}{\eta}\left(\frac{dP_1}{dr}\right)_{r=r_1}, \quad v_{t_2} = -\frac{kA_{p2}}{\eta}\left(\frac{dP_2}{dr}\right)_{r=r_2} \tag{8.175}$$

where A_{p1} is the surface area of pressure injection hole, and A_{p2} is the area of pressure release surface. Pressure gradients in the preceding equations are as follows:

$$\frac{dP_1}{dr} \approx -\frac{1}{r_1}\frac{(P_1 - P_2)}{\ln(r_2/r_1)}, \quad \frac{dP_2}{dr} \approx -\frac{1}{r_2}\frac{(P_2 - P_1)}{\ln(r_2/r_1)} \tag{8.176}$$

Inserting the preceding equation into Equation (8.175) and equalizing the resulting equation to Equation (8.174) yields the following set of equations:

$$\frac{\partial P_1}{\partial t} = -\beta\frac{A_{p1}}{V_1 r_1}\frac{(P_1 - P_2)}{\ln(r_2/r_1)} \tag{8.177}$$

$$\frac{\partial P_2}{\partial t} = \beta\frac{A_{p2}}{V_2 r_2}\frac{(P_1 - P_2)}{\ln(r_2/r_1)} \tag{8.178}$$

where

$$\beta = \frac{k}{c_f \eta}$$

Equation (8.177) can be rearranged as follows:

$$P_2 = P_1 + \frac{V_1 r_1 \ln(r_2/r_1)}{\beta A_{p1}}\frac{\partial P_1}{\partial t} \tag{8.179}$$

Taking the time derivative of the preceding equation, the following expression is obtained:

$$\frac{\partial P_2}{\partial t} = \frac{\partial P_1}{\partial t} + \frac{V_1 r_1 \ln(r_2/r_1)}{\beta A_{p1}}\frac{\partial^2 P_1}{\partial t^2} \tag{8.180}$$

Substituting Equations (8.180) and (8.181) into Equation (8.179) and rearranging the resulting equation yields the following homogeneous differential equation:

$$\frac{\partial^2 P_1}{\partial t^2} + \alpha\frac{\partial P_1}{\partial t} = 0 \tag{8.181}$$

where

$$\alpha = \beta\frac{V_2 r_2 A_{p1} + V_1 r_1 A_{p2}}{\ln(r_2/r_1)V_1 V_2 r_2 r_1}$$

The general solution of this differential equation is:

$$P_1 = C_1 + C_2 e^{-\alpha t} \tag{8.182}$$

Introducing the following initial conditions:

$$t = 0 \quad \text{as} \quad P_1 = P_i, \qquad t = \infty \quad \text{as} \quad P_1 = P_f$$

yields the integration constants C_1 and C_2 as:

$$C_1 = P_f, \quad C_2 = P_i - P_f \tag{8.183}$$

Inserting these integration constants into Equation (8.182) gives the following equation:

$$P_1 = P_f + (P_i - P_f)e^{-\alpha t} \tag{8.184}$$

Taking the time derivative of the preceding equation:

$$\frac{\partial P_1}{\partial t} = -(P_i - P_f)\beta \frac{V_2 r_2 A_{p1} + V_1 r_1 A_{p2}}{V_2 r_2 V_1 r_1 \ln(r_2/r_1)} e^{-\alpha t} \tag{8.185}$$

Substituting Equations (8.184) and (8.185) into Equation (8.179) and rearranging yields the following equation:

$$P_2 = P_f - (P_i - P_f)\frac{V_1 r_1 A_{p2}}{V_2 r_2 A_{p1}} e^{-\alpha t} \tag{8.186}$$

For the following initial condition for P_2:

$$t = 0 \quad \text{as} \quad P_2 = P_0$$

Equation (8.187) takes the following form:

$$(P_i - P_f) = (P_f - P_0)\frac{V_2 r_2 A_{p1}}{V_1 r_1 A_{p2}} \tag{8.187}$$

The preceding equation can be rewritten in a different way for $P_i - P_0$ as follows:

$$(P_i - P_f) = (P_i - P_0)\frac{V_2 r_2 A_{p1}}{V_1 r_1 A_{p2} + V_2 r_2 A_{p1}} \tag{8.188}$$

Inserting this equation into Equation (8.185) and rearranging yields the following:

$$-\alpha t = \ln\left(\frac{P_1 - P_f}{P_i - P_0}\frac{V_1 r_1 A_{p2} + V_2 r_2 A_{p1}}{V_2 r_2 A_{p1}}\right) \tag{8.189}$$

where

$$\alpha = \frac{k}{c_f \eta}\frac{V_2 r_2 A_{p1} + V_1 r_1 A_{p2}}{V_2 r_2 V_1 r_1 \ln(r_2/r_1)}$$

From the preceding equations, one get the following equation to compute permeability:

$$k = \eta c_f \frac{V_2 r_2 V_1 r_1 \ln(r_2/r_1)}{V_2 r_2 A_{p1} + V_1 r_1 A_{p2}}\ln\left(\frac{P_1 - P_f}{P_i - P_0}\frac{V_2 r_2 A_{p1}}{V_2 r_2 A_{p1} + V_1 r_1 A_{p2}}\right)\frac{1}{t} \tag{8.190}$$

When gas is used as a permeation fluid, P_1 and P_2 are replaced with U_1 $(= P_1^2)$ and U_2 $(= P_2^2)$, and permeability can be calculated using the same relation previously given. If the volume of reservoir 2 (V_2) is greater than the volume of reservoir 1 (V_1), $(V_2 ? V_1)$ (for instance, the outer side of the specimen is open to air), then P_0 ve P_f given in the preceding equation will be equal to atmospheric pressure (P_a). For this particular case, the permeability of a specimen can be obtained from the following equation:

$$k = \eta c_f \frac{V_1 r_1 \ln(r_2/r_1)}{A_{p1}} \ln\left(\frac{P_1 - P_f}{P_i - P_0}\right) \frac{1}{t} \tag{8.191}$$

8.3.3.3   Theory of interface or discontinuity permeability in radial flow tests

For the cylindrical coordinate system, the force equilibrium equation for the fluid can be given as follows (Fig. 8.34) (Aydan *et al.*, 1997; Aydan, 2016):

$$\left(\tau_{(z+\Delta z)} - \tau_{(z)}\right) r\Delta r\Delta\theta + \sigma_{r_{(r)}} r\Delta\theta\Delta y - \sigma_{r_{(r+\Delta r)}}(r + \Delta r)\Delta\theta\Delta y + 2\sigma_\theta \Delta r \sin\left(\frac{\Delta\theta}{2}\right) = 0 \tag{8.192}$$

with the use of the following relation:

$$\sin\left(\frac{\Delta\theta}{2}\right) \approx \frac{\Delta\theta}{2}$$

The preceding equation can be rearranged. Then, if Taylor expansion is used for τ and σ_r in Equation (8.192), the following partial differential is obtained:

$$\frac{d\sigma_r}{dr} - \frac{\sigma_r - \sigma_\theta}{r} - \frac{d\tau}{dz} = 0 \tag{8.193}$$

For hydrostatic case ($\sigma_r = \sigma_\theta = p$), Equation (8.193) becomes:

$$\frac{d\sigma_r}{dr} - \frac{d\tau}{dz} = 0 \tag{8.194}$$

Assuming that flow is laminar and the relation between shear force and shear strain rate $\dot{\gamma}$ is linear:

$$\tau = \eta\dot{\gamma}, \quad \dot{\gamma} = \frac{d\dot{u}}{dz} = \frac{dv}{dz}, \quad \sigma_r = p, \quad v = \dot{u} = \frac{du}{dt} \tag{8.195}$$

Substituting the preceding relations into Equation (8.194) yields the following partial differential equation:

$$\frac{dp}{dr} - \eta\frac{d^2v}{dz^2} = 0 \tag{8.196}$$

Integrating the preceding partial differential equation for *y*-direction yields the following expression for flow velocity *v*:

$$v = \frac{1}{\eta}\frac{dp}{dr}\frac{z^2}{2} + C_1 z + C_2 \tag{8.197}$$

Introducing the following boundary conditions in Equation (8.197):

$$z = \frac{h}{2} \quad \text{as} \quad v = 0, \quad z = 0 \quad \text{as} \quad \tau = 0$$

yields the integration constants C_1 and C_2 as:

$$C_1 = 0, \quad C_2 = -\frac{1}{\eta}\frac{dp}{dr}\frac{h^2}{8} \tag{8.198}$$

Total flow rate v_t through the discontinuity at a given time is:

$$v_t = 2\int_0^{2\pi}\int_{z=0}^{z=\frac{h}{2}} vr\,dz\,d\theta \tag{8.199}$$

The explicit form of v_t is obtained as:

$$v_t = -\frac{2\pi r}{\eta}\frac{h^3}{12}\frac{dp}{dr} \tag{8.200}$$

Let us redefine the flow rate v_t in terms of an average velocity $\bar{v}$ over the discontinuity aperture area as:

$$v_t = 2\pi r\bar{v}h \tag{8.201}$$

Inserting this equation into Equation (8.200), the following expression is obtained:

$$\bar{v} = -\frac{1}{\eta}\frac{h^2}{12}\frac{dp}{dr} \tag{8.202}$$

In an analogy to Darcy's law, the preceding equation may be rewritten as:

$$\bar{v} = -\frac{k_d}{\eta}\frac{dp}{dr} \tag{8.203}$$

where

$$k_d = \frac{h^2}{12} \quad \text{or} \quad h = \sqrt{12k_d}$$

and k_d is called intrinsic permeability of discontinuity. Inserting Equation (8.203) into Equation (8.199), the flow rate v_t can be obtained as:

$$v_t = -2\pi rhk_d\frac{dp}{dr} \tag{8.204}$$

The preceding equation can be rewritten in the following form:

$$\frac{dr}{r} = -\frac{2\pi hk_d dp}{v_t} \tag{8.205}$$

The integral form of the preceding equation may be written as:

$$\int_{r_i}^{r_o}\frac{dr}{r} = -\int_{p_i}^{p_o}\frac{2\pi hk_d}{v_t}dp \tag{8.206}$$

If integration is carried out, the permeability of discontinuity (k_d) can be found after some manipulations as follows:

$$k_d = [\frac{v_t\ln(\frac{r_o}{r_i})}{4\pi\sqrt{3}(p_i - p_o)}]^{2/3} \tag{8.207}$$

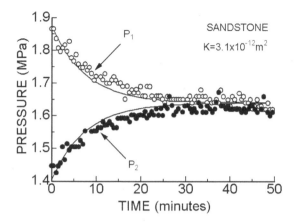

Figure 8.35 Pressure responses of a sandstone sample in a transient pulse test

Figure 8.35 shows pressure responses of reservoirs 1 and 2 in a transient pulse test on a sandstone sample. Despite some scattering of pressure responses, the variations of reservoir pressures tend to decrease with time and become asymptotic to a stabilizing pressure. The permeability of the sandstone sample was 3.1×10^{-12} m^2.

8.4 Analytical solutions for heat flow: temperature distribution in the vicinity of geological active faults

As a first case, the geological fault is assumed to be sandwiched between two nonconductive rock slabs, and closed-form solutions are derived for temperature rises within the fault due to shearing. Then a more general case is considered such that a seismic energy release takes place within the fault, and the adjacent rock is conductive. The solution of the governing equation for this case is solved with the use of the finite element method (Aydan, 2016). Several examples were solved by considering some hypothetical energy release functions, and their implications are discussed.

If a geological fault and its close vicinity may be simplified to a one-dimensional situation, as shown in Figure 8.36, by assuming that mechanical energy release is due purely to shearing with no heat production source. Thus, Equation (4.5) may be reduced to the following form:

$$\rho c \frac{\partial T}{\partial t} = -\nabla q + \tau \dot{\gamma} \tag{8.208}$$

Let us assume that the heat flux obeys Fourier's law, which is given by:

$$q = -k \frac{\partial T}{\partial x} \tag{8.209}$$

Inserting Equation (8.209) into Equation (8.208) yields the following equation:

$$\rho c \frac{\partial T}{\partial t} = k \frac{\partial^2 T}{\partial x^2} + \tau \dot{\gamma} \tag{8.210}$$

The solution of the preceding equation will yield the temperature variation with time.

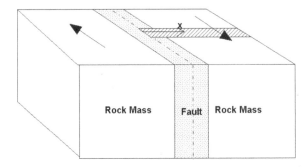

Figure 8.36 Fault model

The energy release during earthquakes is a very complex phenomenon. Nevertheless, some simple forms relevant for overall behavior may be assumed in order to have some insight to the phenomenon. Two energy release rate functions of the following form are assumed:

$$\dot{E} = \tau\dot{\gamma} = Ate^{-\frac{t}{\theta}} \tag{8.211}$$

$$\dot{E} = \tau\dot{\gamma} = A^*e^{-\frac{t}{\theta^*}} \tag{8.212}$$

Constants A and A^* depend on the shear stress and shear strain rate history with time and fault thickness. Constants θ and θ^* are time history constants. For situations illustrated in Figure 8.36, constants A and A^* will take the following forms:For Equation (8.210):

$$A = \frac{\tau_o u_f}{h\theta^2} \tag{8.213}$$

For Equation (8.211):

$$A^* = \frac{\tau_o u_f}{h\theta^*} \tag{8.214}$$

Where u_f, h are the final relative displacement and thickness of the fault. τ_o is the shear stress acting on the fault, and it is assumed to be constant during the motion.

Two specific situations are analysed:

- Creeping fault
- Fault with hill-shaped seismic energy release rate

In the case of creeping fault, the energy release rate is almost constant with time. The geometry of the fault is assumed to be one-dimensional as shown in Figure 8.37. Figures 8.38 and 8.39 show the computed temperature differences at selected locations with time and temperature difference distribution throughout the whole domain at selected time steps. In the computations, the energy release rate is assumed to be taking place within the fault zone only. The increase of temperature difference is parabolic, and they keep increasing as time goes by. Nevertheless, the temperature difference increases are about one-tenth of those of the fault sandwiched between nonconductive rock mass slabs.

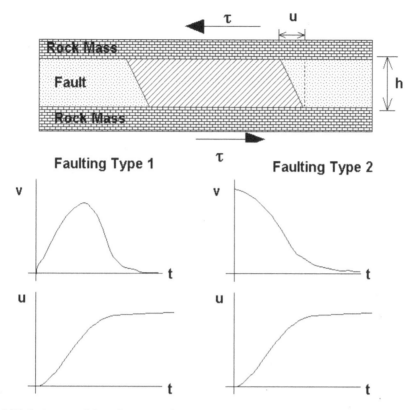

Figure 8.37 Faulting models and energy release types

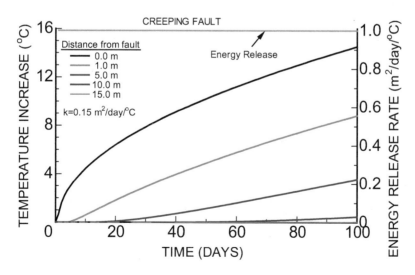

Figure 8.38 Temperature difference variations for a fault sandwiched between conductive rock mass slabs for creeping condition

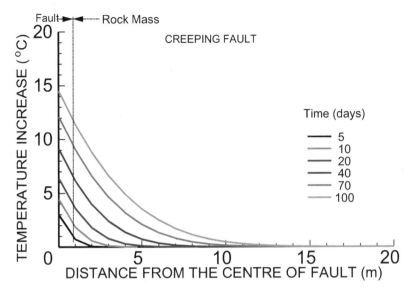

Figure 8.39 Temperature distributions at different time steps for a fault sandwiched between conductive rock mass slabs for creeping condition

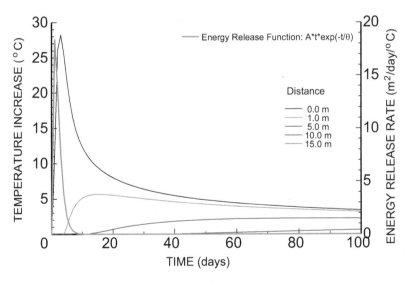

Figure 8.40 Temperature difference variations for a fault sandwiched between conductive rock mass slabs for hill-shaped energy release function

Figures 8.40 and 8.41 show the computed temperature differences at selected locations with time and temperature difference distribution throughout the whole domain at selected time steps for a fault with a hill-like energy release rate. In the computations, the energy release rate is assumed to be taking place within the fault zone only. The increase of

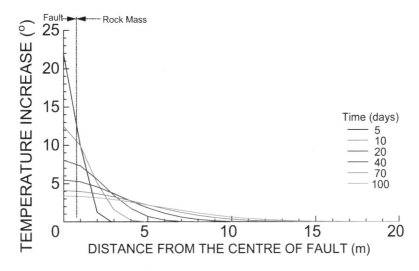

Figure 8.41 Temperature distributions at different time steps for a fault sandwiched between conductive rock mass slabs for hill-shaped energy release function

temperature difference is parabolic. Temperature differences increase at first, and then they tend to decay in a similar manner to the assumed seismic energy release rate function.

This situation will be probably quite similar to the actual situation in nature. The temperature difference increases are about one-tenth of those of the fault sandwiched between nonconductive rock mass slabs. These results indicate that the observation of ground temperatures may be a very valuable source of information in the prediction of earthquakes because atmospheric temperature measurements near the ground surface may be quite problematic in interpreting the observations. However, the observation of hot-spring temperature, which reflects the actual ground temperature, may be very good tool for such measurements without any deep boring.

8.5 Analytical solutions for diffusion problems

8.5.1 *Drying testing procedure*

Let us consider a sample with volume V dried in air with infinite volume as shown in Figure 8.42 (Aydan, 2003). Water-contained Q in a geo-material sample may be given in the following form:

$$Q = \rho_w \theta_w V \tag{8.215}$$

where ρ_w, θ_w and V are water density, water content ratio and volume of sample, respectively. Assuming that water density and sample volume remain constant, the flux q of water content may be written in the following form:

$$q = \frac{dQ}{dt} = -\rho_w V \frac{d\theta_w}{dt} \tag{8.216}$$

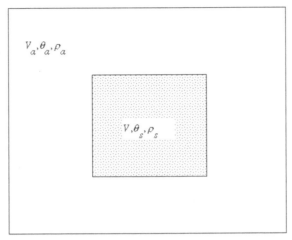

a) Physical Model

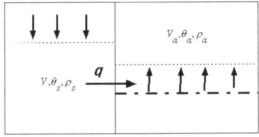

b) Mechanical model

Figure 8.42 Physical and mechanical models for water migration during drying process

Air is known to contain water molecules of 6 g m^{-3} when relative humidity is 100%. When the relative humidity is less than 100%, water is lost from geo-materials to air. If such a situation presents, the water lost from the sample to air may be given in the following form using a concept similar to Newton's cooling law in thermodynamics:

$$q = \rho_w A_s h \Delta\theta = \rho_w A_s h(\theta_w - \theta_a) \tag{8.217}$$

Where h and A_s are the water loss coefficient and surface area of sample. Requiring that the water loss rate of sample should be equal to the water loss into air on the basis of the mass conservation law, one can easily write the following relation:

$$\rho_w A_s h(\theta_w - \theta_a) = -\rho_w V \frac{d\theta_w}{dt} \tag{8.218}$$

The solution of differential Equation (8.217) is easily obtained in the following form

$$\theta_w = \theta_a + Ce^{-\alpha t} \tag{8.219}$$

where

$$\alpha = h\frac{A_s}{V}$$

The integration constant may be obtained from the initial condition, that is:

$$\theta_w = \theta_{w0} \quad \text{at} \quad t = 0 \tag{8.220}$$

as follows:

$$C = \theta_{w0} - \theta_a \tag{8.221}$$

Thus the final expression takes the following form:

$$\theta_w = \theta_a + (\theta_{w0} - \theta_a)e^{-\alpha t} \tag{8.222}$$

If the water content migration is considered a diffusion process, Fick's law in one dimension may be written as follows:

$$q = \rho_w D\frac{\partial \theta_w}{\partial x} \tag{8.223}$$

Requiring that the water loss rate given by Equation (8.222) to be equal to that given by Equation (8.222) yields the following relation:

$$D = h\frac{V}{A_s} \tag{8.224}$$

If surface area A_s and volume V of sample are known, it is easy to determine the water migration diffusion constant D from drying tests, provided that the coefficient α and subsequently h are determined from experimental results fitted to Equation (8.221).

If samples behave linearly, water migration characteristics should remain the same during the swelling and drying processes. Recent technological developments have made it quite easy to measure the weight of samples and the environmental conditions such as temperature and humidity. Figure 8.43 shows an automatic weight and environmental conditions monitoring system developed for such tests. It is also possible to measure the volumetric variations (shrinkage) during the drying process using noncontact-type displacement transducers (i.e. laser transducers).

Physical and mechanical properties of materials can be measured using the conventional testing machines such as wave velocity measurements, uniaxial compression tests, elastic modulus. Tuff samples used in the tests were from Avanos, Ürgüp and Derinkuyu of the Cappadocia region in Turkey and Oya in Japan. The samples from the Cappadocia region are gathered from historical and modern underground rock structures. They represent the rocks in which historical and modern underground structures were excavated. These tuff samples bear various clay minerals as given in Table 8.2 (Temel, 2002; Aydan and Ulusay, 2003). As noted from the table, the clay content is quite high in Avanos tuff, and most of the clay minerals are smectite.

In drying experiments, the samples that underwent swelling were dried in a room with an average temperature of 23°C and relative humidity of 65–70. Figures 8.44, 8.45 and 8.46 show the drying test results for some tuff samples from the Cappadocia region in Turkey. As seen from the figures, it takes a longer time for the tuff sample from Avanos compared with Ürgüp and Derinkuyu samples. The Derinkuyu sample dries much rapidly than the

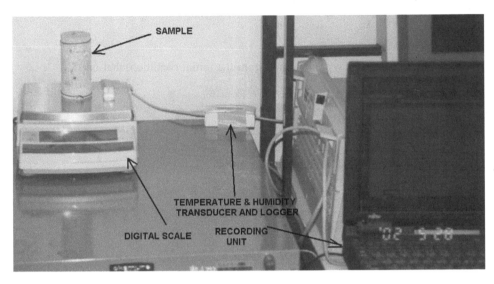

Figure 8.43 Experimental setup for measuring water content during drying

Table 8.2 XRD results from the samples of Ürgüp (Kavak tuff) and Avanos

Specimen Number	Clay Percentage	Clay Fraction		
		Smectite	Kaolin	Illite
UR-1 (Ürgüp)	74	83	14	3
UR-2 (Ürgüp)	60	67	25	8
AV-1 (Avanos)	94	84	13	3
AV-2 (Avanos)	82	95	5	T

T: Trace amount

others. Each sample was subject to drying twice. Once again, it is noted that the drying period increases for Avanos tuff after each run, whereas Derinkuyu tuff tends to dry much rapidly in the second run. From these tests, it may be also possible to determine the diffusion characteristics of each tuff.

The theory derived in the previous section could be applied to the experimental results shown in Figures 8.44, 8.45 and 8.46. To obtain the constants of water migration model, Equation (8.221) may be rewritten as follows:

$$ln\left(\frac{\theta_w - \theta_a}{\theta_{w0} - \theta_a}\right) = -\alpha t \tag{8.225}$$

The plot of experimental results in the semilogarithmic space first yields the constant α, from which constant h and diffusion coefficient D can be computed subsequently.

The results are shown in Figures 8.44, 8.45 and 8.46. The unit of parameters α, h and D are $1\ h^{-1}$, $cm\ h^{-1}$ and $cm^2\ h^{-1}$, respectively. The computed values of parameters α, h and D are also shown in the same figures.

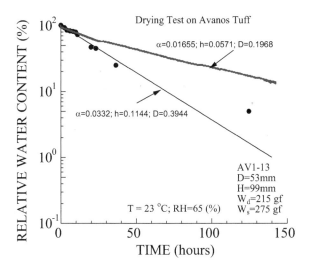

Figure 8.44 Determination of constants for relative water content variation during drying of Avanos tuff

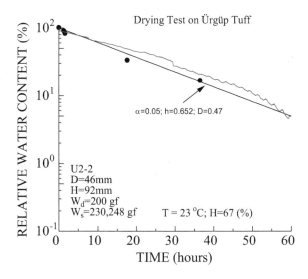

Figure 8.45 Determination of constants for relative water content variation during drying of Ürgüp tuff

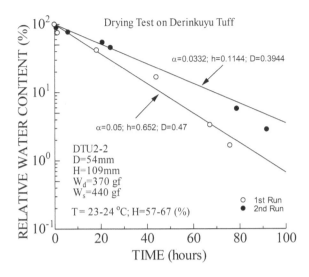

Figure 8.46 Determination of constants for relative water content variation during drying of Derinkuyu tuff

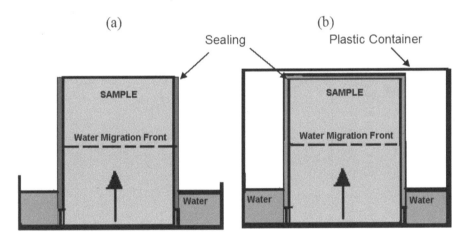

Figure 8.47 Experimental setups: (a) top surface unsealed, (b) top-surface sealed

8.5.2 Saturation testing technique

Initially dry samples can be subjected to saturation, and water migration characteristics may be obtained. The sides of samples can be sealed and subjected to saturation from the bottom. The top surface may be sealed and unsealed, as illustrated in Figure 8.47. Samples can be isolated against water migration from the sides by sealing while the bottom surface of the samples can be exposed to saturation by immersing in water up to a given depth. There may be two conditions at the top surface, which could be either exposed to air directly or sealed. When the top surface is sealed, the boundary value would be changing with time. The water migration coefficient can be determined from the solution of the following diffusion equation:

$$\frac{\partial \theta_w}{\partial t} = D \frac{\partial^2 \theta_w}{\partial x^2} \qquad (8.226)$$

When the top surface is unsealed, the top boundary condition $(x=H)$ is:

$$\theta_w = \theta_a \qquad (8.227)$$

On the other hand, if the top surface is sealed, the boundary condition is time dependent, and it can be estimated from the following condition:

$$q_{x=H} = \hat{q}_n(t) \qquad (8.228)$$

For some simple boundary conditions, the solution of partial differential Equation (8.226) can be easily obtained using the technique of separation of variables (i.e. Keryszig, 1983). In the general case, it would be appropriate to solve it using finite difference technique or finite element method (i.e. Aydan, 2003, 2016).

8.6 Evaluation of creep-like deformation of semi-infinite soft rock layer

The simplified analytical model introduced in this section is based the theoretical model developed by Aydan (1994, 1998). The momentum conservation law for an infinitely small element of ground on a plane with an inclination of α for each respective direction can be written in the following form (Fig. 8.48):

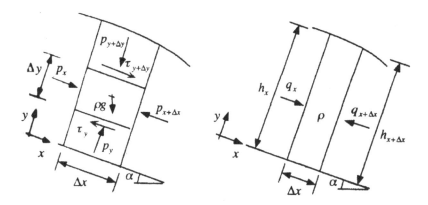

Figure 8.48 Modeling of a layer subjected to shearing
Source: From Aydan, 1994, 1998

x-direction

$$\frac{\partial \tau}{\partial y} = \frac{\partial p}{\partial x} - \rho g \sin \alpha \tag{8.229}$$

y-direction

$$\frac{\partial p}{\partial y} = \rho g \sin \alpha \tag{8.230}$$

where τ, p, ρ, g are shear stress, pressure, density and gravitational acceleration, respectively. The variation of pressure along the *x*-direction is given by:

$$\frac{\partial p}{\partial x} = \rho g \cos \alpha \frac{\partial h}{\partial x} \tag{8.231}$$

If shear stress related to shear strain is linearly as given in the following form:

$$\tau = G\gamma; \gamma = \frac{\partial u}{\partial y}$$

one can easily obtain the solution given as:

$$\tau = \rho g \cos \alpha (\tan \alpha - \frac{\partial h}{\partial x})(h - y) \tag{8.232}$$

If the variation of ground surface height (*h*) is neglected, the resulting equation for shear stress and displacement takes the following form:

$$\tau = \rho g \sin \alpha (h - y); u = \frac{\rho g \sin \alpha}{G} y \left(h - \frac{y}{2} \right) \tag{8.233}$$

As is well-known, rainfall induces groundwater level fluctuations. However, these fluctuations are not as high as presumed in many limiting equilibrium approaches to analyzing the failure of slopes. In other words, the whole body, which is prone to fail, does not become fully saturated. However, the monitoring results indicate that a certain thickness of layer becomes saturated. In view of experimental results, the deformation modulus would become smaller during the saturation process and recover its original value upon drying. The deformation modulus during saturation may be assumed to be the plastic deformation modulus (G_p), and the displacement induced during the saturation period may be viewed as the plastic (irrecoverable) deformation (Fig. 8.49). With the use of this concept and the analytical model previously presented, one can easily derive the following equation for deformation induced by saturation as:

$$u_s = \frac{\rho g \sin \alpha}{G_s} y \left(h - \left(t - \frac{y}{2} \right) \right) \tag{8.234}$$

where *t* is the thickness of saturated zone in a given cycle of saturation drying. The plastic deformation would be the difference between displacements induced under saturated and

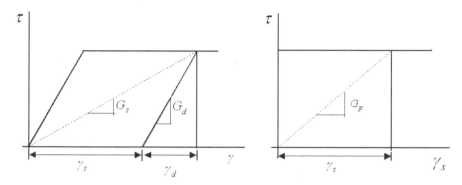

Figure 8.49 Constitutive modeling of cyclic softening-hardening of marl layer

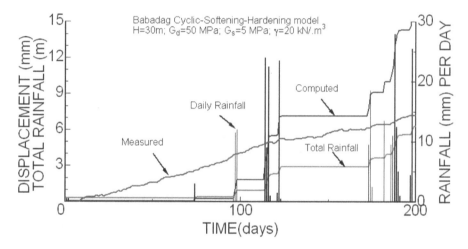

Figure 8.50 Comparison of measured and computed displacements

dry states, and it will take the following form:

$$u_p = \rho g \sin \alpha y \left(\frac{1}{G_s} - \frac{1}{G_d} \right) \cdot \left(h - \left(t - \frac{y}{2} \right) \right) \tag{8.235}$$

where G_d and G_s are shear modulus for dry and saturated states, respectively. Thus the equivalent shear modulus may be called the plastic deformation modulus (G_p) in this chapter and can be written as:

$$G_p = \frac{G_s G_d}{G_d - G_s} \tag{8.236}$$

The time for saturation and drying of marls is very short (say, in hours). With this observational fact and experimental results, the analysis presented is based on the day unit. Figure 8.50 compares the computed displacement and displacement measured at monitoring station No.1 of the Gündoğdu district of Babadağ town with the consideration of thickness of the saturation zone (Kumsar *et al.*, 2016). Despite some differences between computed and measured responses, the analytical model can efficiently explain the overall response of the landslide area of the Gündoğdu district of the town of Babadağ.

References

Aydan, Ö. (1987). Approximate estimation of plastic zones about underground openings. Interim report (unpublished), Nagoya University, 8p.

Aydan, Ö. (1989). The stabilisation of rock engineering structures by rockbolts. Doctorate Thesis, Nagoya University.

Aydan, Ö. (1994) The dynamic shear response of an infinitely long visco-elastic layer under gravitational loading. *Soil Dynamics and Earthquake Engineering*, Elsevier, 13, 181–186.

Aydan, Ö. (1995) Mechanical and numerical modelling of lateral spreading of liquified soil. *The 1st Int. Conf. on Earthquake Geotechnical Engineering, IS-TOKYO'95*, Tokyo. pp. 881–886.

Aydan, Ö. (1997) Dynamic uniaxial response of rock specimens with rate-dependent characteristics. *SARES'97*. pp. 322–331.

Aydan, Ö. (1998). A simplified finite element approach for modelling the lateral spreading of liquefied ground. The 2nd Japan-Turkey Workshop on Earthquake Engineering, Istanbul.

Aydan, Ö. (2003) The moisture migration characteristics of clay-bearing geo-materials and the variations of their physical and mechanical properties with water content. *2nd Asian Conference on Saturated Soils, UNSAT-ASIA*. pp. 383–388.

Aydan, Ö. (2016). *Time Dependency in Rock Mechanics and Rock Engineering*. CRC Press, Taylor and Francis Group, London, 241p.

Aydan, Ö. (2018). *Rock Reinforcement and Rock Support*. CRC Press, Taylor and Francis Group, London, 486p.

Aydan, Ö. & Geniş, M. (2010) Rockburst phenomena in underground openings and evaluation of its counter measures. *Journal of Rock Mechanics*, Turkish National Rock Mechanics Group, (Special Issue 17), 1–62.

Aydan, Ö. & Nawrocki, P. (1998) Rate-dependent deformability and strength characteristics of rocks. *In Proceedings of Symposium on the Geotechnics of Hard Soils-Soft Rock, Napoli*, 1. pp. 403–411.

Aydan, Ö. & Ulusay, R. (2003) Geotechnical and geoenvironmental characteristics of man-made underground structures in Cappadocia, Turkey. *Engineering Geology*, 69, 245–272.

Aydan, Ö., Ersen, A., Ichikawa, Y. & Kawamoto, T. (1985) Temperature and thermal stress distributions in mass concrete shaft and tunnel linings during the hydration of concrete (in Turkish). *The 9th Mining Science and Technology Congress of Turkey*, Ankara. pp. 355–368.

Aydan, Ö., Güloğlu, R. & Kawamoto, T. (1986) Temperature distributions and thermal stresses in tunnel linings due to hydration of cement (in Japanese). *Tunnels and Underground*, 17(2), 29–36.

Aydan, Ö., Akagi, T. & Kawamoto, T. (1993) Squeezing potential of rocks around tunnels: Theory and prediction. *Rock Mechanics and Rock Engineering*, 26(2), 137–163.

Aydan, Ö., Seiki, T., Jeong, G.C. & Tokashiki, N. (1994). Mechanical behaviour of rocks, discontinuities and rock masses. *Int. Symp. Pre-failure Deformation Characteristics of Geomaterials*, Sapporo, 2, 1161–1168.

Aydan, Ö., Akagi, T., Ito, T., Ito, J. & Sato, J. (1995a) Prediction of deformation behaviour of a tunnel in squeezing rock with time-dependent characteristics. *Numerical Models in Geomechanics*, NUMOG, V, 463–469.

Aydan, Ö., Akagi, T., Ito, T. & Sezaki, M. (1995b) The design of supports of tunnels in squeezing rocks (in Japanese). *The 25th Rock Mechanics Symposium of Japan*, JSCE, 51–55.

Aydan, Ö., Akagi, T. & Kawamoto, T. (1996) The squeezing potential of rock around tunnels: theory and prediction with examples taken from Japan. *Rock Mechanics and Rock Engineering*, 29(3), 125–143.

Aydan, Ö., Üçpırtı, H. & Türk, N. (1997) Theory of laboratory methods for measuring permeability of rocks and tests. *Kaya Mekaniği Bülteni*, 13, 19–36.

Aydan, Ö., Watanabe, S. & Tokashiki, N. (2008) The inference of mechanical properties of rocks from penetration tests. 5th Asian Rock Mechanics Symposium (ARMS5), Tehran, 213–220.

Aydan, Ö., Ohta, Y., Tano, H. (2010). Multi-parameter response of soft rocks during deformation and fracturing with an emphasis on electrical potential variations and its implications in geomechanics and geoengineering. The 39th Rock Mechanics Symposium of Japan, Tokyo, 116–121.

Aydan, Ö., Uehara, F. & Kawamoto, T. (2012) Numerical study of the long-term performance of an underground powerhouse subjected to varying initial stress states, cyclic water heads, and temperature variations. *International Journal of Geomechanics*, ASCE, 12(1), 14–26.

Boussinesq, J. (1885) *Applications des potentiels à l'étude de l'équilibre et mouvement des solides élastiques*. Gauthier-Villard, Paris.

Brace, W.F., Walsh, J.B. & Frangos, W.T. (1968) Permeability of granite under high pressure. *Journal of Geophysical Research*, 73, 2225–2236.

Bieniawski, Z.T. (1970) Time-dependent behaviour of fractured rock. *Rock Mechanics*, 2, 123–137.

Brady, B.H.G. & Brown, E.T. (1985) *Rock Mechanics for Underground Mining*. Kluwer Academic Publications, New York, Boston, London, Moscow. 527p.

Detournay, E. (1986) An approximate statical solution of the elastoplastic interface for the problem of Galin with a cohesive-frictional material. *International Journal of Solids and Structures*, Elsevier, 22, 1435–1454.

England, A.H. (1971) *Complex variable methods in elasticity*. Wiley-Interscience, 181p.

Eringen, A.C. (1980) *Mechanics of Continua*. R. E. Krieger Pub. Co., New York.

Fenner, R. (1938) Researches on the notion of ground stress (in German). *Glückauf*, 74, 681–695.

Galin, L.A. (1946) Plane elastic-plastic problem: Plastic regions around circular holes in plates and beams. *Prikladnaia Matematika i Mechanika*, 10, 365–386.

Gerçek, H. (1993) Qualitative prediction of failures around non-circular openings. In: Paşamehmetoğlu, A.G. *et al.* (eds.) *Proc. Int. Symp. on Assessment and Prevention of Failure Phenomena in Rock Engineering*. A.A. Balkema, Rotterdam. pp. 727–732.

Gerçek, H. (1996) Special elastic solutions for underground openings. *Milestones in Rock Engineering: The Bieniawski Jubilee Collection, Balkema, Rotterdam*. pp. 275–290.

Gerçek, H. (1997) An elastic solution for stresses around tunnels with conventional shapes. *International Journal of Rock Mechanics and Mining*. Science, 34(3–4), paper No. 096.

Green, A.E. & Zerna, W. (1968) *Theory of elasticity*, Clarendon Press, Oxford.

Hoek, E. & Brown, E.T. (1980) Underground Excavations in Rock. *Inst. Min. & Metall.*, 251, 21–26. London.

Inglis, C.E. (1913) Stresses in plates due to the presence of cracks and sharp corners. *Transactions of the Institute of Naval Architects*, 55, 219–241.

Jaeger, J.C. & Cook, N.G.W. (1979) *Fundamentals of Rock Mechanics*, 3rd edition. Chapman & Hall, London. pp. 79, 311.

Kastner, H. (1961) *Statik des Tunnel- and Stollenbaues, ("Design of Tunnels")*, 2nd edition. Springer-Verlag, Berlin.

Kirsch, G. (1898) Die theorie der elastizitat und die bedürfnisse der festigkeitslehre. *Veit Ver. Deut. Ing.*, 42, 797–807.

Kolosov, G.V. (1909) An application of the theory of functions of a complex variable to a planar problem in the mathematical theory of elasticity. Dorpat (Yuriev) University, Doctoral Thesis, 187p.

Kreyszig, E. (1983) *Advanced Engineering Mathematics*. John Wiley & Sons, New York.

Kumsar, H., Aydan, Ö., Tano, H., Çelik, S.B. & Ulusay, R. (2016) An integrated geomechanical investigation, multi-parameter monitoring and analyses of Babadağ-Gündoğdu creep-like landslide. *Rock Mechanics and Rock Engineering*, Special Issue on the Deep-seated landslides. DOI:10.1007/s00603-015-0826-7.

Ladanyi, B. (1974) Use of the long-term strength concept in the determination of ground pressure on tunnel linings. *Proc. of 3rd Congr. Int. Soc. Rock Mech.*, Denver, 2B, 1150–1165.

Lama, R.D. & Vutukuri, V.S. (1978) *Handbook on Mechanical Properties of Rocks*. Trans Tech Publications, Clausthal, Germany.

Milne-Thomson, L.M. (1960) *Plane elastic systems*. Springer-Verlag, Berlin, Heidelberg.

Muskhelishvili, N.I. (1962) *Some Basic Problems of the Mathematical Theory of Elasticity*. Noordhoff, Groningen.

Polubarinova-Kochina, P.YA. (1962) *Theory of Groundwater Movement*. Princeton University Press, Princeton.

Sezaki, M., Aydan, Ö. & Yokota, H. (1994) Non-destructive testing of shotcrete for tunnels. *Int. Conf. on Inspection, Appraisal, Repairs & Maintenance of Buildings & Structures*, Bangkok, 209–215.

Snow, D.T. (1965) *A Parallel Plate Model of Fractured Permeable Media*. PhD Dissertation, University of California, Berkeley.

Talobre, J. (1957) *The Mechanics of Rocks, Dunod* (in French), Paris.

Temel, A. (2002) *Personal Communication*. Hacettepe University, Geological Engineering Department, Ankara, Turkey.

Terzaghi, K. (1925) *Erdbaumechanik auf bodenphysikalischer Grundlage*. F. Deuticke's Verlag, Leipzig, Vienna.

Terzaghi, K. (1946) Rock defects and loads on tunnel support. Introduction to rock tunnelling with steel supports. R.V. Proctor & T.L. White (eds.). Commercial Sheering & Stamping Co., Youngstown, Ohio, U.S.A., 271p.

Terzaghi, K. (1960) Stability of steep slopes on hard unweathered rock. *Geotechnique*, 12, 251–270.

Timoshenko, S. & Goodier, J.N. (1951) *Theory of elasticity*. McGraw Hill Book Company, New York, 519p.

Üçpırtı, H. & Aydan, Ö. (1997) An experimental study on the permeability of interface between sealing plug and rock. *The 28th Rock Mechanics Symposium of Japan*. pp. 268–272.

Verruijt, A. (1970) *Theory of Groundwater Flow*. MacMillian, London, UK.

Zachmanoglou, E.C. & Thoe, D.W. (1986) *Introduction to Partial Differential Equations with Applications*. Dover Pub. Inc., New York.

Zeigler, T.W. (1976). Determination of rock mass permeability. U.S. Army Corps of Engineers Waterways Experiment Station, Vicksburg, Miss. Tech. Rept. S-76-2. 112pp.

Zoback, M.D., Tsukahara, H. & Hickman, S.H. (1980) Stress measurements at depth in the vicinity of the San Andreas Fault: Implications for the magnitude of shear stress at depth. *Journal of Geophysical Research*, 85(B11), 6157–6173.

Chapter 9

Numerical methods

In this chapter, the first part is related to the solution of fundamental governing equations using the finite element method. Nevertheless, an illustrative example is given in the introduction to explain the similarity and dissimilarity of various numerical methods as well as exact solutions. Although formulations for multidimensional situations are not presented, they can be easily extended to such situations by just selecting shape functions for multidimensional situations as the general forms of equations would remain the same. In the second part, some numerical procedures developed for rock masses involving discontinuities are presented, and several examples are given.

9.1 Introduction

There are three approximate methods:

- Finite difference method (FDM)
- Finite element method (FEM)
- Boundary element method (BEM)

The characteristics of the closed-form and approximate methods are briefly discussed through solving the following ordinary differential equation:

$$\frac{d^2u}{dx^2} - u = 0 \tag{9.1}$$

The boundary conditions are as follows:

$$
\begin{array}{llll}
u = 0 & \text{at} & x = 0 \\
u = 1 & \text{at} & x = 1
\end{array}
$$

9.1.1 Closed-form solution

If the solution of Equation (9.1) is a series of exponential functions $e^{\lambda x}$, the characteristics equation can be obtained as:

$$\lambda^2 - 1 = 0 \tag{9.2}$$

Hence, the roots are:

$$\lambda_1 = 1, \quad \lambda_2 = -1 \tag{9.3}$$

Thus, the solution is of the following form:

$$u = C_1 e^x + C_2 e^{-x} \tag{9.4}$$

The integration constants are obtained from the boundary conditions as:

$$C_1 = 0.4254589 \quad \text{and} \quad C_2 = -0.4254589$$

9.1.2   Finite Difference Method (FDM)

The finite difference method is the earliest approximate method, and it is called a strong form approximate solution. It utilizes the Taylor expansion of dependent variable to discretize the governing equation. Let us assume that the domain is discretized into n segments with equal interval Δx. Equation (9.1) at a node j may be rewritten as:

$$\left(\frac{d^2 u}{dx^2}\right)_{x=x_j} - u_j = 0 \tag{9.5}$$

The Taylor expansions of function u at nodes i, j and k may be written as:

$$u_i(x_j - \Delta x) = u_j - \left(\frac{du}{dx}\right)_{x=x_j} \frac{\Delta x}{1!} + \left(\frac{d^2 u}{dx^2}\right)_{x=x_j} \frac{\Delta x^2}{2!} - 0^3 \tag{9.6}$$

$$u_j(x_j) = u_j \tag{9.7}$$

$$u_k(x_j + \Delta x) = u_j + \left(\frac{du}{dx}\right)_{x=x_j} \frac{\Delta x}{1!} + \left(\frac{d^2 u}{dx^2}\right)_{x=x_j} \frac{\Delta x^2}{2!} + 0^3 \tag{9.8}$$

From the preceding relations, one gets the following relation:

$$\left(\frac{d^2 u}{dx^2}\right)_{x=x_j} = \frac{u_k - 2u_j + u_i}{\Delta x^2} \tag{9.9}$$

Thus the finite difference form of the preceding equation takes the following form:

$$\left[\frac{1}{\Delta x^2} u_i - \left(\frac{2}{\Delta x^2} + 1\right) u_j + \frac{1}{\Delta x^2} u_k\right] = 0 \tag{9.10}$$

This simultaneous equation system for a domain divided into n segments will result in:

$$[K]\{U\} = \{F\} \tag{9.11}$$

where matrix $[K]$ has $n - 1$ rows and $n + 1$ columns, vector $\{U\}$ has $n + 1$ rows and vector $\{F\}$ has $n - 1$ rows. However, if the boundary conditions are introduced, it yields the following simultaneous equation system:

$$[K^*]\{U^*\} = \{F^*\} \tag{9.12}$$

The resulting matrix $[K^*]$ has $n - 1$ rows and $n - 1$ columns. Similarly, the resulting vectors $\{U^*\}$ and $\{F^*\}$ have $n - 1$ rows. Therefore, it becomes possible to solve this simultaneous equation system.

Example 1: Let us assume that we have two segments and three nodes. Accordingly, $u_1 = 0$, $u_3 = 1$, $\Delta x = 0.5$. From Equation (9.12), we obtain unknown u_2 as:

$$u_2 = 0.444444444444 \tag{9.13}$$

Example 2: Let us assume that we have four segments and five nodes. Accordingly, $u_1 = 0$, $u_5 = 1$, $\Delta x = 0.25$. From Equation (9.12), we get the following equation system for unknown $\{u^*\}$:

$$\begin{bmatrix} -33 & 16 & 0 \\ 16 & -33 & 16 \\ 0 & 16 & -33 \end{bmatrix} \begin{Bmatrix} u_2 \\ u_3 \\ u_4 \end{Bmatrix} = \begin{Bmatrix} 0 \\ 0 \\ -16 \end{Bmatrix} \tag{9.14}$$

The solution of this simultaneous equation system yields the following:

$$u_2 = 0.215114752376, \quad u_3 = 0.443674176776, \quad u_4 = 0.69963237225$$

9.1.3 Finite Element Method (FEM)

The finite element method is relatively new, but it is the most widely used method in engineering and science as compared with FDM or other methods. The governing equation is first integrated over the domain, and then the resulting integral equation is discretized. Therefore, it is called a *weak form solution* as there is a possibility that the solution may be different from the actual one.

(a) Weak formulation

Taking a dot product of Equation (9.1) by a trial function δv and integrating it yields the following:

$$\int_0^1 \delta v \cdot \frac{d^2 u}{dx^2} dx - \int_0^1 \delta v \cdot u \, dx = 0 \tag{9.15}$$

Introducing the integral by parts for the first term gives:

$$\int_0^1 \frac{d\delta v}{dx} \cdot \frac{du}{dx} dx + \int_0^1 \delta v \cdot u \, dx = [\delta v \cdot \hat{t}]_0^1 \tag{9.16}$$

where

$$\hat{t} = \frac{du}{dx} n$$

and n is the unit normal vector. Let us assume that the trial function v is the same as the function u, which is generally called the Galerkin approach in finite element formulation.

(b)   Discretization

The domain is discretized into subdomains called elements. The function u is approximated by a chosen function in an element, and it is summed up for the whole domain. For this particular problem, let us choose a linear function of the following form:

$$u = ax + b \tag{9.17}$$

Let us assume that the function u at nodes i and j are known. Thus we can write the following:

$$\begin{bmatrix} x_i & 1 \\ x_j & 1 \end{bmatrix} \begin{Bmatrix} a \\ b \end{Bmatrix} = \begin{Bmatrix} u_i \\ u_j \end{Bmatrix} \tag{9.18}$$

Taking the inverse of the preceding relation, one gets coefficients a and b. Inserting these coefficients in Equation (9.17) yields the following:

$$u = N_i u_i + N_j u_j \tag{9.19}$$

where

$$N_i = \frac{x_j - x}{x_j - x_i}, \quad N_j = \frac{x - x_i}{x_j - x_i}$$

The preceding equation may be rewritten in a compact form as:

$$u = [N]\{U_e\} \quad \text{or} \quad u = \mathbf{N} U_e \tag{9.20}$$

where $[N] = [N_i, N_j]$, $\{U_e\}^T = \{u_i, u_j\}$.[1] The derivative of the preceding equation takes the following form:

$$\frac{du}{dx} = \frac{dN_i}{dx} u_i + \frac{dN_j}{dx} u_j \tag{9.21}$$

The preceding relation is rewritten in a compact form as:

$$\frac{du}{dx} = [B]\{U_e\} \quad \text{or} \quad \frac{du}{dx} = \mathbf{B} U_e \tag{9.22}$$

where $[B] = [B_i, B_j]$, $B_i = -1/L_e, B_j = 1/L_e, L_e = x_j - x_i$. The dot product of two vectors are presented in the following form in the finite element method:

$$c = \mathbf{a} \cdot b \rightarrow c = \{a\}^T \{b\} \tag{9.23}$$

Equation (9.16), which holds for the whole domain, must also hold for each element as:

$$\int_{x_i}^{x_j} \frac{\delta u}{dx} \cdot \frac{du}{dx} dx + \int_{x_i}^{x_j} \delta u \cdot u dx - [\delta u \cdot \tilde{t}]_{x_i}^{x_j} \tag{9.24}$$

---

1 should be noted that the horizontally written vector is defined as the transpose of the vector.

Inserting relations given by Equations (9.20) and (9.22) into Equation (9.24) and using the finite element convention for dot product (Equation (9.23)) yields the following:

$$\{\delta U_e\} \left(\left[\int_{x_i}^{x_j} [B]^T [B] dx + \int_{x_i}^{x_j} [N]^T [N] dx \right] \{U_e\} - [[N]^T \hat{t}]_{x_i}^{x_j} \right) = 0 \tag{9.25}$$

The preceding relation implies the following:

$$[K_e]\{U_e\} = \{F_e\} \tag{9.26}$$

where

$$[K_e] = \int_{x_i}^{x_j} [B]^T [B] dx + \int_{x_i}^{x_j} [N]^T [N] dx, \quad \{F_e\} = [[N]^T \hat{t}]_{x_i}^{x_j}$$

For a typical element, one gets the preceding relations specifically for a shape function given by Equation (9.17):

$$[K_e] = \frac{1}{L} \begin{bmatrix} 1 & -1 \\ -1 & 1 \end{bmatrix} + \frac{L}{6} \begin{bmatrix} 2 & 1 \\ 1 & 2 \end{bmatrix}, \quad \{U_e\} = \begin{Bmatrix} u_i \\ u_j \end{Bmatrix}, \quad \{F_e\} = \begin{Bmatrix} \hat{t}_i \\ \hat{t}_j \end{Bmatrix}$$

The sum-up of the preceding relation for the whole domain is:

$$[K]\{U\} = \{F\} \tag{9.27}$$

where

$$[K] = \sum_{k=1}^{n} [K_e], \quad \{U\} = \sum_{k=1}^{n} \{U_e\}, \quad \{F\} = \sum_{k=1}^{n} \{F_e\}$$

Example 1: Let us assume that we have two elements and three nodes. Accordingly, $u_1 = 0$, $u_3 = 0$, $x_j - x_i = 0.5$. From Equation (9.27), we obtain unknown u_2 as:

$$u_2 = 0.4423076 \tag{9.28}$$

Example 2: Let us assume that we have four elements and five nodes. Accordingly, $u_1 = 0$, $u_5 = 0$, $x_j - x_i = 0.25$. The solution of the simultaneous equation system (Equation (9.27)) yields the followings:

$$u_2 = 0.214787576025, \quad u_3 = 0.443140650725, \quad u_4 = 0.699481489062$$

9.1.4 Comparisons

Solutions obtained from the approximate methods for the example chosen are compared with that by the closed-form solution (CFS). Figures 9.1(a) and (b) show comparisons of computations for two element (three nodes) and four element (five nodes) discretizations of the domain by the FDM and FEM with that by the CFS, respectively. As seen from both figures, the approximate solutions almost coincide with the exact ones at nodal

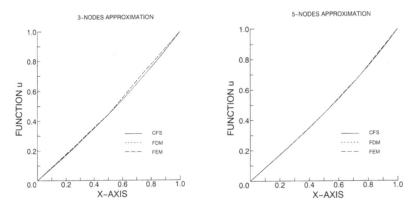

Figure 9.1 Comparison of computations by FDM and FEM with that by CFS

points. Increasing the number of nodes results in better solutions, and errors caused by discretization decreases.

9.2 1-D hyperbolic problem: equation of motion

As shown in Chapter 3, the equation of momentum for 1-D problems can be written as:

$$\frac{\partial \sigma}{\partial x} + b = \rho \frac{\partial^2 u}{\partial t^2} \tag{9.29}$$

Let us assume that this equation is subjected to following boundary and initial conditions

Boundary conditions

$$u(0,t) = 0 \quad \text{as} \quad x = 0$$
$$t(0,t) = t_0 \quad \text{as} \quad x = L \tag{9.30}$$

Initial conditions

$$u(x,0) = 0 \quad \text{as} \quad t = 0$$
$$\dot{u}(x,0) = 0 \quad \text{as} \quad t = 0 \tag{9.31}$$

Let us further assume that the material is of Kelvin-type as given here:

$$\sigma = E\varepsilon + C\dot{\varepsilon} \tag{9.32}$$

Displacement–strain and strain rate are given as:

$$\varepsilon = \frac{\partial u}{\partial x}, \quad \dot{\varepsilon} = \frac{\partial \dot{u}}{\partial x} \tag{9.33}$$

where

$$\dot{u} = \frac{\partial u}{\partial t}$$

We utilize finite element method to solve this equation system in the following section.

9.2.1   Weak form formulation

Taking a variation on displacement field δu, the integral form of Equation (9.29) becomes:

$$\int_V \delta u \cdot \frac{\partial \sigma}{\partial x} dV + \int_V \delta u \cdot b dV = \int_V \rho \delta u \cdot \frac{\partial^2 u}{\partial t^2} dV \tag{9.34}$$

where $dV = dA dx$.

Applying the integral by parts to the first term on the LHS with respect to x yields the following:

$$\int_A [\delta u \cdot t]_0^L dA + \int_V \delta u \cdot b dV = \int_V \frac{\partial \delta u}{\partial x} \cdot \sigma dV + \int_V \rho \delta u \cdot \frac{\partial^2 u}{\partial t^2} dV \tag{9.35}$$

Where $t = \sigma \cdot n$. The preceding equation is called the weak form of Equation (9.29). Inserting the constitutive relation given by Equation (9.32) into Equation (9.35), we obtain the following:

$$\int_A [\delta u \cdot t]_0^L dA + \int_V \delta u \cdot b dV = \int_V E \frac{\partial \delta u}{\partial x} \cdot \frac{\partial u}{\partial x} dV + \int_V C \frac{\partial \delta u}{\partial x} \cdot \frac{\partial u}{\partial x} dV + \int_V \rho \delta u \cdot \frac{\partial^2 u}{\partial t^2} dV \tag{9.36}$$

9.2.2   Discretization

For this particular problem, a linear function of the following form for the space is chosen:

$$u(t) = ax + b \tag{9.37}$$

Let us assume that the function u at nodes i and j are known. Thus we can write:

$$\begin{bmatrix} x_i & 1 \\ x_j & 1 \end{bmatrix} \begin{Bmatrix} a \\ b \end{Bmatrix} = \begin{Bmatrix} u_i \\ u_j \end{Bmatrix} \tag{9.38}$$

Taking the inverse of the preceding relation, one gets coefficients a and b. Inserting these coefficients in Equation (9.37) yields the following:

$$u = N_i u_i + N_j u_j \tag{9.39}$$

where

$$N_i = \frac{x_j - x}{x_j - x_i}, \quad N_j = \frac{x - x_i}{x_j - x_i}$$

The preceding equation may be rewritten in a compact form as:

$$u = [N]\{U_e\} \quad \text{or} \quad u = \mathbf{N}U_e \tag{9.40}$$

where $[N] = [N_i, N_j]$, $\{U_e\}^T = \{u_i, u_j\}$. The derivative of the preceding equation takes the following form:

$$\frac{du}{dx} = \frac{dN_i}{dx}u_i + \frac{dN_j}{dx}u_j \tag{9.41}$$

The preceding relation is rewritten in a compact form as:

$$\frac{du}{dx} = [B]\{U_e\} \quad \text{or} \quad \frac{du}{dx} = \mathbf{B}U_e \tag{9.42}$$

Where $[B] = [B_i, B_j]$, $B_i = -1/L_e$, $B_j = 1/L_e$, $L_e = x_j - x_i$. Equation (9.36), which holds for the whole domain, must also hold for each element:

$$\int_{V_e} \rho \delta u \cdot \frac{\partial^2 u}{\partial t^2} dV + \int_{V_e} C \frac{\partial \delta u}{\partial x} \cdot \frac{\partial u}{\partial x} dV + \int_{V_e} E \frac{\partial \delta u}{\partial x} \cdot \frac{\partial u}{\partial x} dV = \int_{A_e} [\delta u \cdot t]_{x_i}^{x_j} dA + \int_{V_e} \delta u \cdot b_e dV \tag{9.43}$$

The discretised form of the preceding equation becomes:

$$\int_{V_e} \rho [N]^T [N] \{\ddot{U}_e\} dV + \int_{V_e} C[B]^T[B]\{\dot{U}_e\} dV + \int_{V_e} E[B]^T[B]\{U\}_e dV = \tag{9.44}$$

$$\int_{A_e} [N_k]^T t_e]_{x_i}^{x_j} dA + \int_{V_e} [N]^T b_e dV$$

The preceding equation may be written in a compact form as:

$$[M_e]\{\ddot{U}_e\} + [C_e]\{\dot{U}_e\} + [K_e]\{U_e\} = \{F_e\} \tag{9.45}$$

where

$$[M_e] = \int_{V_e} \rho [N]^T[N] dV, \quad [C_e] = \int_{V_e} C[B]^T[B] dV, \quad [K_e] = \int_{V_e} E[B]^T[B] dV,$$

$$\{F_e\} = \int_{A_e} [[N]^T t_e]_{x_i}^{x_j} dA + \int_{V_e} [N]^T b_e dV$$

For the total domain, we have the following:

$$[M]\{\ddot{U}\} + [C]\{\dot{U}\} + [K]\{U\} = \{F\} \tag{9.46}$$

where

$$[M] = \sum_{k=1}^{n} [M_e]_k, \quad [C] = \sum_{k=1}^{n} [C_e]_k, \quad [K] = \sum_{k=1}^{n} [K_e]_k, \quad \{F\} = \sum_{k=1}^{n'} \{F_e\}_k, \quad \{U\} = \sum_{k=1}^{n} \{U_e\}_k$$

Equation (9.46) could not be solved as it is. For a time step m, we can rewrite Equation (9.46) as:

$$[M]\{\ddot{U}\}_m + [C]\{\dot{U}\}_m + [K]\{U\}_m = \{F\}_m \tag{9.47}$$

Therefore, we discretize displacement field $\{U\}$ for time-domain using the Taylor expansion as it is in the finite difference method as:

$$\{U\}_{m-1} = \{U\}_m - \frac{\partial \{U\}_m}{\partial t}\frac{\Delta t}{1!} + \frac{\partial^2 \{U_m\}}{\partial t^2}\frac{\Delta t^2}{2!} - 0^3 \tag{9.48}$$

$$\{U\}_m = \{U\}_m \tag{9.49}$$

$$\{U\}_{m+1} = \{U\}_m + \frac{\partial \{U_m\}}{\partial t}\frac{\Delta t}{1!} + \frac{\partial^2 \{U_m\}}{\partial t^2}\frac{\Delta t^2}{2!} + 0^3 \tag{9.50}$$

From the preceding relations, one easily gets the following:

$$\{\dot{U}\}_m = \frac{1}{\Delta t}(\{U\}_{m+1} - \{U\}_{m-1}) \tag{9.51}$$

$$\{\ddot{U}\}_m = \frac{1}{\Delta t^2}(\{U\}_{m+1} - 2\{U\}_m + \{U\}_{m-1}) \tag{9.52}$$

Inserting these relations into Equation (9.47), we get the following:

$$[M^*]\{U\}_{m+1} = \{F^*\}_{m+1} \tag{9.53}$$

where

$$[M^*] = \left[\frac{1}{\Delta t^2}[M] + \frac{1}{\Delta t}[C]\right]$$

$$\{F^*\}_{m+1} = \left[\frac{2}{\Delta t^2}[M] - [K]\right]\{U\}_m - \left[\frac{1}{\Delta t^2}[M] - \frac{1}{\Delta t}[C]\right]\{U\}_{m-1} + \{F\}_m$$

9.2.3 Specific example

For a typical two-noded element, the followings are obtained:

$$[M_e] = \frac{\rho L_e A_e}{6}\begin{bmatrix} 2 & 1 \\ 1 & 2 \end{bmatrix}, \quad [C_e] = \frac{CA_e}{L}\begin{bmatrix} 1 & -1 \\ -1 & 1 \end{bmatrix}, \quad [K_e] = \frac{EA_e}{L}\begin{bmatrix} 1 & -1 \\ -1 & 1 \end{bmatrix},$$

$$\{F_e\} = \begin{Bmatrix} t_i \\ t_j \end{Bmatrix} + \frac{b}{2}\begin{Bmatrix} 1 \\ 1 \end{Bmatrix}, \quad \{U_e\} = \begin{Bmatrix} u_i \\ u_j \end{Bmatrix}$$

If the space is discretized into two elements, we have the following simultaneous equation system:

$$
\begin{bmatrix}
(K_{11}^*)^1 & (K_{12}^*)^1 & 0 \\
(K_{21}^*)^1 & (K_{22}^*)^1 + (K_{11}^*)^2 & (K_{12}^*)^2 \\
0 & (K_{21}^*)^2 & (K_{22}^*)^2
\end{bmatrix}
\begin{Bmatrix}
U_1 \\ U_2 \\ U_3
\end{Bmatrix}_{m+1}
=
\begin{Bmatrix}
F_1^* \\ F_2^* \\ F_3^*
\end{Bmatrix}_{m+1}
$$

9.2.4   1-D Parabolic problem: creep problem

If the inertia term in Equation (9.29) is negligible, and the constitutive law is of the Kelvin type, then the finite element form of Equation (9.29) becomes:

$$
[C]\{\dot{U}\} + [K]\{U\} = \{F\}
\tag{9.54}
$$

For a time step m, we get the following equation using the Taylor expansion:

$$
[C^*]\{U\}_{m+1} = \{F^*\}_{m+1}
\tag{9.55}
$$

where

$$
[C^*] = \frac{1}{\Delta t}[C], \quad \{F^*\}_{m+1} = \left[\frac{1}{\Delta t}[C] - [K]\right]\{U\}_m + \{F\}_m
$$

9.2.5   1-D elliptic problem: static problem

If the inertia term in Equation (9.29) is negligible, and the constitutive law is of the Hookean type, then the finite element form of Equation (9.29) becomes:

$$
[K]\{U\} = \{F\}
\tag{9.56}
$$

9.2.6   Computational examples

In the first example, the dynamic response of a layer of infinite length and 1 m thick is analysed. The body force of the layer is assumed to be applied suddenly, and the selected viscosity coefficient (V) is 0.2 and 0.5. Figure 9.2 shows the computed displacement response of some points with time. It is interesting to note that fluctuations occur as the viscosity coefficient decreases in magnitude.

In the second example, the same problem is analysed using hyperbolic and parabolic formulations. The results are shown in Figure 9.3 As seen from the figure, the computed responses of both hyperbolic and parabolic selected points converge to those, which could be obtained from the elliptical formulation.

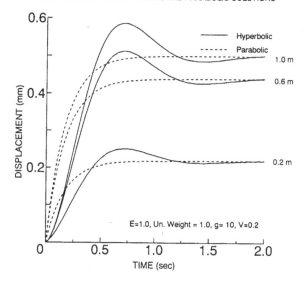

Figure 9.2 Comparison of hyperbolic and parabolic solutions (V = 0.2)

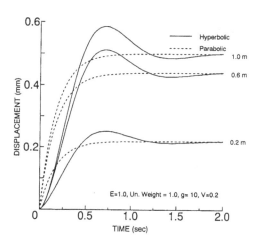

Figure 9.3 Comparison of hyperbolic and parabolic solutions (V = 0.5)

9.3   Parabolic problems: heat flow, seepage and diffusion

9.3.1   Introduction

The governing equation for heat flow, fluid flow, seepage and diffusion problems takes exactly the same form except the physical meaning of variables are different. In the followings, a finite element formulation of such a governing equation and its discretisation are given. Furthermore, some sample computations are carried out.

9.3.2   Governing equation

As shown in Chapter 4, the laws of mass conservation law and heat flow of nonconvective or nonadvective type take the following form for 1-D problems:

$$-\frac{\partial q}{\partial x} + g = \rho c \frac{\partial T}{\partial t} \tag{9.57}$$

where T can be temperature, water head or mass concentration. Let us assume that this equation is subjected to following boundary and initial conditions:

Boundary conditions

$$\begin{aligned} T(0,t) &= 0 \quad \text{at} \quad x = 0 \\ q_n(0,t) &= q_0 \quad \text{at} \quad x = L \end{aligned} \tag{9.58}$$

Initial conditions

$$\begin{aligned} T(x,0) &= 0 \quad \text{at} \quad t = 0 \\ \dot{T}(x,0) &= 0 \quad \text{at} \quad t = 0 \end{aligned} \tag{9.59}$$

Let us further assume that the material obeys a linear type of constitutive law between flux q and dependent variable T:

$$q = -k \frac{\partial T}{\partial x} \tag{9.60}$$

9.3.3   Weak form formulation

Taking a variation on variable δT, the integral form of Equation (9.57) becomes:

$$-\int_V \delta T \cdot \frac{\partial q}{\partial x} dV + \int_V \delta T \cdot g dV = \int_V \rho c \delta T \cdot \frac{\partial T}{\partial t} dV \tag{9.61}$$

where $dV = dA dx$.

Applying the integral by parts to the first term on the LHS with respect to x yields the following:

$$-\int_A [\delta T \cdot q_n]_0^L dA + \int_V \delta T \cdot g dV = -\int_V \frac{\partial \delta T}{\partial x} \cdot q dV + \int_V \rho c \delta T \cdot \frac{\partial T}{\partial t} dV \tag{9.62}$$

where $q_n = q \cdot n$. The preceding equation is called the weak form of Equation (9.57). Inserting the constitutive relation given by Equation (9.60) into Equation (9.62), we obtain the following:

$$-\int_A [\delta T \cdot q_n]_0^L dA + \int_V \delta T \cdot g dV = \int_V k \frac{\partial \delta T}{\partial x} \cdot \frac{\partial T}{\partial x} dV + \int_V \rho c \delta T \cdot \frac{\partial T}{\partial t} dV \qquad (9.63)$$

9.3.4 Discretization

For this particular problem, let us choose a linear function of the following form for the space:

$$T(t) = ax + b \qquad (9.64)$$

Let us assume that the function T at nodes i and j are known. Thus we can write the following:

$$\begin{bmatrix} x_i & 1 \\ x_j & 1 \end{bmatrix} \begin{Bmatrix} a \\ b \end{Bmatrix} = \begin{Bmatrix} T_i \\ T_j \end{Bmatrix} \qquad (9.65)$$

Taking the inverse of the preceding relation, one gets coefficients a and b. Inserting these coefficients in Equation (9.64) yields the following:

$$T = N_i T_i + N_j T_j \qquad (9.66)$$

where

$$N_i = \frac{x_j - x}{x_j - x_i}, \quad N_j = \frac{x - x_i}{x_j - x_i}$$

The preceding equation may be rewritten in a compact form as:

$$T = [N]\{T_e\} \quad \text{or} \quad T = \mathbf{N}T_e \qquad (9.67)$$

where $[N] = [N_i, N_j], [N] \{T_e\}^T = \{T_i, T_j\}$. The derivative of the preceding equation takes the following form:

$$\frac{dT}{dx} = \frac{dN_i}{dx} T_i + \frac{dN_j}{dx} T_j \qquad (9.68)$$

The preceding relation is rewritten in a compact form as:

$$\frac{dT}{dx} = [B]\{T_e\} \quad \text{or} \quad \frac{dT}{dx} = \mathbf{B}T_e \qquad (9.69)$$

where $[B] = [B_i, B_j]$, $B_i = -1/L_e$, $L_e = x_j - x_i$. Equation (9.63), which holds for the whole domain, must also hold for each element:

$$\int_{V_e} \rho c \delta T \cdot \frac{\partial T}{\partial t} dV + \int_{V_e} k \frac{\partial \delta T}{\partial x} \cdot \frac{\partial T}{\partial x} dV = \int_{A_e} [\delta T \cdot q_n^e]_{x_i}^{x_j} dA + \int_{V_e} \delta T \cdot g_e dV \qquad (9.70)$$

The discretized form of the preceding equation becomes:

$$\int_{V_e} \rho c [N]^T [N] \{\dot{T}_e\} dV + \int_{V_e} k [B]^T [B] \{U\}_e dV = \int_{A_e} [[N_k]^T q_n^e]_{x_i}^{x_j} dA + \int_{V_e} [N]^T g_e dV \qquad (9.71)$$

The preceding equation may be written in a compact form as:

$$[M_e]\{\dot{T}_e\} + [K_e]\{T_e\} = \{F_e\} \qquad (9.72)$$

where

$$[M_e] = \int_{V_e} \rho c [N]^T [N] dV, \quad [K_e] = \int_{V_e} k [B]^T [B] dV, \{F_e\} = \int_{A_e} [[N]^T q_n^e]_{x_i}^{x_j} dA + \int_{V_e} [N]^T g_e dV.$$

For the total domain, we have:

$$[M]\{\dot{T}\} + [K]\{T\} = \{F\} \qquad (9.73)$$

where

$$[M] = \sum_{k=1}^{n} [M_e]_k, \quad [K] = \sum_{k=1}^{n} [K_e]_k, \quad \{F\} = \sum_{k=1}^{n} \{F_e\}_k, \quad \{T\} = \sum_{k=1}^{n} \{T_e\}_k$$

Equation (9.73) could not be solved as it is. For a time step m, we can rewrite Equation (9.73) as:

$$[M]\{\dot{T}\}_{(m+\theta)} + [K]\{T\}_{(m+\theta)} = \{F\}_{(m+\theta)} \qquad (9.74)$$

Therefore, we discretisz dependent variable {T} for time-domain for using the Taylor expansion as it is in the finite difference method:

$$\{T\}_m = \{T\}_{(m+\theta)-\theta} = \{T\}_{(m+\theta)} - \frac{\partial\{T\}_{(m+\theta)}}{\partial t}\frac{\theta\Delta t}{1!} + \frac{\partial^2\{T\}_{(m+\theta)}}{\partial t^2}\frac{\theta^2\Delta t^2}{2!} - 0^3 \qquad (9.75)$$

$$\{T\}_{m+1} = \{T\}_{(m+\theta)+(1-\theta)} = \{T\}_{(m+\theta)} + \frac{\partial\{T\}_{(m+\theta)}}{\partial t}\frac{(1-\theta)\Delta t}{1!} + \frac{\partial^2\{T\}_{(m+\theta)}}{\partial t^2}\frac{(1-\theta)^2\Delta t^2}{2!} + 0^3 \qquad (9.76)$$

From preceding relations, one easily gets:

$$\{T\}_{(m+\theta)} = \theta\{T\}_{m+1} + (1-\theta)\{T\}_m \quad \{\dot{T}\}_{(m+\theta)} = \frac{\{T\}_{m+1} - \{T\}_m}{\Delta t} \qquad (9.77)$$

$$\{F\}_{(m+\theta)} = \theta\{F\}_{m+1} + (1-\theta)\{F\}_m \qquad (9.78)$$

Inserting these relations into Equation (9.74), we get:

$$[M^*]\{T\}_{m+1} = \{F^*\}_{m+1} \qquad (9.79)$$

where

$$[M^*] = \left[\frac{1}{\Delta t}[M] + \theta[K]\right]$$

$$\{F^*\}_{m+1} = \left[\frac{1}{\Delta t}[M] - (1-\theta)[K]\right]\{T\}_m + \theta\{F\}_{m+1} + (1-\theta)\{F\}_m$$

9.3.5 Steady-state problem

When the time variation of dependent variable T is negligible, then the problem is called a steady-state problem. This type of special case corresponds to the elliptical problem. The final finite element form of the discretized governing equation becomes:

$$[K]\{T\} = \{F\} \tag{9.80}$$

where

$$[K] = \sum_{k=1}^{n}[K_e]_k, \quad \{F\} = \sum_{k=1}^{n}\{F_e\}_k, \quad \{T\} = \sum_{k=1}^{n}\{T_e\}_k$$

9.3.6 Specific example

For a typical two noded element, the followings are obtained:

$$[M_e] = \frac{\rho c L_e A_e}{6}\begin{bmatrix} 2 & 1 \\ 1 & 2 \end{bmatrix} \quad [K_e] = \frac{kA_e}{L}\begin{bmatrix} 1 & -1 \\ -1 & 1 \end{bmatrix}$$

If the space is discretized into two elements, we have the following simultaneous equation system:

$$\begin{bmatrix} (K_{11}^*)^1 & (K_{12}^*)^1 & 0 \\ (K_{21}^*)^1 & (K_{22}^*)^1 + (K_{11}^*)^2 & (K_{12}^*)^2 \\ 0 & (K_{21}^*)^2 & (K_{22}^*)^2 \end{bmatrix}\begin{Bmatrix} T_1 \\ T_2 \\ T_3 \end{Bmatrix}_{m+1} = \begin{Bmatrix} F_1^* \\ F_2^* \\ F_3^* \end{Bmatrix}_{m+1}$$

9.3.7 Example: simulation of a solid body with heat generation

A specific example is given here by simulating the temperature variation in a solid body for the following three different conditions:

1 Heat generation only
2 Heat flux input only
3 Heat generation + heat flux

The heat generation function is assumed to be of the following form:

$$g = At \exp - t/\tau \tag{9.81}$$

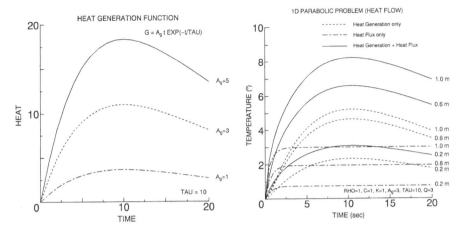

Figure 9.4 Comparison of temperature responses of some selected points for various conditions

where A and τ are physical constants determined from heat generation tests of a solid, and t is time. Figure 9.4 compares the results of computations for three different conditions. When the *heat generation only* condition is considered, temperature first increases and then decreases in a similar manner to the heat generation function. When the *heat flux only* condition is considered, temperature increases monotonically and tends to be asymptotic. When the *heat generation + heat flux* condition is considered, temperature first increases and then decreases. Finally, it becomes asymptotic to those computed for the *heat flux only* *condition*.

9.4 Finite element method for 1-D pseudo-coupled parabolic problems: heat flow and thermal stress; swelling and swelling pressure

9.4.1 Introduction

In the following sections, a finite element formulation for a coupled heat flow and stress problem (thermo-mechanics) and its discretisation are given. Although the stress field is coupled with the heat flow, the heat flow field is uncoupled. Therefore, such a problem may be called as a pseudo-coupled problem.

9.4.2 Governing equations

9.4.2.1 Governing equation for a heat flow

As shown in Chapter 4, the laws of mass conservation law and heat flow of nonconvective or nonadvective type takes the following form for 1-D problems:

$$-\frac{\partial q}{\partial x} + g = \rho c \frac{\partial T}{\partial t} \tag{9.82}$$

where T is temperature. Let us assume that this equation is subjected to following boundary and initial conditions:

Boundary conditions

$$T(0,t) = 0 \quad \text{as} \quad x = 0$$
$$q_n(0,t) = q_0 \quad \text{as} \quad x = L$$

$$(9.83)$$

Initial conditions

$$T(x,0) = 0 \quad \text{as} \quad t = 0$$
$$\dot{T}(x,0) = 0 \quad \text{as} \quad t = 0$$

$$(9.84)$$

Let us further assume that the material obeys a linear type of constitutive law between flux q and dependent variable T:

$$q = -k\frac{\partial T}{\partial x} \tag{9.85}$$

9.4.2.2 Governing equation for stress field

As shown in Chapter 4, the equation of momentum for 1-D problems without inertia term can be written as:

$$\frac{\partial \sigma}{\partial x} + b = 0 \tag{9.86}$$

Let us assume that this equation is subjected to the following boundary conditions:

$$u(0,t) = 0 \quad \text{as} \quad x = 0$$
$$t(0,t) = t_0 \quad \text{as} \quad x = L$$

$$(9.87)$$

Let us further assume that the material is of the Hookean type:

$$\sigma = E\varepsilon \tag{9.88}$$

The displacement–strain relation is given as:

$$\varepsilon = \frac{\partial u}{\partial x} \tag{9.89}$$

The incremental form of Equation (9.86) takes the following form if the body force remains constant with time:

$$\frac{\partial \dot{\sigma}}{\partial x} = 0 \tag{9.90}$$

Similarly, the constitutive law and displacement rate and strain rate relations may be rewritten as:

$$\dot{\sigma} = E\dot{\varepsilon} \tag{9.91}$$

$$\dot{\varepsilon} = \frac{\partial \dot{u}}{\partial x} \tag{9.92}$$

9.4.3 Coupling of heat and stress fields

If the mechanical energy rate is slow, the effect of a heat field on stress field is coupled through volumetric strain caused by temperature variation. For one-dimensional situations, this is written as:

$$\dot{\varepsilon}_t = \dot{\varepsilon} - \dot{\varepsilon}_0 \tag{9.93}$$

where $\dot{\varepsilon}_0 = \alpha \Delta T$, α is the thermal expansion coefficient, and ΔT is the temperature variation of a given point per unit time.

9.4.4 Weak form formulation

9.4.4.1 Weak formulation for heat flow field

Taking a variation on variable δT, the integral form of Equation (9.4.1) becomes:

$$-\int_V \delta T \cdot \frac{\partial q}{\partial x} dV + \int_V \delta T \cdot g dV = \int_V \rho c \delta T \cdot \frac{\partial T}{\partial t} dV \tag{9.94}$$

where $dV = dA dx$.

Applying the integral by parts to the first term on the LHS with respect to x yields:

$$-\int_A [\delta T \cdot q_n]_0^L dA + \int_V \delta T \cdot g dV = -\int_V \frac{\partial \delta T}{\partial x} \cdot q dV + \int_V \rho c \delta T \cdot \frac{\partial T}{\partial t} dV \tag{9.95}$$

where $q_n = q \cdot n$. The preceding equation is called the weak form of Equation (9.82). Inserting the constitutive relation given by Equation (9.85) into Equation (9.95), we obtain the following:

$$-\int_A [\delta T \cdot q_n]_0^L dA + \int_V \delta T \cdot g dV = \int_V k \frac{\partial \delta T}{\partial x} \cdot \frac{\partial T}{\partial x} dV + \int_V \rho c \delta T \cdot \frac{\partial T}{\partial t} dV \tag{9.96}$$

9.4.4.2 Weak formulation for stress field

Taking a variation on displacement rate field $\delta \dot{u}$, the integral form of Equation (9.4.9) becomes:

$$\int_V \delta \dot{u} \cdot \frac{\partial \dot{\sigma}}{\partial x} dV = 0 \tag{9.97}$$

where $dV = dA dx$.

Applying the integral by parts to the LHS with respect to x yields:

$$\int_A [\delta \dot{u} \cdot \dot{t}]_0^L dA = \int_V \frac{\partial \delta \dot{u}}{\partial x} \cdot \dot{\sigma} dV \tag{9.98}$$

where $i = \dot{\sigma} \cdot n$. The preceding equation is called the weak form of Equation (9.90). Inserting the constitutive relation given by Equation (9.91) and Equation (9.93) into Equation (9.98), we obtain:

$$\int_A [\delta \dot{u} \cdot i]_0^L dA + \int_V E\alpha \frac{\partial \delta \dot{u}}{\partial x} \Delta T_e dV = \int_V E \frac{\partial \delta \dot{u}}{\partial x} \cdot \frac{\partial \dot{u}}{\partial x} dV \tag{9.99}$$

9.4.5 Discretization

9.4.5.1 Discretization of heat flow field

For this particular problem, let us choose a linear function of the following form for the space:

$$T(t) = ax + b \tag{9.100}$$

Let us assume that the function T at nodes i and j are known. Thus we can write:

$$\begin{bmatrix} x_i & 1 \\ x_j & 1 \end{bmatrix} \begin{Bmatrix} a \\ b \end{Bmatrix} = \begin{Bmatrix} T_i \\ T_j \end{Bmatrix} \tag{9.101}$$

Taking the inverse of the preceding relation, one gets coefficients a and b. Inserting these coefficients in Equation (9.100) yields:

$$T = N_i T_i + N_j T_j \tag{9.102}$$

where

$$N_i = \frac{x_j - x}{x_j - x_i}, \quad N_j = \frac{x - x_i}{x_j - x_i}$$

The preceding equation may be rewritten in a compact form as:

$$T = [N]\{T_e\} \quad \text{or} \quad T = \mathbf{N}T_e \tag{9.103}$$

where $[N] = [N_i, N_j]$, $\{T_e\}^T = \{T_i, T_j\}$. The derivative of the preceding equation takes the following form:

$$\frac{dT}{dx} = \frac{dN_i}{dx} T_i + \frac{dN_j}{dx} T_j \tag{9.104}$$

The preceding relation is rewritten in a compact form as:

$$\frac{dT}{dx} = [B]\{\dot{T}_e\} \quad \text{or} \quad \frac{dT}{dx} = \mathbf{B}\dot{T}_e \tag{9.105}$$

where $[B] = [B_i, B_j]$, $B_i = -1/L_e$, $B_j = 1/L = x_j - x_i$. Equation (9.106), which holds for the whole domain, must also hold for each element:

$$\int_{V_e} \rho c \delta T \cdot \frac{\partial T}{\partial t} dV + \int_{V_e} k \frac{\partial \delta T}{\partial x} \cdot \frac{\partial T}{\partial x} dV = \int_{A_e} [\delta T \cdot q_n^e]_{x_i}^{x_j} dA + \int_{V_e} \delta T \cdot g_e dV \tag{9.106}$$

The discretized form of the preceding equation becomes:

$$\int_{V_e} \rho c [N]^T [N] \{\dot{T}_e\} dV + \int_{V_e} k[B]^T [B] \{U\}_e dV = \int_{A_e} [[N]^T q_n^e]_{x_i}^{x_j} dA + \int_{V_e} [N]^T g_e dV$$

(9.107)

The preceding equation may be written in a compact form:

$$[M_e]\{\dot{T}_e\} + [K_e]\{T_e\} = \{F_e\}$$

(9.108)

where

$$[M_e] = \int_{V_e} \rho c [N]^T [N] dV, \quad [K_e] = \int_{V_e} k[B]^T [B] dV, \quad \{F_e\} = \int_{A_e} [[N_k]^T q_n^e]_{x_i}^{x_j} dA + \int_{V_e} [N]^T g_e dV$$

For the total domain, we have the following:

$$[M]\{\dot{T}\} + [K]\{T\} = \{F\}$$

(9.109)

where

$$[M] = \sum_{k=1}^{n} [M_e]_k, \quad [K] = \sum_{k=1}^{n} [K_e]_k, \quad \{F\} = \sum_{k=1}^{n} \{F_e\}_k, \quad \{T\} - \sum_{k=1}^{n} \{T_e\}_k$$

Equation (9.109) could not be solved as it is. For a time step m, we can rewrite Equation (9.4.28) as:

$$[M]\{\dot{T}\}_{(m+\theta)} + [K]\{T\}_{(m+\theta)} = \{F\}_{(m+\theta)}$$

$\cdot$ (9.110)

Therefore, we discretise dependent variable {T} for time-domain for using the Taylor expansion as it is in the finite difference method:

$$\{T\}_m = \{T\}_{(m+\theta)-\theta} = \{T\}_{(m+\theta)} - \frac{\partial \{T\}_{(m+\theta)}}{\partial t} \frac{\theta \Delta t}{1!} + \frac{\partial^2 \{T\}_{(m+\theta)}}{\partial t^2} \frac{\theta^2 \Delta t^2}{2!} - 0^3$$

(9.111)

$$\{T\}_{m+1} = \{T\}_{(m+\theta)+(1-\theta)} = \{T\}_{(m+\theta)} + \frac{\partial \{T\}_{(m+\theta)}}{\partial t} \frac{(1-\theta)\Delta t}{1!} + \frac{\partial^2 \{T\}_{(m+\theta)}}{\partial t^2} \frac{(1-\theta)^2 \Delta t^2}{2!} + 0^3$$

(9.112)

From preceding relations, one easily gets the following:

$$\{T\}_{(m+\theta)} = \theta \{T\}_{m+1} + (1-\theta)\{T\}_m, \quad \{\dot{T}\}_{(m+\theta)} = \frac{\{T\}_{m+1} - \{T\}_m}{\Delta t}$$

(9.113)

$$\{F\}_{(m+\theta)} = \theta \{F\}_{m+1} + (1-\theta)\{F\}_m$$

(9.114)

Inserting these relations into Equation (9.110), we get the following:

$$[M^*]\{T\}_{m+1} = \{F^*\}_{m+1}$$

(9.115)

where

$$[M^*] = \left[\frac{1}{\Delta t}[M] + \theta[K]\right], \quad \{F^*\}_{m+1} = \left[\frac{1}{\Delta t}[M] - (1-\theta)[K]\right]\{T\}_m + \theta\{F\}_{m+1} + (1-\theta)\{F\}_m$$

9.4.5.2   Discretization of stress field

For this particular problem, let us choose a linear function of the following form for the space:

$$\dot{u} = ax + b \tag{9.116}$$

Let us assume that the function $\dot{u}$ at nodes i and j are known. Thus we can write the following:

$$\begin{bmatrix} x_i & 1 \\ x_j & 1 \end{bmatrix} \begin{Bmatrix} a \\ b \end{Bmatrix} = \begin{Bmatrix} \dot{u}_i \\ \dot{u}_j \end{Bmatrix} \tag{9.117}$$

Taking the inverse of the preceding relation, one gets coefficients a and b. Inserting these coefficients in Equation (9.116) yields the following:

$$\dot{u} = N_i\dot{u}_i + N_j\dot{u}_j \tag{9.118}$$

where

$$N_i = \frac{x_j - x}{x_j - x_i}, \quad N_j = \frac{x - x_i}{x_j - x_i}$$

The preceding equation may be rewritten in compact form as:

$$\dot{u} = [N]\{\dot{U}_e\} \quad \text{or} \quad \dot{u} = \mathbf{N}\dot{\mathbf{U}}_e \tag{9.119}$$

where $[N] = [N_i, N_j]$, $\{\dot{U}_e\}^T = \{\dot{u}_i, \dot{u}_j\}$. The derivative of the preceding equation takes the following form:

$$\frac{d\dot{u}}{dx} = \frac{dN_i}{dx}\dot{u}_i + \frac{dN_j}{dx}\dot{u}_j \tag{9.120}$$

The preceding relation is rewritten in a compact form:

$$\frac{d\dot{u}}{dx} = [B]\{\dot{U}_e\} \quad \text{or} \quad \frac{d\dot{u}}{dx} = \mathbf{B}\dot{\mathbf{U}}_e \tag{9.121}$$

where $[B] = [B_i, B_j], B_i = -1/L_e, B_j = 1/L_e, L_e = x_j - x_i$. Equation (9.119), which holds for the whole domain, must also hold for each element:

$$\int_{V_e} E \frac{\partial \delta \dot{u}}{\partial x} \cdot \frac{\partial \dot{u}}{\partial x} dV = \int_{V_e} E\alpha \frac{\partial \delta \dot{u}}{\partial x} \Delta T_e dV + \int_{A_e} [\delta \dot{u} \cdot \dot{t}]_{x_i}^{x_j} dA \tag{9.122}$$

The discretized form of the preceding equation becomes:

$$\int_{V_e} E[B]^T[B]\{\dot{U}\}_e dV = \int_{V_e} E\alpha[B]^T \Delta T_e dV + \int_{A_e} [N]^T \dot{t}_e]_{x_i}^{x_j} dA \tag{9.123}$$

The preceding equation may be written in compact form as:

$$[K_e]\{\dot{U}_e\} = \{\dot{F}_e\} \tag{9.124}$$

where

$$[K_e] = \int_{V_e} E[B]^T[B]dV, \quad \{\dot{F}_e\} = \int_{A_e} [N_k]^T \dot{t}_e]_{x_i}^{x_j} dA + \int_{V_e} E\alpha[B]^T \Delta T_e dV$$

For the total domain, we have:

$$[K]\{\dot{U}\} = \{\dot{F}\} \tag{9.125}$$

where

$$[K] = \sum_{k=1}^{n} [K_e]_k, \quad \{\dot{F}\} = \sum_{k=1}^{n} \{\dot{F}_e\}_k, \quad \{\dot{U}\} = \sum_{k=1}^{n} \{\dot{U}_e\}_k$$

9.4.6   Specific example

For a typical two-noded element for heat flow field, the following is obtained:

$$[M_e] = \frac{\rho c L_e A_e}{6}\begin{bmatrix} 2 & 1 \\ 1 & 2 \end{bmatrix}, \quad [K_e] = \frac{kA_e}{L}\begin{bmatrix} 1 & -1 \\ -1 & 1 \end{bmatrix}, \{F_e\} = \begin{Bmatrix} q_i \\ q_j \end{Bmatrix} + \frac{g_e}{2}\begin{Bmatrix} 1 \\ 1 \end{Bmatrix}, \quad \{T_e\} = \begin{Bmatrix} T_i \\ T_j \end{Bmatrix}$$

Similarly for a typical two-noded element for stress field, the following is obtained:

$$[K_e] = \frac{EA_e}{L}\begin{bmatrix} 1 & -1 \\ -1 & 1 \end{bmatrix}, \quad \{\dot{F}_e\} = \begin{Bmatrix} \dot{t}_i \\ \dot{t}_j \end{Bmatrix} + \frac{E\alpha A_e \Delta T_e}{L}\begin{Bmatrix} -1 \\ 1 \end{Bmatrix}, \quad \{\dot{U}_e\} = \begin{Bmatrix} \dot{u}_i \\ \dot{u}_j \end{Bmatrix}$$

9.4.7   Example: simulation of heat generation and associated thermal stress

A specific example is given herein by simulating the temperature variation in a solid with a length of 1 m and associated stress field for the following conditions.

(a)   Heat generation + heat flux

The heat generation function is assumed to be of the following form:

$$g = A_g t e^{-t/\tau_g} \tag{9.126}$$

where A_g and τ_g are physical constants determined from heat generation tests, and t is time. Furthermore, the variation elastic modulus of hardening solid with time is assumed to be of the following form:

$$E(t) = A_e(1 - e^{-t/\tau_e}) \tag{9.127}$$

where A_e and τ_e are physical constants determined from uniaxial tests of hardening solid, and t is time.

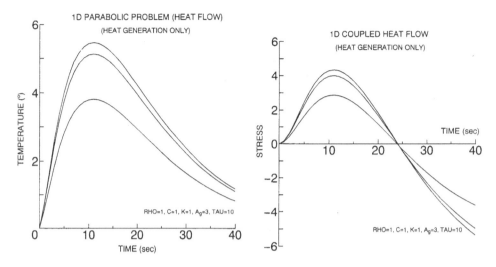

Figure 9.5 Temperature and stress response of some selected points

Figure 9.5 shows the variation of temperature and associated stress of some selected points.

9.5 Hydromechanical coupling: seepage and effective stress problem

9.5.1 Introduction

In the following sections, a finite element formulation for a coupled seepage and stress problem and its discretization are given. This is a fully coupled problem, and it is generally called a consolidation problem in the geotechnical engineering field.

9.5.2 Governing equations

9.5.2.1 Governing equation for seepage field

As shown in Chapter 4, the volumetric variation of porous media takes the following form for 1-D problems:

$$\frac{\partial \varepsilon}{\partial t} = -\frac{\partial v}{\partial x} \tag{9.128}$$

where v is relative velocity. Let us assume that this equation is subjected to the following boundary and initial conditions:

Boundary conditions

$$v(0,t) = 0 \quad \text{at} \quad x = 0$$
$$q_n(0,t) = q_0 \quad \text{at} \quad x = L \tag{9.129}$$

Initial conditions

$$p(x,0) = 0 \quad \text{at} \quad t = 0$$
$$\dot{p}(x,0) = 0 \quad \text{at} \quad t = 0 \tag{9.130}$$

Let us further assume that the seepage obeys a linear type of constitutive law (Darcy's law) between velocity v and dependent variable pressure p:

$$v = -k\frac{\partial p}{\partial x} \tag{9.131}$$

9.5.2.2 Governing equation for stress field

As shown in Chapter 4, the equation of momentum for 1-D problems without the inertia term can be written as:

$$\frac{\partial \sigma}{\partial x} + b = 0 \tag{9.132}$$

Let us assume that this equation is subjected to the following boundary conditions:

$$u(0,t) = 0 \quad \text{at} \quad x = 0$$
$$t(0,t) = t_0 \quad \text{at} \quad x = L \tag{9.133}$$

Displacement–strain relation is given as:

$$\varepsilon = \frac{\partial u}{\partial x} \tag{9.134}$$

Let us assume that the total stress σ may be related to the effective stress law of Terzaghi through the following relation:

$$\sigma = \sigma' - p \tag{9.135}$$

The incremental form of Equation (9.132) together with the effective stress law takes the following form if the body force remains constant with time:

$$\frac{\partial \dot{\sigma}\prime}{\partial x} - \frac{\partial \dot{p}}{\partial x} = 0 \tag{9.136}$$

Let us further assume that the material is of Hookean type:

$$\sigma' = E\varepsilon \tag{9.137}$$

Similarly, the constitutive law and displacement rate and strain rate relations may be rewritten as:

$$\dot{\sigma}' = E\dot{\varepsilon} \tag{9.138}$$

$$\dot{\varepsilon} = \frac{\partial \dot{u}}{\partial x} \tag{9.139}$$

9.5.3 Weak form formulation

9.5.3.1   Weak form formulation for seepage field

Taking a variation on variable δP, the integral form of Equation (9.128) becomes:

$$\int_V \delta p \cdot \frac{\partial \varepsilon}{\partial t} dV = -\int_V \delta p \cdot \frac{\partial v}{\partial x} dV \tag{9.140}$$

where $dV = dAdx$.

Applying the integral by parts to the RHS with respect to x yields:

$$\int_V \delta p \cdot \frac{\partial \varepsilon}{\partial t} dV - \int_V \frac{\partial \delta p}{\partial x} \cdot v dV = -\int_A [\delta p \cdot q_n]_0^L dA \tag{9.141}$$

where $q_n = v \cdot n$. The preceding equation is called the weak form of Equation (9.128). Inserting the constitutive relation given by Equation (9.121) into Equation (9.131), we obtain the following:

$$\int_V \delta p \cdot \frac{\partial p}{\partial t} dV + \int_V k \frac{\partial \delta p}{\partial x} \cdot \frac{\partial p}{\partial x} dV = -\int_A [\delta T \cdot q_n]_0^L dA \tag{9.142}$$

9.5.3.2   Weak form formulation for stress field

Taking a variation on displacement rate field $\delta \dot{u}$, the integral form of Equation (9.136) becomes:

$$\int_V \delta \dot{u} \cdot \frac{\partial(\dot{\sigma}' - \dot{p})}{\partial x} dV = 0 \tag{9.143}$$

where $dV = dAdx$.

Applying the integral by parts to the LHS with respect to x yields:

$$\int_A [\delta \dot{u} \cdot \dot{t}]_0^L dA = \int_V \frac{\partial \delta \dot{u}}{\partial x} \cdot (\dot{\sigma}' - \dot{p}) dV \tag{9.144}$$

where $i = (\acute{\sigma}\prime - \dot{p}) \cdot n$. The preceding equation is called the weak form of Equation (9.135). Inserting the constitutive relation given by Equation (9.137) and Equation (9.138) into Equation (9.144), we obtain:

$$\int_A [\delta\dot{u} \cdot \dot{t}]_0^L dA = \int_V E\frac{\partial\delta\dot{u}}{\partial x} \cdot \frac{\partial\dot{u}}{\partial x}dV - \int_V E\frac{\partial\delta\dot{u}}{\partial x} \cdot \dot{p}dV \tag{9.145}$$

9.5.4 Discretization

9.5.4.1 Discretization for physical space

(A) INTERPOLATION (SHAPE) FUNCTION FOR PRESSURE FIELD

For this particular problem, let us choose a linear function of the following form for the space in a local coordinate system whose origin is at a nodal point t:

$$p(t) = a\xi + b, \quad \xi = x - x_i, \quad d\xi = dx \tag{9.146}$$

Let us assume that the function P at nodes i and k are known. Thus we can write:

$$\begin{bmatrix} 0 & 1 \\ L & 1 \end{bmatrix}\begin{Bmatrix} a \\ b \end{Bmatrix} = \begin{Bmatrix} P_i \\ P_k \end{Bmatrix} \tag{9.147}$$

Taking the inverse of the preceding relation, one gets coefficients a and b. Inserting these coefficients in Equation (9.146) yields the following:

$$p = N_i P_i + N_k P_k \tag{9.148}$$

where

$$N_i = 1 - \frac{\xi}{L}, \quad N_k = \frac{\xi}{L}$$

The preceding equation may be rewritten in a compact form:

$$p = [N]_p\{P_e\} \quad \text{or} \quad p = \mathbf{N}_p\mathbf{P}_e \tag{9.149}$$

where $[N]_p = [N_i, N_k]$, $\{P_e\}^T = \{P_i, P_k\}$. The derivative of the preceding equation takes the following form:

$$\frac{dp}{dx} = \frac{dp}{d\xi} = \frac{dN_i}{d\xi}P_i + \frac{dN_k}{d\xi}P_k \tag{9.150}$$

The preceding relation is rewritten in a compact form:

$$\frac{dp}{dx} = \frac{dp}{d\xi} = [B]_p\{P_e\} \quad \text{or} \quad \frac{dp}{dx} = \frac{dp}{d\xi} = \mathbf{B}_p\mathbf{P}_e \tag{9.151}$$

where $[B] = [B_i, B_k]$, $B_i = -1/L, B_k = 1/L, L = x_k - x_i$.

(B) INTERPOLATION (SHAPE) FUNCTION FOR DISPLACEMENT FIELD

For this particular problem, we have to choose a quadratic function of the following form for the space discretisation of displacement field if the shape function for the pressure field is linear in a local coordinate system whose origin is at a nodal point i:

$$u(t) = a + b\xi + c\xi^2, \quad \xi = x - x_i, \quad d\xi = dx \quad (9.152)$$

Let us assume that the function u at nodes i, j and k are known. Thus we can write the following:

$$\begin{bmatrix} 1 & 0 & 0 \\ 1 & \dfrac{L}{2} & \dfrac{L^2}{4} \\ 1 & L & L^2 \end{bmatrix} \begin{Bmatrix} a \\ b \\ c \end{Bmatrix} = \begin{Bmatrix} U_i \\ U_j \\ U_k \end{Bmatrix} \quad (9.153)$$

Taking the inverse of the preceding relation, one gets coefficients a, b and c. Inserting these coefficients in Equation (9.152) yields the following:

$$u = N_i U_i + N_j U_j + N_k U_k \quad (9.154)$$

where

$$N_i = (1 - \frac{2\xi}{L})(1 - \frac{\xi}{L}), \quad N_j = \frac{4\xi}{L}(1 - \frac{\xi}{L}), \quad N_k = -\frac{\xi}{L}(1 - \frac{2\xi}{L})$$

The preceding equation may be rewritten in a compact form:

$$u = [N]_u \{U_e\} \quad \text{or} \quad u = \mathbf{N}_u \mathbf{U}_e \quad (9.155)$$

where $[N]_u = [N_i, N_j, N_k]$, $\{U_e\}^T = \{U_i, U_j, U_k\}$. The derivative of the preceding equation takes the following form:

$$\frac{du}{dx} = \frac{du}{d\xi} = \frac{dN_i}{d\xi} U_i + \frac{dN_j}{d\xi} U_j + \frac{dN_k}{d\xi} P_k \quad (9.156)$$

The preceding relation is rewritten in a compact form:

$$\frac{du}{dx} = \frac{du}{d\xi} = [B]_u \{U_e\} \quad \text{or} \quad \frac{du}{dx} = \frac{du}{d\xi} = \mathbf{B}_u \mathbf{U}_e \quad (9.157)$$

where $[B] = [B_i, B_j, B_k]$,

$$B_i = \frac{1}{L^2}(4\xi - 3L), \quad B_j = \frac{4}{L^2}(L - 2\xi), \quad B_k = \frac{1}{L^2}(4\xi - L), \quad L = x_k - x_i$$

Equation (9.142) for seepage field, which holds for the whole domain, must also hold for each element:

$$\int_{V_e} \delta p \cdot \frac{\partial \varepsilon}{\partial t} dV + \int_{V_e} k \frac{\partial \delta p}{\partial x} \cdot \frac{\partial p}{\partial x} dV = -\int_{A_e} [\delta p \cdot q_n^e]_0^L dA \quad (9.158)$$

The discretized form of the preceding equation becomes:

$$\int_{V_e} [N]_p^T [B]_u dV \{\dot{U}_e\} + \int_{V_e} k[B]_p^T [B]_p dV \{P\}_e dV = -\int_{A_e} [[\bar{N}]_p^T q_n^e]_0^L dA \tag{9.159}$$

The preceding equation may be written in a compact form:

$$[C_e]_{pu} \{\dot{U}_e\} + [K_e]_{pp} \{P_e\} = \{Q_e\} \tag{9.160}$$

where

$$[C_e]_{pu} = \int_{V_e} [N]_p^T [B]_u dV, \quad [K_e]_{pp} = \int_{V_e} k[B]_p^T [B]_p dV, \quad \{Q_e\} = -\int_{A_e} [[\bar{N}]_p^T q_n^e]_0^L dA$$

For the total domain, we have:

$$[C]_{pu} \{\dot{U}\} + [K]_{pp} \{P\} = \{Q\} \tag{9.161}$$

where

$$[C]_{pu} = \sum_{k=1}^{n} [C_e]_{pu}^k, \quad [K]_{uu} = \sum_{k=1}^{n} [K_c]_{pp}^k, \quad \{Q\} = \sum_{k=1}^{n} \{Q_e\}_k, \quad \{T\} = \sum_{k=1}^{n} \{T_e\}_k$$

Similarly Equation (9.145) for stress field, which holds for the whole domain, must also hold for each element:

$$\int_{V_e} E \frac{\partial \delta \dot{u}}{\partial x} \cdot \frac{\partial \dot{u}}{\partial x} dV + \int_{V_e} E \frac{\partial \delta \dot{u}}{\partial x} \cdot \dot{p} dV = \int_{A_e} [\delta \dot{u} \cdot \dot{t}]_0^L dA \tag{9.162}$$

The discretised form of the preceding equation becomes:

$$\int_{V_e} E[B]_u^T [B]_u dv \{\dot{U}\}_e \int_{V_e} [B]_u^T [N]_p dv \{\dot{P}\}_e = \int_{A_e} [[\bar{N}]^T \dot{t}_e]_0^L dA \tag{9.163}$$

The preceding equation may be written in a compact form:

$$[K_e]_{uu} \{\dot{U}_e\} - [C_e]_{up} \{\dot{P}_e\} = \{\dot{F}_e\} \tag{9.164}$$

where

$$[K_e]_{uu} = \int_{V_e} E[B]_u^T [B]_u dV, \{\dot{F}_e\} = \int_{A_e} [\bar{N}]^T \dot{t}_e]_0^L dA$$

For the total domain, we have:

$$[K]_{uu} \{\dot{U}\} - [C]_{up} \{\dot{P}\} = \{\dot{F}\} \tag{9.165}$$

where

$$[K]_{uu} = \sum_{k=1}^{n}[K_e]_{uu}^k, \quad [C]_{up} = \sum_{k=1}^{n}[C_e]_{up}^k, \quad \{F\} = \sum_{k=1}^{n}\{F_e\}_k, \quad \{U\} = \sum_{k=1}^{n}\{U_e\}_k, \quad \{P\} = \sum_{k=1}^{n}\{P_e\}_k$$

Above equations (9.161) and (9.145) can be written in a compact form:

$$\begin{bmatrix} \mathbf{K}_{uu} & -\mathbf{C}_{up} \\ \mathbf{C}_{pu} & \mathbf{0} \end{bmatrix} \begin{Bmatrix} \dot{\mathbf{U}} \\ \dot{\mathbf{P}} \end{Bmatrix} + \begin{bmatrix} \mathbf{0} & \mathbf{0} \\ \mathbf{0} & \mathbf{K}_{pp} \end{bmatrix} \begin{Bmatrix} \mathbf{U} \\ \mathbf{P} \end{Bmatrix} = \begin{Bmatrix} \dot{\mathbf{F}} \\ \mathbf{Q} \end{Bmatrix} \tag{9.166}$$

The preceding equation may be rewritten in a more compact form:

$$[M]\{\dot{T}\} + [H]\{T\} = \{Y\} \tag{9.167}$$

where

$$[M] = \begin{bmatrix} \mathbf{K}_{uu} & -\mathbf{C}_{up} \\ \mathbf{C}_{pu} & \mathbf{0} \end{bmatrix}, \quad [H] = \begin{bmatrix} \mathbf{0} & \mathbf{0} \\ \mathbf{0} & \mathbf{K}_{pp} \end{bmatrix}, \quad \{\dot{T}\} = \begin{bmatrix} \dot{\mathbf{U}} \\ \dot{\mathbf{P}} \end{bmatrix}, \quad \{T\} \begin{bmatrix} \mathbf{U} \\ \mathbf{P} \end{bmatrix}, \quad [Y] = \begin{bmatrix} \dot{\mathbf{U}} \\ \dot{\mathbf{P}} \end{bmatrix}$$

9.5.4.2 Discretization for time domain

Equation (9.167) could not be solved as it is. For a time step m, we can rewrite Equation (9.167):

$$[M]\{\dot{T}\}_{(m+\theta)} + [H]\{T\}_{(m+\theta)} = \{Y\}_{(m+\theta)} \tag{9.168}$$

Therefore, we discretize dependent variable {T} for time-domain by using the Taylor expansion as it is in the finite difference method:

$$\{T\}_m = \{T\}_{(m+\theta)-\theta} = \{T\}_{(m+\theta)} - \frac{\partial\{T\}_{(m+\theta)}}{\partial t}\frac{\theta\Delta t}{1!} + 0^2 \tag{9.169}$$

$$\{T\}_{m+1} = \{T\}_{(m+\theta)+(1-\theta)} = \{T\}_{(m+\theta)} + \frac{\partial\{T\}_{(m+\theta)}}{\partial t}\frac{(1-\theta)\Delta t}{1!} + 0^2 \tag{9.170}$$

From preceding relations, one easily gets the following:

$$\{T\}_{(m+\theta)} = \theta\{T\}_{m+1} + (1-\theta)\{T\}_m, \quad \{\dot{T}\}_{(m+\theta)} = \frac{\{T\}_{m+1} - \{T\}_m}{\Delta t} \tag{9.171}$$

$$\{Y\}_{(m+\theta)} = \theta\{Y\}_{m+1} + (1-\theta)\{Y\}_m \tag{9.172}$$

Inserting these relations into Equation (9.168), we get the following

$$[M^*]\{T\}_{m+1} = \{Y^*\}_{m+1} \tag{9.173}$$

where

$$[M^*] = \left[\frac{1}{\Delta t}[M] + \theta[H]\right]$$

$$\{Y^*\}_{m+1} = \left[\frac{1}{\Delta t}[M] - (1-\theta)[H]\right]\{T\}_m + \theta\{Y\}_{m+1} + (1-\theta)\{Y\}_m$$

9.5.5   Specific example

For a typical element, the followings are obtained:

$$[M_e] = A_e \begin{bmatrix} \dfrac{7E}{3L} & -\dfrac{8E}{3L} & \dfrac{E}{3L} & \dfrac{5}{6} & \dfrac{1}{6} \\[2mm] -\dfrac{8E}{3L} & \dfrac{16E}{3L} & -\dfrac{8E}{3L} & \dfrac{2}{3} & \dfrac{2}{3} \\[2mm] \dfrac{E}{3L} & -\dfrac{8E}{3L} & \dfrac{7E}{3L} & -\dfrac{1}{6} & -\dfrac{5}{6} \\[2mm] -\dfrac{5}{6} & \dfrac{2}{3} & \dfrac{1}{6} & 0 & 0 \\[2mm] -\dfrac{1}{6} & -\dfrac{2}{3} & \dfrac{5}{6} & 0 & 0 \end{bmatrix}, \quad [H_e] = \frac{kA_e}{L}\begin{bmatrix} 0 & 0 & 0 & 0 & 0 \\ 0 & 0 & 0 & 0 & 0 \\ 0 & 0 & 0 & 0 & 0 \\ 0 & 0 & 0 & 1 & -1 \\ 0 & 0 & 0 & -1 & 1 \end{bmatrix}$$

$$\{Y_e\} = \begin{Bmatrix} \dot{F}_i \\ \dot{F}_j \\ \dot{F}_k \\ Q_i \\ Q_k \end{Bmatrix}, \quad \{\dot{T}_e\} = \begin{Bmatrix} \dot{T}_i \\ \dot{T}_j \\ \dot{T}_k \\ \dot{P}_i \\ \dot{P}_k \end{Bmatrix}, \quad \{T_e\} = \begin{Bmatrix} T_i \\ T_j \\ T_k \\ P_i \\ P_k \end{Bmatrix}$$

9.5.6   Example: simulation of settlement under sudden loading

A specific example is given herein by simulating the settlement and pore pressure variation in the ground by considering the order of approximation function for pressure and displacement field. Figures 9.6 and 9.7 show the variation of settlement and pore pressure at some selected nodes. The results for both situations are almost exactly the same.

9.6   Biot problem: coupled dynamic response of porous media

9.6.1   Introduction

In the following subsections, a finite element formulation for a coupled seepage and stress problem for dynamic responses saturated porous media and its discretization is given. This is a fully coupled problem, generally called Biot's problem in the geotechnical engineering field.

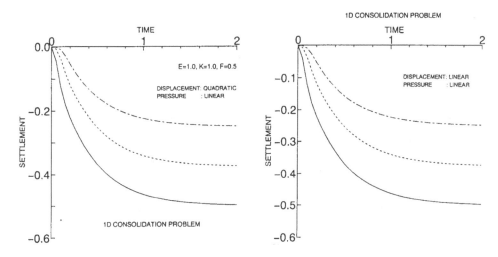

Figure 9.6 Settlement of ground under rapid load for different shape functions

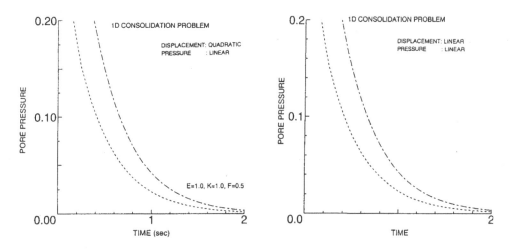

Figure 9.7 Pore pressure of ground under rapid load for different shape functions

9.6.2   Governing equations

9.6.2.1   Governing equation for fluid phase

As shown in Chapter 4, the equation of motion for fluid phase of porous media takes the following form for 1-D problems:

$$\frac{\partial \sigma_f}{\partial x} + \rho_f g = \rho_f \frac{\partial^2 u_s}{\partial t^2} + \frac{\rho_f}{n}\frac{\partial^2 w}{\partial t^2} + \frac{1}{K}\frac{\partial w}{\partial t} \tag{9.174}$$

where w is relative displacement. Let us assume that this equation is subjected to following boundary and initial conditions:

Boundary conditions

$$w(0,t) = 0 \quad \text{at} \quad x = 0$$
$$t_n^f(0,t) = t_0^f \quad \text{at} \quad x = L \tag{9.175}$$

Initial conditions

$$w(x,0) = 0 \quad \text{at} \quad t = 0$$
$$\dot{w}(x,0) = 0 \quad \text{at} \quad t = 0 \tag{9.176}$$

Let us further assume that the fluid obeys a linear type of constitutive law (Biot's law) in terms of fluid and skeleton strains and dependent variable fluid pressure p:

$$-p = \sigma_f = M(\alpha\varepsilon_s + \varepsilon_f) \tag{9.177}$$

9.6.2.2 Governing equation for total stress system

As shown in Chapter 4, the equation of motion for 1-D problems takes the following form:

$$\frac{\partial\sigma}{\partial x} + \rho g = \rho\frac{\partial^2 u_s}{\partial t^2} + \rho_f\frac{\partial^2 w}{\partial t^2} \tag{9.178}$$

Let us assume that this equation is subjected to the following boundary conditions:

$$u_s(0,t) = 0 \quad \text{at} \quad x = 0$$
$$t^s(0,t) = t_0^s \quad \text{at} \quad x = L \tag{9.179}$$

Displacement–strain relation is given as:

$$\varepsilon_s = \frac{\partial u_s}{\partial x}, \quad \varepsilon_f = \frac{\partial w}{\partial x} \tag{9.180}$$

Let us assume that the total stress σ may be related to the effective stress law of Biot through the following relation:

$$\sigma = \sigma' - \alpha p \tag{9.181}$$

Let us further assume that the skeleton obeys the following constitutive law:

$$\sigma = (2\mu + \lambda)\varepsilon_s - \alpha p = (2\mu + \lambda + \alpha^2 M)\varepsilon_s + \alpha M\varepsilon_f \tag{9.182}$$

9.6.3 Weak form formulation

9.6.3.1 Weak form formulation for fluid phase

Taking a variation on variable δw, the integral form of Equation (9.169) becomes:

$$\int_V \delta w \cdot \rho_f \frac{\partial^2 u_s}{\partial t^2} dV + \int_V \delta w \cdot \frac{\rho_f}{n} \frac{\partial^2 w}{\partial t^2} dV + \int_V \delta w \cdot \frac{1}{K} \frac{\partial w}{\partial t} dV$$
$$= -\int_V \delta w \cdot \frac{\partial p}{\partial x} dV + \int_V \delta w \cdot \rho_f g dV \tag{9.183}$$

where $dV = dAdx$.

Applying the integral by parts to the RHS with respect to x yields the following:

$$\int_V \delta w \cdot \rho_f \frac{\partial^2 u_s}{\partial t^2} dV + \int_V \delta w \cdot \frac{\rho_f}{n} \frac{\partial^2 w}{\partial t^2} dV + \int_V \delta w \cdot \frac{1}{K} \frac{\partial w}{\partial t} dV - \int_V \frac{\partial \delta w}{\partial x} \cdot p dV =$$
$$- \int_A [\delta w \cdot t_n^f]_0^L dA + \int_V \delta w \cdot \rho_f g dV \tag{9.184}$$

where $t_n^f = p \cdot n$. The preceding equation is called the weak form of Equation (9.184). Inserting the constitutive relation given by Equation (9.177) into Equation (9.184), we obtain:

$$\int_V \delta w \cdot \rho_f \frac{\partial^2 u_s}{\partial t^2} dV + \int_V \delta w \cdot \frac{\rho_f}{n} \frac{\partial^2 w}{\partial t^2} dV + \int_V \delta w \cdot \frac{1}{K} \frac{\partial w}{\partial t} dV + \int_V \alpha M \frac{\partial \delta w}{\partial x} \cdot \frac{\partial u_s}{\partial x} dV +$$
$$\int_V M \frac{\partial \delta w}{\partial x} \cdot \frac{\partial w}{\partial x} dV = -\int_A [\delta w \cdot t_n^f]_0^L dA + \int_V \delta w \cdot \rho_f g dV \tag{9.185}$$

9.6.3.2 Weak form formulation for total stress field

Taking a variation on displacement rate field δu_s, the integral form of Equation (9.6.5) becomes:

$$\int_V \delta u_s \cdot \frac{\partial \sigma}{\partial x} dV + \int_V \delta u_s \cdot \rho g dV = \int_V \delta u_s \cdot \rho \frac{\partial^2 u_s}{\partial t^2} dV + \int_V \delta u_s \cdot \rho_f \frac{\partial^2 w}{\partial t^2} dV \tag{9.186}$$

where $dV = dAdx$.

Applying the integral by parts to the LHS with respect to x yields the following:

$$\int_A [\delta u_s \cdot t_n]_0^L dA + \int_V \delta u_s \cdot \rho g dV = \int_V \frac{\partial \delta u_s}{\partial x} \cdot \sigma dV + \int_V \delta u_s \cdot \rho \frac{\partial^2 u_s}{\partial t^2} dV + \int_V \delta u_s \cdot \rho_f \frac{\partial^2 w}{\partial t^2} dV \tag{9.187}$$

where $t^s = \sigma_s \cdot n$. The preceding equation is called the weak form of Equation (9.178). Inserting the constitutive relation given by Equation (9.182) into Equation (9.187), we obtain the following:

$$\int_A [\delta u_s \cdot t]_0^L dA + \int_V \delta u_s \cdot \rho g dV = (2\mu + \lambda + \alpha^2 M) \int_V \frac{\partial \delta u_s}{\partial x} \cdot \frac{\partial u_s}{\partial x} dV$$

$$+ \alpha M \int_V \frac{\partial \delta u_s}{\partial x} \cdot \frac{\partial w}{\partial x} dV + \int_V \delta u_s \cdot \rho \frac{\partial^2 u_s}{\partial t^2} dV + \int_V \delta u_s \cdot \rho_f \frac{\partial^2 w}{\partial t^2} dV \qquad (9.188)$$

9.6.4 Discretization

9.6.4.1 Discretization for physical space

(A)   INTERPOLATION (SHAPE) FUNCTION FOR RELATIVE DISPLACEMENT FIELD

For this particular problem, let us choose a linear function of the following form for the space in a local coordinate system whose origin is at a nodal point i:

$$w(t) = a\xi + b, \quad \xi = x - x_i, \quad d\xi = dx \qquad (9.189)$$

Let us assume that the function w at nodes i and j are known. Thus we can write:

$$\begin{bmatrix} 0 & 1 \\ L & 1 \end{bmatrix} \begin{Bmatrix} a \\ b \end{Bmatrix} = \begin{Bmatrix} w_i \\ w_j \end{Bmatrix} \qquad (9.190)$$

Taking the inverse of the preceding relation, one gets coefficients a and b. Inserting these coefficients in Equation (9.189) yields the following:

$$w = N_i W_i + N_j W_j \qquad (9.191)$$

where

$$N_i = 1 - \frac{\xi}{L}, \quad N_j = \frac{\xi}{L}$$

The preceding equation may be rewritten in a compact form as

$$w = [N]\{W_e\} \quad \text{or} \quad w = \mathbf{N}\mathbf{W}_e \qquad (9.192)$$

where $[N] = [N_i, N_j]$, $\{W_e\}^T = \{W_i, W_j\}$. The derivative of the preceding equation takes the following form:

$$\frac{dw}{dx} = \frac{dw}{d\xi} = \frac{dN_i}{d\xi} W_i + \frac{dN_j}{d\xi} W_j \qquad (9.193)$$

The preceding relation is rewritten in a compact form:

$$\frac{dw}{dx} = \frac{dw}{d\xi} = [B]p\{W_e\} \quad \text{or} \quad \frac{dw}{dx} = \frac{dw}{d\xi} = \mathbf{B}\mathbf{W}_e \qquad (9.194)$$

where $[B] = [B_i, B_j]$, $B_i = -1/L, B_j = 1/L, L = x_j - x_i$.

(B) INTERPOLATION (SHAPE) FUNCTION FOR SKELETON DISPLACEMENT FIELD

As for displacement field of skeleton, we chose also a linear interpolation function. The resulting expressions will be similar so that it will not be presented herein.

Equation (9.180) for fluid phase, which holds for the whole domain, must also hold for each element:

$$\int_{V_e} \delta w \cdot \rho_f \frac{\partial^2 u_s}{\partial t^2} dV + \int_{V_e} \delta w \cdot \frac{\rho_f}{n} \frac{\partial^2 w}{\partial t^2} dV + \int_{V_e} \delta w \cdot \frac{1}{K} \frac{\partial w}{\partial t} dV + \int_{V_e} \alpha M \frac{\partial \delta w}{\partial x} \cdot \frac{\partial u_s}{\partial x} dV$$

$$+ \int_{V_e} M \frac{\partial \delta w}{\partial x} \cdot \frac{\partial w}{\partial x} dV = -\int_{A_e} [\delta w \cdot t_n^f]_0^L dA + \int_{V_e} \delta w \cdot \rho_f g dV \qquad (9.195)$$

The discretised form of the preceding equation becomes:

$$\int_{V_e} \rho_f [N]^T [N] \{\ddot{U}_e\} dV + \int_{V_e} \frac{\rho_f}{n} [N]^T [N] \{\ddot{W}\}_e dV + \int_{V_e} \frac{1}{K} [N]^T [N] \{\dot{W}\}_e dV$$

$$+ \int_{V_e} \alpha M [B]^T [B] \{U\} dV + \int_{V_e} M [B]^T [B] \{W\} dV = -\int_{A_e} [[\bar{N}]^T t_n^e]_0^L dA + \int_{V_e} \rho_f g [N]^T dV$$

$$(9.196)$$

The preceding equation may be written in a compact form:

$$[M_e]_1^s \{\ddot{U}_e\} + [M_e]_1^f \{\ddot{W}_e\} + [C_e]_1^f \{\dot{W}_e\} + [K_e]_1^s \{U_e\} + [K_e]_1^f \{W_e\} + = \{F_e\}_1 \qquad (9.197)$$

where

$$[M_e]_1^s = \int_{V_e} \rho [N]^T [N] dV, \quad [M_e]_1^f = \int_{V_e} \frac{\rho_f}{n} [N]^T [N] dV, \quad [C_e]_1^f = \int_{V_e} \frac{1}{K} [N]^T [N] dV,$$

$$[K_e]_1^s = \alpha M \int_{V_e} [B]^T [B] dV, \quad [K_e]_1^f = M \int_{V_e} [B]^T [B] dV$$

$$\{F_e\}_1 = -\int_{A_e} [[\bar{N}]^T t_n^e]_0^L dA + \int_{V_e} \rho_f g [N]^T dV$$

For the total domain, we have:

$$[M]_1^s \{\ddot{U}\} + [M]_1^f \{\ddot{W}\} + [C]_1^f \{\dot{W}\} + [K]_1^s \{U\} + [K]_1^f \{W\} + = \{F\}_1 \qquad (9.198)$$

where

$$[M]_1^s = \sum_{k=1}^n [M_e]_1^s, \quad [M]_1^f = \sum_{k=1}^n [M_e]_1^f, \quad [C]_1^f = \sum_{k=1}^n [C_e]_1^f,$$

$$[K]_1^s = \sum_{k=1}^n [K_e]_1^s, \quad [K]_1^f = \sum_{k=1}^n [K_e]_1^f, \quad \{F\}_1 = \sum_{k=1}^n \{F_e\}_1$$

Similarly Equation (9.183) for stress field, which holds for the whole domain, must also hold for each element:

$$\int_{A_e} [\delta u_s \cdot t]_0^L dA + \int_{V_e} \delta u_s \cdot \rho g dV = (2\mu + \lambda + \alpha^2 M) \int_{V_e} \frac{\partial \delta u_s}{\partial x} \cdot \frac{\partial u_s}{\partial x} dV$$

$$+ \alpha M \int_{V_e} \frac{\partial \delta u_s}{\partial x} \cdot \frac{\partial w}{\partial x} dV + \int_{V_e} \delta u_s \cdot \rho \frac{\partial^2 u_s}{\partial t^2} dV + \int_{V_e} \delta u_s \cdot \rho_f \frac{\partial^2 w}{\partial t^2} dV \qquad (9.199)$$

The discretized form of the preceding equation becomes:

$$\int_{V_e} \rho [N]^T [N] dv \{\ddot{U}\}_e \int_{V_e} \rho_f [N]^T [N] dv \{\ddot{W}\}_e + \int_{V_e} (2\mu + \lambda + \alpha^2 M) [B]^T [B] \{U\} dV$$

$$+ \int_{V_e} \alpha M [B]^T [B] \{W\} dV = \int_{A_e} [[\bar{N}]^T i_e]_0^L dA + \int_{V_e} \rho g [N]^T dV \qquad (9.200)$$

The preceding equation may be written in a compact form:

$$[M_e]_2^s \{\ddot{U}\} + [M_e]_2^f \{\ddot{W}\} + [K_e]_2^s \{U\} + [K_e]_2^f \{W\} + = \{F\}_2 \qquad (9.201)$$

where

$$[M_e]_2^s = \int_{V_e} \rho [N]^T [N] dV, \quad [M_e]_2^f = \int_{V_e} \rho_f [N]^T [N] dV,$$

$$[K_e]_2^s = (2\mu + \lambda + \alpha^2 M) \int_{V_e} [B]^T [B] dV, \quad [K_e]_2^f = \alpha M \int_{V_e} [B]^T [B] dV,$$

$$\{F_e\}_1 = -\int_{A_e} [[\bar{N}]^T t_n^e]_0^L dA + \int_{V_e} \rho_f g [N]^T dV$$

For the total domain, we have the following:

$$[M]_2^s \{\ddot{U}\} + [M]_2^f \{\ddot{W}\} + [K]_2^s \{U\} + [K]_2^f \{W\} + = \{F\}_2 \qquad (9.202)$$

where

$$[M]_2^s = \sum_{k=1}^n [M_e]_2^s, \ [M]_2^f = \sum_{k=1}^n [M_e]_2^f, \ [K]_2^s = \sum_{k=1}^n [K_e]_2^s, \ [K]_2^f = \sum_{k=1}^n [K_e]_2^f, \{F\}_2 = \sum_{k=1}^n \{F_e\}_2$$

Equations (9.193) and (9.197) can be written in a compact form:

$$\begin{bmatrix} \mathbf{M}_1^s & \mathbf{M}_1^f \\ \mathbf{M}_2^s & \mathbf{M}_2^f \end{bmatrix} \begin{Bmatrix} \ddot{\mathbf{U}} \\ \ddot{\mathbf{W}} \end{Bmatrix} + \begin{bmatrix} 0 & \mathbf{C}_1^f \\ 0 & 0 \end{bmatrix} \begin{Bmatrix} \dot{\mathbf{U}} \\ \dot{\mathbf{W}} \end{Bmatrix} + \begin{bmatrix} \mathbf{K}_1^s & \mathbf{K}_1^f \\ \mathbf{K}_2^s & \mathbf{K}_2^f \end{bmatrix} \begin{Bmatrix} \mathbf{U} \\ \mathbf{W} \end{Bmatrix} = \begin{Bmatrix} \dot{\mathbf{F}}_1 \\ \mathbf{F}_2 \end{Bmatrix} \qquad (9.203)$$

The preceding equation may be rewritten in a more compact form:

$$[M]\{\ddot{T}\} + [C]\{\dot{T}\} + [H]\{T\} = \{Y\} \tag{9.204}$$

where

$$[M] = \begin{bmatrix} \mathbf{M}_1^s & \mathbf{M}_1^f \\ \mathbf{M}_2^s & \mathbf{M}_2^f \end{bmatrix}, \quad [C] = \begin{bmatrix} \mathbf{0} & \mathbf{C}_1^f \\ \mathbf{0} & \mathbf{0} \end{bmatrix}, \quad [H] = \begin{bmatrix} \mathbf{K}_1^s & \mathbf{K}_1^f \\ \mathbf{K}_2^s & \mathbf{K}_2^f \end{bmatrix}$$

$$\{\ddot{T}\} = \begin{Bmatrix} \ddot{\mathbf{U}} \\ \ddot{\mathbf{W}} \end{Bmatrix}, \quad \{\dot{T}\} = \begin{Bmatrix} \dot{\mathbf{U}} \\ \dot{\mathbf{W}} \end{Bmatrix}, \quad \{T\} = \begin{Bmatrix} \mathbf{U} \\ \mathbf{W} \end{Bmatrix}, \quad \{Y\} = \begin{Bmatrix} \mathbf{F}_1 \\ \mathbf{F}_2 \end{Bmatrix}$$

9.6.4.2   Discretization for time domain

Equation (9.199) could not be solved as it is. For a time step m, we can rewrite Equation (9.199):

$$[M]\{\ddot{T}\}_m + [C]\{\dot{T}\}_m + [H]\{T\}_m = \{Y\}_m \tag{9.205}$$

Therefore, we discretize field $\{T\}$ for time-domain using the Taylor expansion as it is in the finite difference method:

$$\{T\}_{m-1} = \{T\}_m - \frac{\partial \{T\}_m}{\partial t} \frac{\Delta t}{1!} + \frac{\partial^2 \{T\}_m}{\partial t^2} \frac{\Delta t^2}{2!} - 0^3 \tag{9.206}$$

$$\{T\}_m = \{T\}_m \tag{9.207}$$

$$\{T\}_{m+1} = \{T\}_m + \frac{\partial \{T\}_m}{\partial t} \frac{\Delta t}{1!} + \frac{\partial^2 \{T\}_m}{\partial t^2} \frac{\Delta t^2}{2!} + 0^3 \tag{9.208}$$

From preceding relations, one easily gets the following:

$$\{\dot{T}\}_m = \frac{1}{\Delta t}(\{T\}_{m+1} - \{T\}_{m-1}) \tag{9.209}$$

$$\{\ddot{T}\}_m = \frac{1}{\Delta t^2}(\{T\}_{m+1} - 2\{T\}_m + \{T\}_{m-1}) \tag{9.210}$$

Inserting these relations into Equation (9.200), we get the following:

$$[M^*]\{T\}_{m+1} = \{Y^*\}_{m+1} \tag{9.211}$$

where

$$[M^*] = \left[\frac{1}{\Delta t^2}[M] + \frac{1}{\Delta t}[C]\right]$$

$$\{Y^*\}_{m+1} = \left[\frac{2}{\Delta t^2}[M] - [H]\right]\{T\}_m - \left[\frac{1}{\Delta t^2}[M] - \frac{1}{\Delta t}[C]\right]\{T\}_{m-1} + \{Y\}_m$$

9.6.5 Specific example

For a typical element, the following is obtained:

$$[M_e] = A_e \begin{bmatrix} \dfrac{\rho L}{3} & \dfrac{\rho L}{6} & \dfrac{\rho_f L}{3} & \dfrac{\rho_f L}{6} \\ \dfrac{\rho L}{6} & \dfrac{\rho L}{3} & \dfrac{\rho_f L}{6} & \dfrac{\rho_f L}{3} \\ \dfrac{\rho_f L}{3} & \dfrac{\rho_f L}{6} & \dfrac{\rho_f L}{3n} & \dfrac{\rho_f L}{6n} \\ \dfrac{\rho_f L}{6} & \dfrac{\rho_f L}{3} & \dfrac{\rho_f L}{6n} & \dfrac{\rho_f L}{3n} \end{bmatrix}, \quad [C_e] = \dfrac{A_e L}{6K}\begin{bmatrix} 0 & 0 & 2 & 1 \\ 0 & 0 & 1 & 2 \\ 0 & 0 & 0 & 0 \\ 0 & 0 & 0 & 0 \end{bmatrix},$$

$$[H_e] = A_e \begin{bmatrix} \dfrac{\alpha M}{L} & -\dfrac{\alpha M}{L} & \dfrac{M}{L} & -\dfrac{M}{L} \\ -\dfrac{\alpha M}{L} & \dfrac{\alpha M}{L} & -\dfrac{M}{L} & \dfrac{M}{L} \\ \dfrac{2\mu + \lambda + \alpha^2 M}{L} & -\dfrac{2\mu + \lambda + \alpha^2 M}{L} & \dfrac{\alpha M}{L} & -\dfrac{\alpha M}{L} \\ -\dfrac{2\mu + \lambda + \alpha^2 M}{L} & \dfrac{2\mu + \lambda + \alpha^2 M}{L} & -\dfrac{\alpha M}{L} & \dfrac{\alpha M}{L} \end{bmatrix},$$

$$\{Y_e\} = \begin{Bmatrix} \dot{F}_{1i} \\ F_j^1 \\ F_i^1 \\ F_i^2 \\ F_j^2 \end{Bmatrix} \quad 2mm\{\ddot{T}_e\} = \begin{Bmatrix} \ddot{U}_i \\ \ddot{U}_j \\ \ddot{W}_i \\ \ddot{W}_j \end{Bmatrix} \{\dot{T}_e\} = \begin{Bmatrix} \dot{U}_i \\ \dot{U}_j \\ \dot{W}_i \\ \dot{W}_j \end{Bmatrix} \{T_e\} = \begin{Bmatrix} U_i \\ U_j \\ W_i \\ W_j \end{Bmatrix}$$

9.6.6 Example: simulation of dynamic response of saturated porous media

A specific example is given herein by simulating the response of a half space under sinusoidal cyclic loading. No drainage is allowed at the bottom, and the fluid phase is free of traction at the ground surface. Figure 9.8 shows the variation of displacement, velocity and acceleration responses of some selected nodes.

9.7 Introduction of boundary conditions in simultaneous equation system

9.7.1 Formulation

The simultaneous equation system results in a $[m{\times}m]$ square matrix:

$$[K]\{U\} = \{F\} \tag{9.212}$$

It should be noticed that matrix $[K]$ is a square symmetric matrix and that its determinant $|K|$ is zero. Thus, its inverse is not possible due to the singularity problem. The solution becomes possible if the boundary conditions are introduced. The introduction of boundary conditions associated with vectors $\{U\}$ and $\{F\}$ can be partitioned into two unknown and known parts:

$$\{U\} = \left\{ \begin{array}{c} \{U\}_u \\ \{U\}_n \end{array} \right\}, \text{ and } \{F\} = \left\{ \begin{array}{c} \{F\}_n \\ \{F\}_u \end{array} \right\} \tag{9.213}$$

Accordingly, Equation (9.207) may be rewritten as:

$$\begin{bmatrix} [K]_{uu} & [K]_{un} \\ [K]_{nu} & [K]_{nn} \end{bmatrix} \left\{ \begin{array}{c} \{U\}_u \\ \{U\}_n \end{array} \right\} = \left\{ \begin{array}{c} \{F\}_n \\ \{F\}_u \end{array} \right\} \tag{9.214}$$

As the solution of $\{U\}_u$ is required, one can write the following equation:

$$[K]_{uu}\{U\}_u + [K]_{un}\{U\}_n = \{F\}_n \tag{9.215}$$

Rearranging Equation (9.215), we have the following relations

$$[K]_{uu}\{U\}_u = \{F\}_n - [K]_{un}\{U\}_n \tag{9.216a}$$

or

$$\{U\}_u = [K]_{uu}^{-1}(\{F\}_n - [K]_{un}\{U\}_n) \tag{9.216b}$$

If the unknown part of vector $\{F\}$ is required, Equation (9.216) results in the following relation:

$$[K]_{nu}\{U\}_u + [K]_{nn}\{U\}_n = \{F\}_u \tag{9.217}$$

One can easily obtain $\{F\}_u$ from Equation (9.217) by inserting unknown vector $\{U\}_u$ obtained from Equation (9.216) in the previous stage.

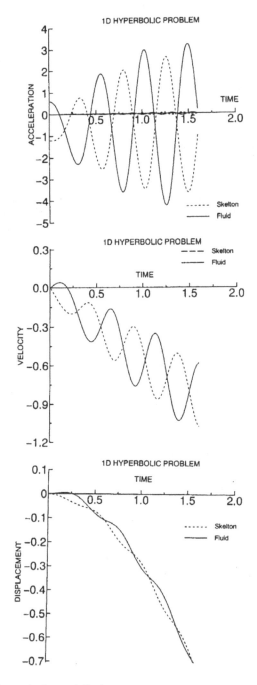

Figure 9.8 Acceleration, velocity, and displacement responses

9.7.2 *Actual implementation and solution of Equation (9.211b)*

Rearranging Equation (9.207) in the form of Equation (9.208) and solving using the relation given by (9.211b) require some extra computational time. Instead of rearrangement in the form of Equation (9.208), the following procedure is implemented. For each boundary condition U_k, the force vector is modified by changing i from 1 to m except row i:

$$F_k^* = F_k - K_{ki}U_k$$

Then 1 is assigned to K_{kk}, and the other components of the row and column matrix [K] are assigned to 0. The value of vector F_k^* is assigned to U_k. The actual implementation of this procedure in the FEM program coded in True BASIC programming language is given here:

```
*************************************************************************
!
! ************* Displacement Boundary condition is implemented ***************
!
FOR IB=1 TO NBS
  IBNI=IBS(IB)
  FOR I=1 TO NODE
    GFS(I)=GFS(I)-GKS(IBNI,I)* UDIS(IB)
  IF I<>IBNI THEN
    GKS(IBNI,I)=0.
    GKS(I,IBNI)=0.
  ELSE
    GKS(IBNI,IBNI)=1.0
    GFS(IBNI)=UDIS(IB)
  END IF
  NEXT I
NEXT IB
!
*************************************************************************
```

9.8 Rayleigh damping and its implementation

The final forms of the discretized form of the equation of motion (Equation 9.47) irrespective of method of solution (FDM, FEM, BEM) and continuum or discontinuum, depending upon the character of governing equation, may be written in the following form:

$$[M]\{\ddot{\varphi}\} + [C]\{\dot{\varphi}\} + [K]\{\varphi\} = \{F\} \tag{9.218}$$

The specific forms of matrices [M], [C], [K] and vector {F} in the preceding equation will only differ depending upon the method of solution chosen and dimensions of physical space. Viscosity matrix [C] is associated with the rate dependency of geomaterials. However, in many dynamic solution schemes, viscosity matrix [C] is expressed in the following form using Rayleigh damping approach

$$[C] = \alpha[M] + \beta[K] \tag{9.219}$$

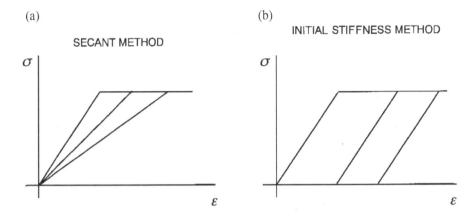

Figure 9.9 Illustration of numerical techniques to deal with nonlinearity

where α, β are called proportionality constants. This approach becomes very convenient in large-scale problems if central finite difference technique and mass lumping are used. However, it should be also noted that it is very difficult to determine these parameters from experiments. Again, in nonlinear problems, the deformation moduli of rocks are reduced in relation to straining using an approach commonly used in soil dynamics. In such approaches, the reduction of moduli is determined from cyclic tests. Nevertheless, it should be noted that the validity of such an approach for rock and rock masses is quite questionable.

9.9 Nonlinear problems

If material behavior involves nonlinearity, the preceding equation system must be solved iteratively with the implementation of required conditions associated with the constitutive law chosen. The iteration techniques may be broadly classified as the initial, secant or tangential stiffness method (e.g. Owen and Hinton, 1980). See Figure 9.9.

9.10 Special numerical procedures for rock mass having discontinuities

The existence of discontinuities in rock mass has special importance on the stability of rock engineering structures, directional seepage, diffusion or heat transport, and its treatment in any analysis requires a special attention. Various types of finite element methods with joint or interface elements, discrete element method (DEM), displacement discontinuity analysis (DDA), discrete finite element method (DFEM) and displacement discontinuity method (DDM) have been developed so far. Although these methods are mostly concerned with the solution of equation of motion, they can be used for seepage, heat transport or diffusion problems. The fundamental features of the available methods are described in a recent review on these methods by Kawamoto and Aydan (1999).

(a) No-tension finite element method

The no-tension finite element method is proposed by Valliappan in 1968 (Zienkiewicz *et al.*, 1968). The essence of this method lies with the assumption of no tensile strength for rock mass since it contains discontinuities. In the finite element implementation, the tensile strength of media is assumed to be nil. It behaves elastically when all principal stresses are compressive. The excess stress is redistributed to the elastically behaving media using a similar procedure adopted in the finite element method with the consideration of elastic-perfectly plastic behavior.

(b) Pseudo-discontinuum finite element method

This method was first proposed by Baudendistel *et al.* in 1970. In this method, the effect of discontinuities in the finite element method is considered through the introduction of directional yield criterion in elasto-plastic behavior. Its effect on the deformation characteristics of the rock mass is not taken into account. If there is any yielding in a given element, the excess stress is computed, and the iteration scheme for elastic-perfectly plastic behavior is implemented. If there is more than one discontinuity set, the excess stress is computed for the discontinuity set that yields the largest value.

(c) Smeared crack element

The smeared crack element method within the finite element method was initially proposed by Rashid (1968) and adopted by Pietruszczak and Mroz (1981) in media having weakness planes or developing fracture planes. This method evaluates the equivalent stiffness matrix of the element and allows the directional plastic yielding within it.

(d) Discrete finite element method (DFEM)

Finite element techniques using contact, joint or interface elements have been developed for representing discontinuities between blocks in rock masses. The simplest approach for representing joints is the contact element, which was originally developed for bond problems between steel bars and concrete. The contact element is a two-noded element having normal and shear stiffnesses. This model has been recently used to model block systems by Aydan and Mamaghani (e.g. Aydan *et al.*, 1996) by assigning a finite thickness to the contact element and employing an updated Lagrangian scheme to deal with large block movements. The contact element can easily deal with sliding and separation movements.

(e) Finite element method with joint or interface element (FEM-J)

Goodman *et al.* (1968) proposed a four-noded joint element for joints. This model is simply a four-noded version of the contact element of Ngo and Scordelis (1967), and it has the following characteristics. In a two-dimensional domain, joints are assumed to be tabular with zero thickness. They have no resistance to the net tensile forces in the normal direction, but they have high resistance to compression. Joint elements may deform under normal pressure, especially if there are crushable asperities. The shear strength is presented by a bilinear Mohr-Coulomb envelope. The joint elements are designed to be compatible with solid elements. Ghaboussi *et al.* (1973) proposed a four-noded interface element for joints. This model is a further improvement of the joint element by assigning a finite thickness to joints.

(f) Displacement discontinuity method (DDM)

This technique is generally used together with the boundary element method (BEM). The discontinuities are modeled as a finite length segment in an elastic medium with a relative displacement. In other words, the discontinuities are treated as internal boundaries with prescribed displacements. As an alternative approach to the technique of Crouch and Starfield (1983), Crotty and Wardle (1985) use interface elements to model discontinuities, and the domain is discretized into several subdomains.

(g) Discontinuous deformation analysis method (DDA)

Shi (1988) proposed a method called discontinuous deformation analysis (DDA). Intact blocks were assumed to be deformable and are subjected to constant strain and stress due to the order of the interpolation functions used for the displacement field of the blocks. In the original model, the inertia term was neglected so that damping becomes unnecessary. For dynamic problems, although damping is not introduced into the system, large time steps are used in the numerical integration in time-domain results in artificial damping. It should be noted that this type of damping is due to the integration technique for time-domain and has nothing to do with the mechanical characteristics of rock masses (i.e. frictional properties). Although the fundamental concept is not very different from Cundall's model, the main difference results from the solution procedure adopted in both methods. In other words, the equation system of blocks and its contacts are assembled into a global equation system in Shi's approach. Recently Ohnishi *et al.* (1995) introduced an elasto-plastic constitutive law for intact blocks and gave an application of this method to rock engineering structures.

(h) Discrete element method (DEM)

Distinct element method (rigid block models) for jointed rocks was developed by Cundall in 1971. In Cundall's model, problems are treated as dynamic ones from the very beginning of formulation. It is assumed that the contact force is produced by the action of springs that are applied whenever a corner penetrates an edge. Normal and shear stiffness were introduced between the respective forces and displacements in his original model. Furthermore, to account for slippage and separation of block contacts, he also introduced the law of plasticity. For the simplicity of calculation of contact forces due to the overlapping of the block, he assumed that the blocks do not change their original configurations. To solve the equations of the whole domain, he never assembled the equilibrium equations of blocks into a large equation system but solved them through a step-by-step procedure, which he called marching scheme. His solution technique has two main merits:

1 The storage memory of computers can be small (note that computer technology was not so advanced during the late 1960s); therefore, it could run on a microcomputer.
2 The separation and slippage of contacts can be easily taken into account since the global matrix representing block connectivity is never assembled. If a large assembled matrix is used, such a matrix will result in zero or very nearly zero diagonals, subsequently causing singularity or ill conditioning of the matrix system.

As the governing equation is of the hyperbolic type, the system could not become stabilized even for static cases unless damping is introduced into the equation system. In recent years, he has improved the original model by considering the deformability of intact blocks and their elasto-plastic behavior. Cundall's model has been actively used in rock engineering structures design by the NGI group in recent years (e.g. Barton *et al.*, 1986).

References

Aydan, Ö., Mamaghani, I.H.P. & Kawamoto, T. (1996) Application of discrete "finite element method (DFEM) to rock engineering". *North American Rock Mechanics Symp.*, Montreal, 2, 2039–2046.

Barton, N., Harvik, L., Christianson, M. & Vik, G. (1986) Estimation of joint deformations, potential leakage and lining stresses for a planned urban road tunnel. *Int. Symp. on Large Rock Caverns.* Helsinki, pp. 1171–1182.

Baudendistel, M., Malina, H. & Müller, L. (1970) Einfluss von Discontinuitaten auf die Spannungen und Deformationen in der Umgebung einer Tunnelröhre. *Rock Mechanics*, Vienna, 2, 17–40.

Crotty, J.M. & Wardle, L.J. (1985) Boundary integral analysis of piece-wise homogenous media with structural discontinuities. *International Journal. Rock Mechanics and Mining Science*, 22(6), 419–427.

Crouch, S.L. & Starfield, A.M. (1983) *Boundary Element Methods in Solid Mechanics.* Allen & Unwin, London.

Cundall, P.A. (1971) *The Measurement and Analysis of Acceleration in Rock slopes.* PhD Thesis, University of London, Imperial College.

Ghaboussi, J., Wilson, E.L. & Isenberg, J. (1973) Finite element for rock joints and interfaces. *Journal Soil Mechanics and Foundation Engineering. Division*, ASCE, 99(SM10), 833–848.

Goodman, R.E., Taylor, R. & Brekke, T.L. (1968) A model for the mechanics of jointed rock. *Journal Soil Mechanics and Foundation Engineering. Division*, ASCE, 94(SM3), 637–659.

Kawamoto, T. & Aydan, Ö. (1999) A review of numerical analysis of tunnels in discontinuous rock masses. *International Journal of Numerical and Analytical Methods in Geomechanics*, 23, 1377–1391.

Ngo, D. & Scordelis, A.C. (1967) Finite element analysis of reinforced concrete beams. *Journal of American Concrete Institute*, 152–163.

Ohnishi, Y., Sasaki, T. & Tanaka, M. (1995) Modification of the DDA for elasto-plastic analysis with illustrative generic problems. *35th US Rock Mechanics Symposium*, Lake Tahoe. pp. 45–50.

Owen, D.R.J. & Hinton, E. (1980) *Finite Element in Plasticity: Theory and Practice.* Pineridge Press Ltd, Swansea.

Pietruszczak, S. & Mroz, Z. (1981) Finite element analysis of deformation of strain-softening materials. *International Journal of Numerical Methods in Engineering*, 17, 327–334.

Rashid, Y.R. (1968) Ultimate strength analysis of prestresses concrete pressure vessels. *Nuclear Engineering and Design*, 7, 334–344.

Shi, G.H. (1988) *Discontinuous Deformation Analysis: A New Numerical Model for the Statics and Dynamics of Block Systems*. PhD Thesis, Department of Civil Engineering, University of California, Berkeley. 378p.

Zienkiewicz, O.C., Valliappan, S. & King, I.P. (1968) Stress analysis of rock as a "no-tension" materials. *Geotechnique*, 18, 56–66.

Chapter 10

Ice mechanics and glacial flow

Ice can be treated as a rock-like material, and some sessions at the Second ISRM Congress in Belgrade in 1970 were devoted to it. However, almost no papers appeared in the following congresses and events on ice and its mechanics. This chapter is devoted to ice, its mechanics and related engineering problems.

10.1 Physics of ice

All materials in nature can be in solid, fluid or gaseous form depending upon the temperature and pressure. Ice is the solidified phase of water under a temperature less than 0°C. Ice-sheets cover the north and south polar regions of the Earth and mountains and plateaus greater than 2500 m above the sea level in other regions. The thickness of ice-sheets is measured to be more than 2700 m in Greenland, and the thickness varies depending upon location. Figures 10.1–10.4 show views of several ice-sheets and glaciers in Northern Hemisphere.

Figure 10.1 Views of ice-sheets and glaciers in Alaska

Figure 10.2 Views of ice-sheets and glaciers in Canada

Figure 10.3 Views of ice-sheets and glaciers in Norway

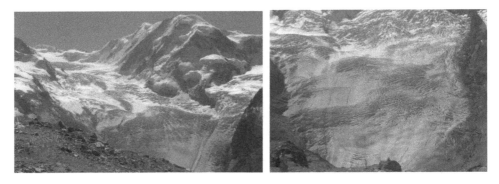

Figure 10.4 Views of ice-sheets and glaciers in Switzerland

10.2 Mechanical properties of ice

Polycrystalline ice is generally treated as an isotropic solid. The p-wave and s-wave velocities of ice are more than 3890 m s^{-1} and 1800 m s^{-1}. However, the wave velocities may be slightly anisotropic. The Young modulus and shear modulus of ice are 9.5 and 3.5 GPa, while the Poisson ratio is about 0.33 (i.e. Truffer, 2013). Ice is generally viewed as a visco-plastic material in terms of constitutive law. As the overall unit weight of ice is less than water, it floats in water.

10.3 Glaciers and ice domes/sheets

Besides the southern and northern poles, glaciers are found in mountainous regions of the Earth such as the Alps and Scandinavian Peninsula in Europe, the Andes in South America, the Rockies in North America, the Taurus, North Anatolian and Elbruz mountains in West Asia, the Himalayas in Asia and even in New Zealand. There are also traces of glaciers even in England and Turkey.

A typical glacier is visualized to be consisted of two zones: an accumulation zone and an ablation zone, as illustrated in Figure 10.5 (e.g. Bennet and Glasser, 2009; Truffer, 2013). They are mainly driven by gravitational forces, and they flow like a visco-plastic fluid.

Ice domes/sheets are found on relatively flat lithospherical areas and mainly in the north and south poles such as Greenland and Antarctica. The motion of the ice domes/sheets is also gravity driven from the center toward their flanks. Nevertheless, their motion entirely depends upon pressure distribution related to the geometry of ice domes/sheets (Fig. 10.6). When they reach open seas or are floating in the seas, they may be broken into pieces called icebergs, and their motion would then be governed by Coriolis forces resulting from the rotation of the Earth.

Velocity and shear stress distribution of a typical longitudinal section of a glacier are illustrated in Figure 10.7. If the ice is modeled as a Newtonian material, the velocity (v) and shear stress distribution (τ) are represented by the following equations:

$$v = v_o + \frac{g}{2\eta} z \sin \alpha \left(H - \frac{z}{2} \right); \tau = gH \sin \alpha \left(1 - \frac{z}{H} \right) \tag{10.1}$$

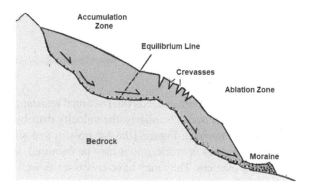

Figure 10.5 Illustration of main components of a glacier

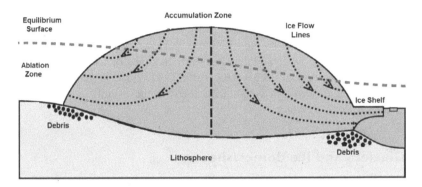

Figure 10.6 Illustration of an ice dome/sheet and its motion

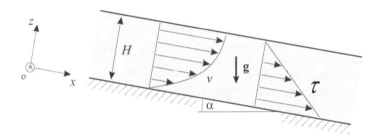

Figure 10.7 Velocity and shear stress distribution along a longitudinal section of a glacier

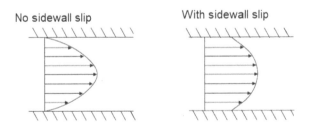

Figure 10.8 Velocity distributions in a glacier in a channel

However, if the shear stress at the base exceeds the frictional resistance between glacier and bedrock, some relative slip will occur. Similarly, the velocity distributions in the direction of the flow will appear as shown in Figure 10.8 for no-slip and slip conditions.

Ice in the upper section of glaciers is brittle, and it may be fractured depending upon the geometry and thickness of the glaciers. They may have crevasses as well as fractures going through its entire thickness. Figure 10.9 shows examples of crevasses and fractures in the Athabasca glacier in Canada. In the ablation zone, some surface and undersurface

Figure 10.9 Crevasses and fractures in the ablation zone of Athabasca glacier in Canada

Figure 10.10 Surface rivers and a sinkhole in Gorner glacier near Zermatt in Switzerland

rivers may form. These rivers may also lead to sinkholes within the glaciers, as seen in Figure 10.10.

Striations on bedrocks are caused by glacier flow in relation to their direction of motion. Figure 10.11 shows an example of striations on the bedrock observed in Athabasca glacier.

10.4   Cliff and slope failures induced by glacial flow

It is well-known that the glaciers apply shear stresses on the bedrock of the channel. As a result, erosion takes place at contact areas, which may result in overhanging configurations on the sidewalls of the channel. Depending upon the resulting geometry and

Figure 10.11 Striations observed on the bedrock of the Athabasca glacier

Figure 10.12 Toe erosions of sidewalls at Gorner glacier in Switzerland

discontinuities and characteristics of rock mass, cliff and/or slope failure may occur as shown in Figure 10.12. As the climatic conditions can be quite harsh and sedimentary rocks are quite vulnerable to degradation, slope failure may take place from time to time. The governing mechanism is quite similar to the cliff failures caused by the toe erosion phenomenon. Figure 10.13 shows an example of a slope failure in the ablation zone at the northern side of the Athabasca glacier in Canada. Rocks were shale, sandstone and mudstone at this locality.

Ice-sheets may also slide down especially during thawing seasons. These failures may be similar to planar slide initially. Figure 10.14 shows examples of ice-sheet slides in the ablation zones.

Huge avalanches of ice-sheets, together with debris material, may occur when they are subjected to seismic forces. Some examples of such failures have been reported in the past. The 1962 and 1970 Huascaran snow, ice and rock avalanches in Peru, Mt. Cook ice-rock avalanche in New Zealand, and 2015 Langtang ice-rock avalanche in Nepal due to the 2015

Figure 10.13 Slope failure at the northern side of the Athabasca glacier in Canada

Gorkha earthquake were all quite devastating. Figure 10.15 shows the satellite images of the Langtang ice-rock avalanche before and after the event.

The Gorkha earthquake also caused the failure of an ice-cliff between Lingtren and Pumori peaks with Khumbu glacier on the mountain of Everest, which resulted in 18 casualties in a base camp about 800 m below. The ice-sheet failure occurred on the western side of the glacier at an elevation of approximately 6300–6400 m.

The receding glaciers reduce the pressure at the sidewalls and base. The variation of normal pressure at the sidewalls particularly induces stress relaxation in the rock mass of slopes, which results in the opening and/or propagation of existing discontinuities and in the increase of permeability and seepage characteristics of the rock mass. These changes result in instability problems at the sidewall slopes of glaciers. For example, a huge rock slope failure occurred in Randa near Zermatt in Switzerland on April 18, 1991 (Götz and Zimmermann, 1993). The Randa slope failure interrupted the railway line connecting Zermatt to the Rhône Valley (Fig. 10.16). The railway line was buried for 800 m and the road for 200 m. The fallen rock mass dammed the Vispa river. A channel was excavated before a potential catastrophic failure of the dam.

A huge rockfall occurred on the east face of Eiger mountain in Grindelwald glacier. In early June of 2006, an 18 cm wide crack appeared at the top of the rock block, and the crack grew at a rate of 90 cm a day. It finally failed when the crack was roughly 4.8 m feet (Fig. 10.17).

(a) Gorner Glacier

(b) Aletsch Glacier

(c) Athabasca

(d) Norway

Figure 10.14 Examples of ice-sheet slides and falls

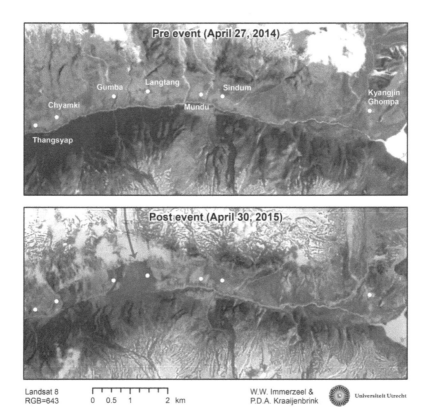

Landsat 8
RGB=643

0 0.5 1   2 km

W.W. Immerzeel &
P.D.A. Kraaijenbrink

Universiteit Utrecht

Figure 10.15 Satellite images of Langtang ice-rock avalanche before and after the event caused by the
2015 Gorkha earthquake

Source: Processed by Immerzel and Kraaijenbrink, 2015

Figure 10.16 Views of Randa slope failure

Figure 10.17 Views of rock slope failure on the east face of Eiger mountain

Figure 10.18 Formation and cracking of glacier caves

10.5　Glacial cave failures

Glacial caves develop within glaciers due to seepage of melting water through crevasses (Fig. 10.18). Most glacier caves are located near the contact zone with bedrock. These caves may cause some sinkholes within the glacier, as seen in Figure 10.10. However, the most catastrophic failures may occur at the tip of the glaciers in the ablation zones when they meet glacial lakes or open seas. The failure of caves may result in some tsunami-like events in the lakes and open seas.

10.6　Moraine lakes and lake burst

Glaciers create moraine lakes in the vicinity of their tips. The material originates from debris from the failure of sidewall slopes of the channels and the erosion of bedrock. The receding glaciers also result in increases in the size of lakes. The lakes may fail due to piping phenomenon or overtopping. These may expose great danger to settlements and engineering facilities downstream of the valleys. See Figure 10.19.

Figure 10.19 Examples of moraine lakes

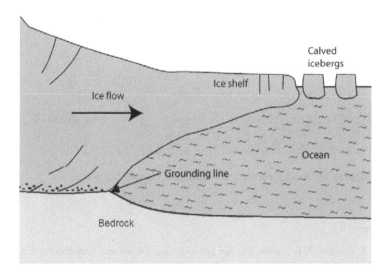

Figure 10.20 Ice thinning and calving.

10.7 Calving and iceberg formation

When ice domes, ice-sheets or glaciers reach to open seas, they start to break down into blocks, which is known as calving, as illustrated in Figures 10.20 and 10.21. As a result of calving, icebergs are formed. These icebergs float in the seas and move mainly as a result of Coriolis forces due to the Earth's rotation. Ice calving results in the toppling of ice blocks into the sea, which may create some small-scale tsunamis (Fig. 10.22). They may expose great danger to nearby settlements as well as ship traffic and offshore platforms. The famous Titanic ship accident was due to the collision of the ship with an iceberg.

Figure 10.21 Calving of glaciers

Figure 10.22 Toppling of calved ice-slabs and subsequent tsunami-like waves

References

Bennet, M.R. & Glasser, N.F. (2009) *Glacial Geology: Ice Sheets and Landforms*. Wiley-Backwell, Chichester, West Sussex, 385p.

Götz, A. & Zimmermann, M. (1993) The 1991 rock slides in Randa: Causes and consequences. *Landslide News*, 7(3), 22–25.

Immerzel, W.W. & Kraaijenbrink, P.D.A. (2015) *Landsat 8 Reveals Extent of Earthquake Disaster in Langtang Valley*. http://mountainhydrology.org/nepal-quake/.

Truffer, M. (2013) *Ice Physics*. University of Alaska, Fairbanks. 120p.

Chapter 11

Extraterrestrial rock mechanics and rock engineering

Mankind is now exploring the ways to find out the characteristics of other planets and the possibility of exploiting their mineral resources. The United States and Russia (former USSR) sent several spacecrafts to the Moon, Venus and Mars, and some of them landed on the Moon, Venus and Mars (NASA, 2008). Some of most impressive images from the Apollo Program of NASA is that of a human next to a lunar rock mass and hitting lunar rock with a geologist's hammer (Figure 11.1). The images from recent Mars exploration rovers showed striking similarities between rocks on Earth and those of Mars. One can easily notice layered rock masses, jointing, weathering effects and some toppling-type rock slope stability problems (Fig. 11.2). This chapter is entirely based on information obtained from images mostly released by NASA and the interpretations of the author, together with some further information and interpretations by others.

Although the environmental conditions on Mars and other planets are different from those on Earth, the principles governing mechanical and engineering aspects of rocks on other planets should be quite similar to those developed for the rocks of Earth. Therefore, the next generations of our discipline would definitely see its extension to the rocks of other planets.

(a) Broken rock mass next to mankind   (b) Geologist hitting lunar rock by a hammer

Figure 11.1 Images from the Moon involving humankind from the Earth
Source: Images by NASA

(a) Layered rock mass on Mars   (b) Toppling of a rock slope on Mars

Figure 11.2 Some images of rocks and rock slope on Mars
Source: Arranged from images of NASA

Figure 11.3 Comparison of Venus, Earth and Mars

11.1 Solar system

The solar system consists of eight planets: Mercury, Venus, Earth, Mars, Saturn, Jupiter, Uranus, Neptune (Watters, 1995). On the basis of computations, the existence of a planet between Mars and Jupiter was estimated. At this location, an asteroid belt now exists that may have corresponded to the missing planet. Initially, Pluto was found in 1930, and it was considered to be the ninth planet until 2006. On the basis of the definition of planets of our solar system, it was then disqualified as the ninth planet. The densities of planets differ as a function of distance from the Sun and decrease as the distance increases. Venus, Earth and Mars are considered to be quite similar to one another in terms of size and geology (Fig. 11.3). Venus has no natural satellites, while the Earth has the Moon as a natural satellite. Mars has two natural satellites called Deimos and Phobos.

11.2   Moon

The Moon is about 284400 km away from the Earth and is 3474.2 km in diameter. The average density is 3.34 g cm^{-3} with a gravitational acceleration of 1.62 m s^{-2} at the surface, which is about one-sixth that of the Earth. The Moon is assumed to be a part of the Earth initially. The Moon is the most extensively investigated solar body by the United States and Russia. Recently, Japan and China sent some explorers to the Moon. However, the most extensive explorations and investigations have been carried out by NASA of the United States so far.

11.2.1   Surface structure

The surface topography of the Moon has been measured by the methods of laser altimetry and stereo image analysis as shown in Figure 11.4. The surface topography of the Moon on the far side is very different from that on the near side. The far side of the Moon is about 1.9 km higher than the near side. The surface of the Moon has been greatly shaped by the impact of meteorites. The south pole Aitken basin has the lowest elevation while the highest elevations are found just to the northeast of this basin, and it has been suggested that this area might represent thick ejecta deposits on the Moon. The difference between lowest and highest elevations is about 13 km.

Although there are no tectonic plates in the Moon, huge rift valleys are found, and they are covered by numerous ancient volcanic plains.

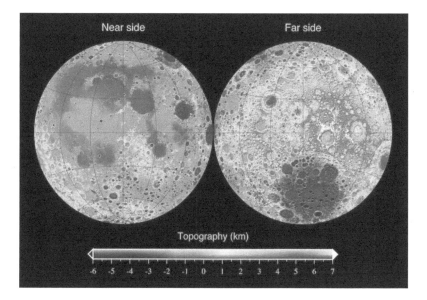

Figure 11.4 Surface topography of the Moon
Source: From NASA

11.2.2 Inner structures

During the NASA Apollo program, some passive-type seismometers were installed on the Moon. During this program from 1969 to 1972, Apollo 12, 14, 15 and 16 seismometers were installed (Fig. 11.5). These seismometers recorded moonquakes, which were initially unexpected (Fig. 11.6). The records from the moonquakes revealed an inner structure of the Moon like that of the Earth (e.g. Irvine, 2002; Latham *et al.*, 1971; Nakamura, 2003; Toksöz *et al.*, 1977). It was found that the earthquakes were categorized as deep earthquakes resulting from the tidal force interaction between the Earth and the Moon, shallow earthquakes due to tectonic forces and thermal stress variations and other earthquakes due to

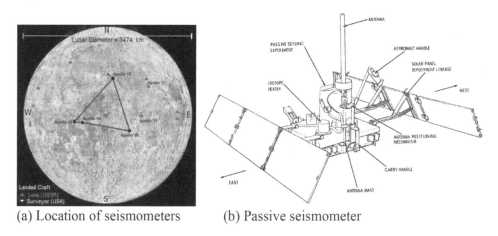

(a) Location of seismometers   (b) Passive seismometer

Figure 11.5 Seismometers and lunar seismic network

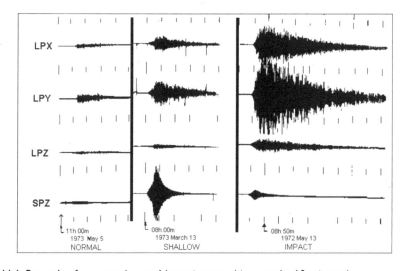

Figure 11.6 Records of moonquakes and lunar impacts (time marks 10 minutes)

meteorite and lunar module impacts. The s of the shallow earthquakes could be up to 5.5, while the magnitudes of the deep earthquakes were much smaller, generally less than 2. The shallow earthquakes may pose a hazard to a lunar habitat, and they may last for more than 10 minutes. The deep earthquakes occur at depths of 700 to 1000 km and occur periodically at 14 and 206 days, implying the causes of the events are terrestrial and solar tidal forces. It was found that the vibrations caused by impacts can last for more than 1 hour while deeper earthquakes last only 10–20 minutes. The reason for such a long seismic vibration is considered to be due to the existence of a thick elastic crust (Fig. 11.7). As the viscosity of the crust is small, the vibrations last longer for surface impacts.

On the basis of these investigations, the inner structure of the Moon is well understood, and it resembles that of the Earth except for the thickness of each layer (Fig. 11.8). Particularly, the outer core is quite thin, and inner core is small.

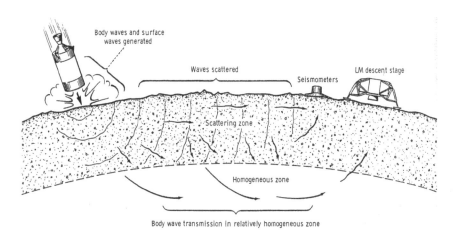

Figure 11.7 Illustration of the response of the Moon crust to impacts

Source: Irvine, 2002

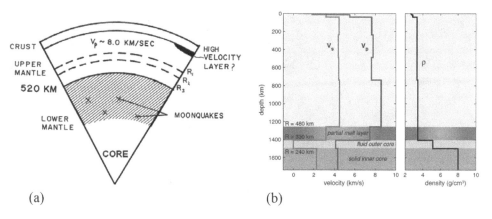

(a)   (b)

Figure 11.8 (a) Inner structure, (b) seismic wave velocity and density distributions

11.2.3 Geology and rocks of the Moon

The geological history of the Moon has been organized into six major epochs starting about 4.5 billion years ago. It was initially in molten state. On the basis of the knowledge of the Earth, it was assumed that olivine and pyroxene formed first, followed by crystallized anorthositic plagioclase feldspar, forming an anorthositic crust about 50 km in thickness. From the Apollo missions, 382 kg rock samples were brought to the Earth from the Moon. Figure 11.9 shows views of several examples of Moon rocks such as olivine, basalt and anorthositic plagioclase. Other rock samples are tuff breccias associated probably with meteorite impacts. Figure 11.10 shows larger rock blocks. One can easily notice bedding planes and discontinuities and fracture planes, which are quite similar to those of the Earth.

The mean lunar surface temperature varies from 107°C at day down to −153°C at night. The temperature difference is thus 260°C. This large temperature difference is very likely to create large thermal stresses, which may impose cyclic contraction–expansion as well as new crack extension in the upper crust of the Moon. It is noted that some thrust-type, extension-type huge cracks/fractures and lava flows exist on the surface of the Moon (Fig. 11.11). The active seismic experiments on the Moon during the Apollo program of NASA revealed that the top 20 km of the Moon's crust is highly heterogeneous and disturbed.

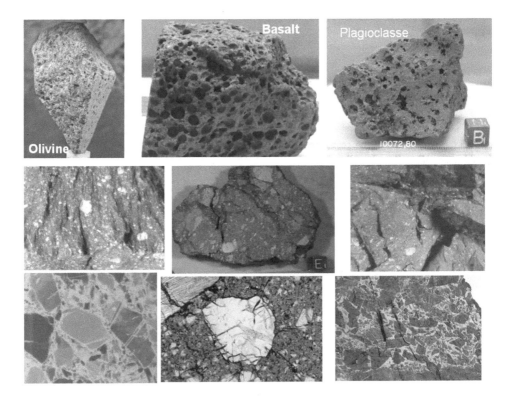

Figure 11.9 Views of some Moon rocks
Source: Original images from NASA

Figure 11.10 Large-scale *in-situ* rock blocks of Moon

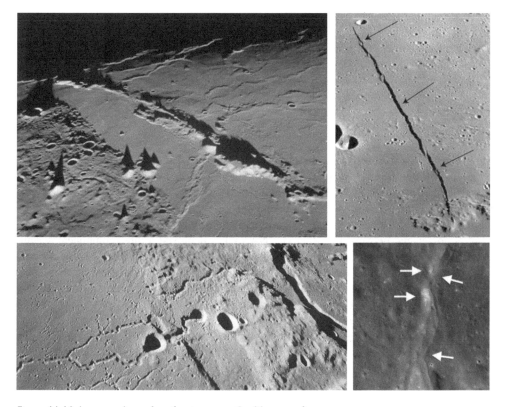

Figure 11.11 Large-scale surface fractures on the Moon surface

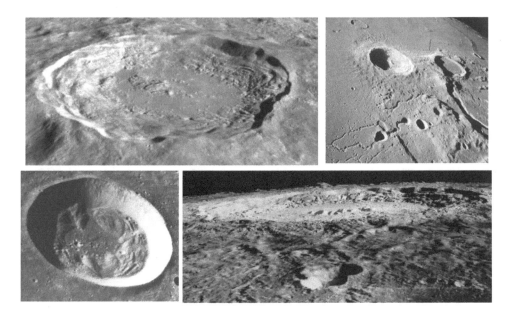

Figure 11.12 Craters and large-scale mass movements on the Moon

The Moon has been heavily bombarded by meteorites of different sizes, creating craters (Fig. 11.12). The surface topography of the Moon is greatly affected and shaped by the impacts of meteorites. In some of large craters, melted rocky material has been noted. In addition, large-scale mass movements exist within the craters.

11.3   Mars

11.3.1   Inner structure

Many models for the inner structure and temperature distribution of Mars have been proposed by various researchers (e.g. Sohl and Spohn, 1997; Johnston and Toksöz, 1977; Toksöz and Hsui, 1978; Steinberger *et al.*, 2010). Particularly, Sohl and Spohn (1997) utilized two meterorites from Mars for estimating the internal structure of that planet. Although all models are based on concepts developed for Earth, they are fundamentally estimations as there is no seismic network for Mars yet. NASA has deployed a three-component seismometer through the Viking program in 1976 (Anderson *et al.*, 1977). In 2018, the second seismometer was installed on Mars as a part of an InSight instrument, which has been equipped with many sensors. The Viking seismometer recorded numerous quakes during its operation. The InSight seismometer recorded the first marsquake on April 6, 2019. Nevertheless, the models for the inner structure of Mars still needs some validation.

The radius of Mars is about 3389 km with a crustal thickness of 10–50 km. The mantle is divided mainly into two layers, which may be called the upper and lower mantles. The core and mantle interface is estimated to be at a depth of 2000 km with a 150 km transition zone. Mars is estimated to be chemically composed of a silicate crust, a Fe-Mg silicate mantle and a metallic Fe core. The average density is about 3.933 g cm^{-3}. The gravitational

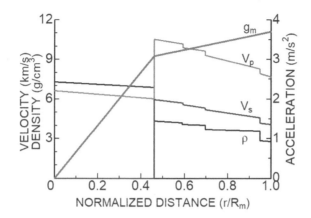

Figure 11.13 Comparison of the distributions of density, wave velocities and gravity of Mars

acceleration is 0.377 g. Figure 11.13 shows the estimated density, wave velocities and gravitational acceleration of Mars.

11.3.2   Geology and tectonics

The surface geology of Mars was mapped by the United States Geological Survey (USGS) and Scott and Carr (1978), utilizing also Mars Orbiter Laser Altimeter (MOLA) data of NASA. The surface geology fundamentally involves volcanics, basalt, breccia, sedimentary deposits (Aydan, 2016). The tectonic activity of Mars is not very active compared to that of Earth. Figure 11.14 shows a map of the tectonics of Mars prepared by the USGS. Nevertheless, a very large-scale Vallis rift zone, Tharsis volcanic chain, depression zones, fracture zones, faults, folding, metamorphism, discordant sedimentation, columnar jointing are found on Mars (Figs. 11.15–11.17).

11.3.3   Rocks of Mars

It is very likely that the rocks of Mars would be quite close to those of Earth. Therefore, the classification of rocks should be igneous, metamorphic and sedimentary on the basis of the images from many NASA Mars explorers. The images and drilling operations performed by Mars explorers (Opportunity, Sprit, Curiosity) on some selected rocks and rock blocks have indicated that the rocks are quite similar to those of the Earth, although it still needs some clarification whether rocks seen in Earth fully exist on Mars. However, the sedimentary rocks (such as conglomerate, sandstone, siltstone, mudstone, sulphate deposits), metamorphic rocks (phyllite, shale, schists) and extrusive rocks (basalt, breccia) of Mars are clearly similar to those of Earth (Figs. 11.18–11.20). Furthermore, flow planes, bedding planes and schistosity planes, which are intrinsic to each rock class, are distinctly observed.

11.3.3.1   Rock discontinuities of Mars

As explained in Chapter 2, there are various causes for the formation of discontinuities in rocks. The causes can be grouped into (1) tension discontinuities, (2) shear discontinuities,

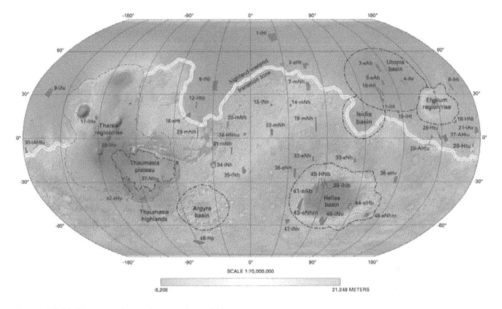

Figure 11.14 Topography and tectonics of Mars

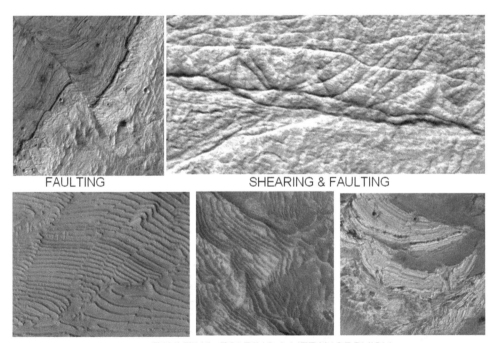

FAULTING

SHEARING & FAULTING

FAULTING, FOLDING & METAMORPHISM

Figure 11.15 Large-scale faulting, shearing, folding, and metamorphism structures

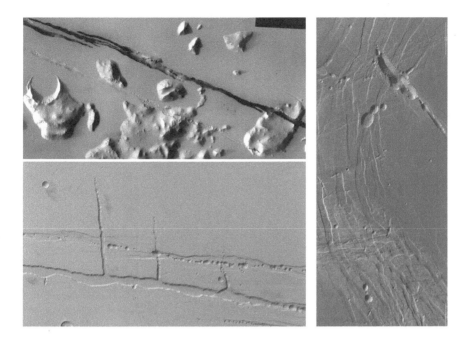

Figure 11.16 Large-scale fractures, rift zones

Sedimentation and Tilting

Sedimentation and Discordance

Volcanic Activity & Columnar Jointing

Folding and metamorphism

Figure 11.17 Sedimentation, tilting, folding and volcanic eruptions and columnar jointing

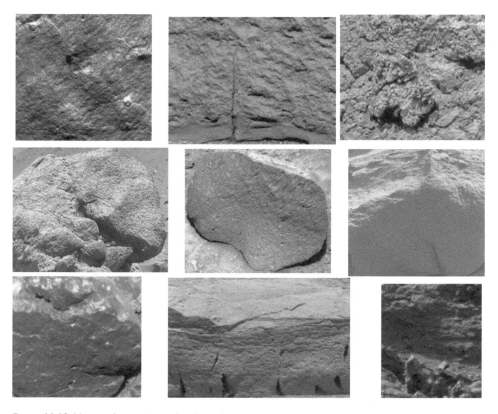

Figure 11.18 Views of extrusive volcanic rocks

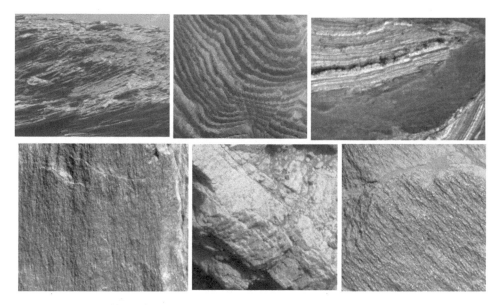

Figure 11.19 Views of metamorphic rocks

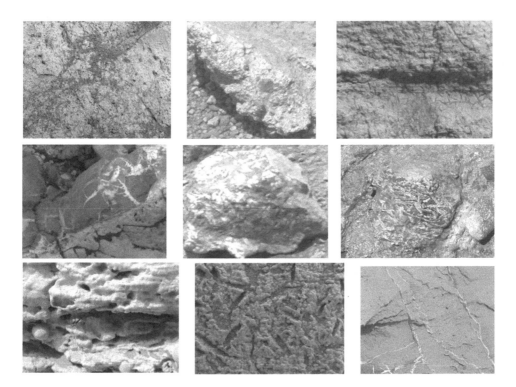

Figure 11.20 Views of sedimentary rocks

(3) discontinuities due to intrinsic properties of rocks (sedimentation, metamorphism, magma flow). Without any exception, discontinuities in the rocks/rock masses of Mars are fundamentally similar to those observed in Earth as seen in Figure 11.21 (Aydan, 2016). Furthermore, the surface morphology and filling situations of rock discontinuities are quite similar to those seen in Earth (Fig. 11.22).

11.3.3.2   Weathering of rocks of Mars

From the surface morphology of Mars, it is well-known that the environment of Mars was quite different from the present. There were lakes, rivers and oceans, and climatic conditions were different from today. Images from Mars explorers clearly show that the weathering of the rocks of Mars is quite similar to that on Earth, as seen in Figure 11.23. As seen in the images of Figure 11.23, differential weathering, solution and oxidation-type weathering can be easily noticed.

Compared to the daily temperature difference on the Moon, the daily temperature difference in Mars is about 80°C. If the rock is not saturated, such a temperature difference would not be of great significance except for soft rocks, as seen in Figure 11.23.

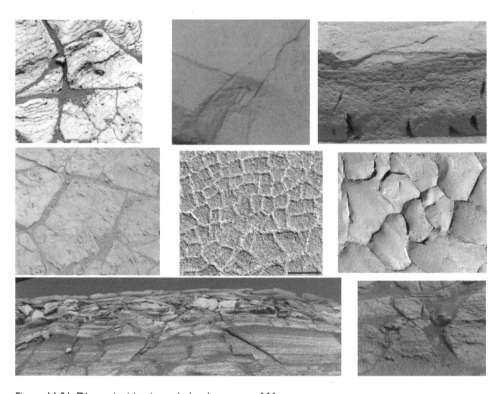

Figure 11.21 Discontinuities in rocks/rock masses of Mars

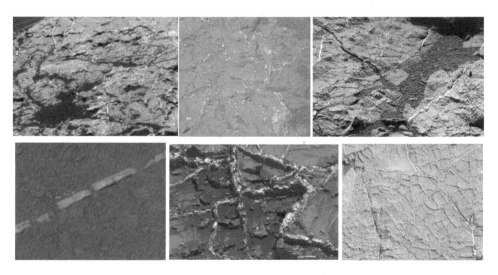

Figure 11.22 Some images of discontinuity filling of rocks on Mars

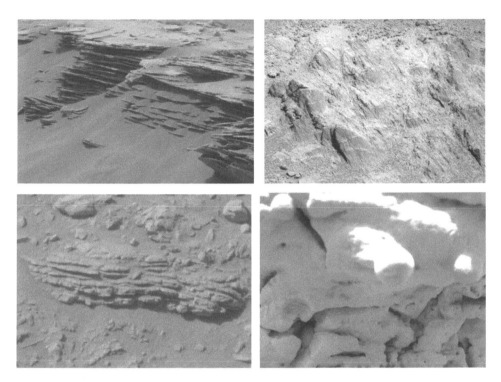

Figure 11.23 Weathering of rocks of Mars

11.4   Venus

Surface observations of Venus have been difficult in the past, due to its extremely dense atmosphere, which is composed primarily of carbon dioxide with a small amount of nitrogen. The first attempts to explore Venus were carried out by the Soviets in the 1960s through the Venera Program. Venera lander missions took place until the early 1980s. Venera 13 and Venera 14 landed on the planet and sent the first color photographs of the surface. Venera 15 and Venera 16 conducted mapping of the Venusian terrain with synthetic aperture radar.

The United States launched the Mariner 1 in 1962 and subsequently NASA's Magellan spacecraft in 1989 to map the surface of Venus with radar. The Magellan provided the most high-resolution images to date of the planet and was able to map 98% of the surface and 95% of its gravity field (Fig. 11.24).

Venus, like Uranus, rotates clockwise, which is known as a retrograde rotation. Due to the slow rotation on its axis, it takes 243 Earth-days to complete one rotation. The orbit of the planet takes 225 Earth-days.

Venus is totally enshrouded by clouds of more than 50 km in height. Due to the continuous global cloud cover and greenhouse effect, surface temperatures on Venus approach 735 K (462°C), and it is the hottest planet in our solar system. As a result of these high temperatures, water is vaporized. The atmospheric pressure at the surface of Venus is 90 times that found at the surface of Earth.

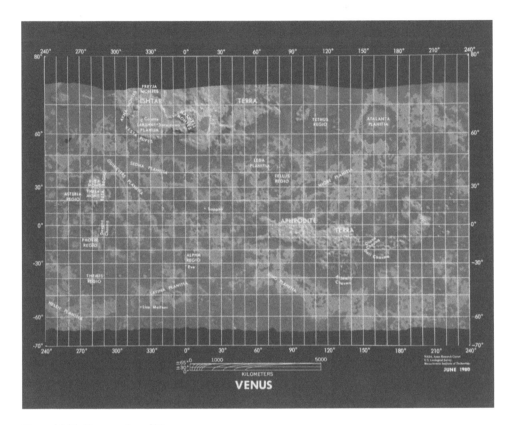

Figure 11.24 Topography of Venus

11.4.1 Interior

It is quite likely that Earth and Venus have similar concentrations of the major elements such as iron, magnesium, calcium, silicon and aluminum. This notion is supported by density measurements determined from the motions of passing spacecraft. The bulk density of Venus (5.24 g/cm^3) is only slightly less than that of Earth (5.52 g cm^{-3}). Venus has a metallic core composed predominantly of iron, a mantle of dense iron and magnesium silicates, and a crust of lighter silicate minerals possibly enriched in aluminum, alkalis, and the radioactive elements uranium and thorium.

Similar to the other terrestrial planets, Venus's interior is essentially composed of three layers: a crust, a mantle and a core. It is believed that Venus's crust is 8–40 km thick depending upon the assumed model, its mantle 3000 km thick, and the core is about 3000 km thick. However, there is an argument about the Venusian interior as to whether or not the planet's core is liquid or solid, as the planet's lacks a substantial magnetic field due to its slow rotation.

Venus is presumed to have emerged about 4.6 billion years ago by the accretion of tiny objects. During the bombardment phase, enough heat was produced to melt the whole proto planet. After a certain cooling period, the molten mass developed a crust, a mantle and a core. Convection in the mantle caused by internal heating deformed the outer crust, which is thinner in low-lying areas (corresponding to seas on the Earth) and thicker in the highlands (corresponding to continents on the Earth). High mountain regions developed by uplifting and outflowing.

11.4.2 Tectonics

The tectonic activity of Venus is quite different from that of Earth (Fig. 11.25). As the surface temperature of Venus is greater than 460°C, the crustal rocks are likely to be softer and weaker than those of Earth and to behave in a ductile manner. As a result of these features of crustal rocks, the stress buildup in the crust and upper mantle of the Venus would be much smaller than that in the Earth. Consequently, the subduction of the soft plastic crust into the upper mantle is not deep. Compared with the subduction of Earth's crust into the upper mantle, which is up to 670 km in depth, the subduction of Venus's crust is estimated to be 250–300 km deep. In view of these facts, the size and the drag of the mantle convection in Venus should be weaker than on Earth so that the tectonic motion of Venus is less pronounced.

Venus's surface appears to have been shaped by extensive volcanic activity. Venus also has 167 large volcanoes that are over 100 km across. The highest point on Venus is Maxwell Montes. Magellan gathered much finer details of the surface topography of Venus, and approximately 1000 impact craters were found. Interestingly, none of the craters seen were smaller than 2 km in diameter. This suggests that any meteoroid small enough to create a crater having a diameter of less than 2 km would have broken apart and burned up during its passage through the dense Venusian atmosphere.

The Ishtar Terra mountains formed as a result of the uplift of the plateau and formation of the Maxwell mountains, which rise 6 km above Lakshmi Planum. Low-lying areas and craters became filled with lava. The shield in the Beta Regis area was formed before the mountain ridges along the fault that broke the surface of the shield (Fig. 11.23).

The surface of Venus has shown that a wide variety of tectonic features are ubiquitous on the planet. Its surface is strongly deformed at a variety of scales. Its lithosphere has been extended and compressed, domed and depressed. Broad crustal domes and rift valleys are common tectonic features on Venus. Beta Regio is a large domical upland about 2500 km

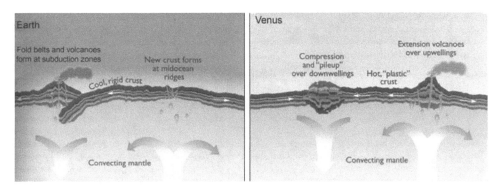

Figure 11.25 Comparison of tectonic model of Venus with that of Earth

Source: From Christiansen and Spilker, 2018

across that is crisscrossed by many faults. The gentle rise is about 4 km high and is crossed by a central trough. A multitude of nearly parallel linear scarps show that the depression is a fault-bounded rift valley, formed as the dome was pulled apart by the extension of the lithosphere. Extensional tectonism has produced long belts of deformation marked by abundant fractures and grabens. These belts persist over hundreds of kilometers. The region around Beta Regio and Atla Regio has fracture belts of diverse orientation. Because of the extensional nature and abundance in the uplands, it is inferred that they formed in response to the uplift of the lithosphere over mantle plumes.

11.4.3   Rocks

The Soviets have successfully landed several spacecraft in the plains regions. Seven of the landers conducted chemical analyses of rocks, which indicate a composition similar to that of terrestrial basaltic volcanic rocks. Venera Spacecraft landed on Venus and took some images of rocks of Venus (Fig. 11.26). The thin, plate-like slabs of rock could be due to molten lava that cooled and cracked. The composition and texture of these rocks is similar to terrestrial basaltic lava.

The weathering processes of rocks on Venus are due to the extremely high temperature and pressure at the surface, as well as the composition of the atmosphere. Each of these factors exerts some control on chemical reactions between gas and rock. Minerals that crystallized at high temperatures in lava flows and that are exposed to CO_2 and SO_2 gas in the Venusian atmosphere are inherently unstable and decompose to form new minerals. The weathered zone probably consists of a mixture of incompletely reacted minerals and newly formed weathering products. For example, it is predicted that, as weathering

Figure 11.26 Images of rocks and soil in landing sites of USSR Venera spacecraft landed on Venus

decomposes basaltic lavas, iron oxides and sulfur-rich minerals will form as iron silicates are destroyed. However, carbonate minerals, which could remove significant amounts of carbon dioxide from the atmosphere, are not stable on the hot, dry surface of Venus.

Weathering varies with altitude on Venus. At high altitudes and low temperatures, sulfur in the atmosphere reacts with lava flows to form the mineral pyrite (FeS_2). Pyrite is highly reflective of radar waves and may explain why mountain peaks and high plateaus of Venus are radar bright. At lower altitudes and higher temperatures, magnetite (Fe_3O_4) and anhydrite ($CaSO_4$) may be the stable minerals produced by weathering. Magnetite is not as reflective as pyrite.

11.5   Issues of rock mechanics and rock engineering on the Moon, Mars and Venus

11.5.1   In-situ *stress*

In-situ stress state is one of the most important items if rock mechanics and rock engineering are to be utilized on the Moon and other planets. Particularly, the engineering design requires data on *in-situ* stress states as well as quake-related topics. Figure 11.27 shows the density, gravity and hydrostatic pressure distributions with depth in Venus and Mars together with those of Earth. The *in-situ* stress measurements and estimations by one or several empirical, analytical and numerical methods can be used as explained in Chapters 7, 8 and 9. However, proper knowledge and information on the characteristics of materials constituting extraterrestrial objects are necessary. For example, the elastic constants obtained from wave velocities, density and gravitational distributions may be used for stress state in the extraterrestrial object. Such a preliminary elastic analysis is carried out for the stress state of Mars under spherical symmetric conditions, and results are shown in Figure 11.28. However, it should be noted that the actual stress conditions are likely to be different from the actual ones as it is very likely that the behavior would involve thermo-elasto-plastic behavior rather than elastic behavior. In other words, experiments on rocks and materials constituting the other extraterrestrial objects are definitely necessary for better evaluations of their *in-situ* stress states.

11.5.2   Slope stability

As on Earth, slope stability issues would be of great significance in other extraterrestrial objects. Although the gravitational acceleration and environmental conditions are different from those of Earth, slope stability issues are likely to be similar to those of Earth. Figure 11.29 shows some examples of slope failures observed on Mars. The slope failures in Mars involve curved shear failure, planar/wedge sliding, toppling failure, bending failure, although their scales may be different. Similar types of slope failures are also noted in the Moon. Nevertheless, such failures are likely to be seen in other planets such as Venus, as seen in Figure 11.26. Figure 11.30 shows a classification of failure modes of rock slopes on Earth (Aydan, 1989). Therefore, the methods developed for rock slopes can be utilized with due considerations of environmental and gravitational differences using the principles of rock mechanics.

Figure 11.31 shows an application of the integrated slope stability assessment proposed by Aydan *et al.* (1991) to rock slopes of Mars. Lines in the figure represent the relations

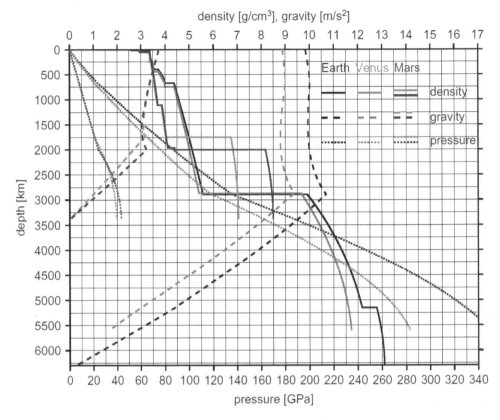

Figure 11.27 Comparison of the density, gravity and hydrostatic pressure distributions with depth in Venus and Mars together with those of Earth

Source: From Steinberger *et al.*, 2010

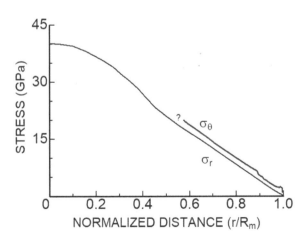

Figure 11.28 Preliminary elastic analysis carried out for the stress state of Mars under spherical symmetric condition

Figure 11.29 Views of slope failures observed in Mars

between the bedding plane and lower slope angle for the stability of rock slopes for jointed rock mass. The friction angles used in this graph are for those of basaltic rock joints on Earth.

11.5.3   Impact- and vibration-induced mass movements

Including Earth, planets and the Moon are bombarded by meteorites of various sizes from time to time. Furthermore, quakes (earthquakes, moonquakes, marsquakes) occur from time to time. These events result in shock waves and vibrations. Figure 11.32 show some examples of recent impacts on Mars and the displaced rock blocks in the Moon and Mars. A rock block displaced on Mars shown in Figure 11.33 is considered, and the conditions for its motion are analyzed herein.

The travel path length was 675 cm, and the inclination of the path was almost 23.5 degrees. As rocks at the site were inferred to be basaltic, the result of a dynamic friction experiment on a saw-cut discontinuity of basalt from Mt. Fuji was utilized (Fig. 11.34).

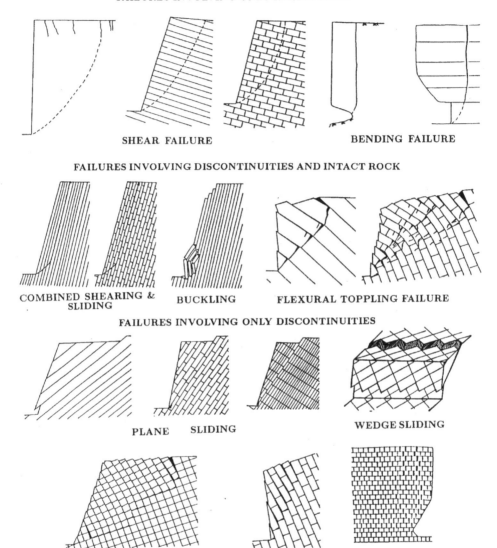

FAILURES INVOLVING ONLY INTACT ROCK

SHEAR FAILURE   BENDING FAILURE

FAILURES INVOLVING DISCONTINUITIES AND INTACT ROCK

COMBINED SHEARING & SLIDING   BUCKLING   FLEXURAL TOPPLING FAILURE

FAILURES INVOLVING ONLY DISCONTINUITIES

PLANE   SLIDING   WEDGE SLIDING

TOPPLING

Figure 11.30 Classification of rock slope failures
Source: From Aydan, 1989

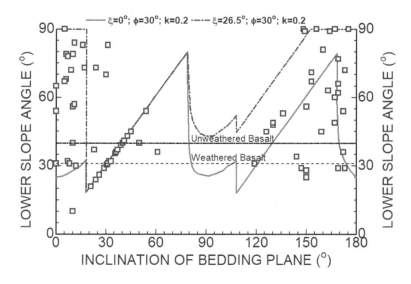

Figure 11.31 Relation between inclination of bedding plane and lower slope angle for rock slopes on Mars

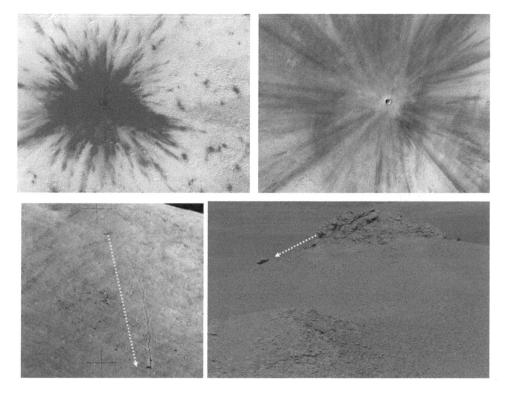

Figure 11.32 Images of impacts and rock block movements

Figure 11.33 Displaced rock block on Mars

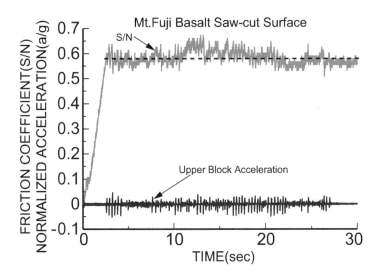

Figure 11.34 Dynamic friction test on a basaltic saw-cut discontinuity

The simple mechanical considerations imply that the maximum acceleration and velocity to displace the rock block may be derived as given here:

- Maximum acceleration

$$a > g_m \tan(\varphi - \alpha) \qquad (11.1)$$

- Maximum velocity

$$v_{max} > \sqrt{2g_m \tan(\varphi - \alpha)\delta} \qquad (11.2)$$

As seen from Figure 11.34, the friction angle of basaltic saw-cut surface from Mt. Fuji is greater than 30 degrees. If these values are used, the conditions to displace the rock block shown Figure 11.33 are estimated to be as:

- Maximum acceleration: 42.13 cm s^{-2}
- Maximum velocity: 238.5 cm s^{-1}

As understood from this simple example, rock dynamics would be necessary for the Moon and other planets.

11.5.4   Properties of rocks and discontinuities

As discussed in previous subsections, rocks from the Moon and other planets are quite similar to those of Earth. In rock mechanics and rock engineering, numerous experiments on rock, rock discontinuities and rock masses have been carried out, and many mechanical and engineering properties have been determined. Nevertheless, it is needless to say that we need to carry out similar experiments for extraterrestrial rocks probably in their environment. During the Apollo program, rock samples of more than 300 kg have been brought to Earth. However, no mechanical experiments have been carried out yet. The Mars rovers provided many images of rocks, rock masses and discontinuities. Back analyses of many images may provide some information on the mechanical analyses. For example, one may easily infer the frictional properties of stable and unstable blocks shown in Figure 11.35 with their base rocks. Furthermore, the back analyses of failed rock slopes and rock cliffs may further provide information on the strength properties of rock masses. Some of these analyses are reported by several researchers (e.g. Brunetti, 2014; Conway et al., 2011; Crosta et al., 2014; Lucchitta, 1979; Schultz, 2002; Neuffer and Schultz, 2006). The force and rate of advance during drilling operations by the Curiosity may also provide other data on the mechanical properties of intact rocks.

11.5.5   Sinkholes

Sinkholes are a severe problem on Earth. This problem has been often observed in karstic terrains, evaporitic formations and volcanic regions on Earth. The images from the Moon and Mars, as seen in Figure 11.36, clearly showed that sinkholes also formed on both the Moon and Mars. These sinkholes have been found in basaltic flow areas. Some of sinkholes may also be formed by impacts of meteorites.

Figure 11.35 Some images to infer frictional properties of discontinuities on Mars

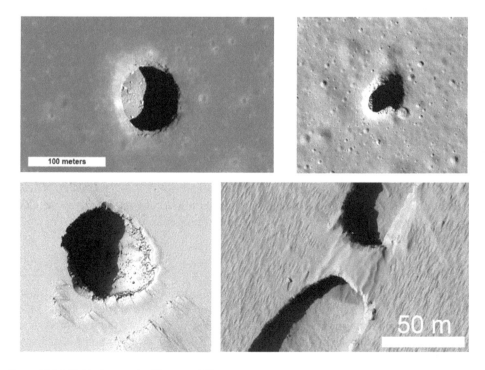

Figure 11.36 Sinkholes on the Moon and Mars

11.6 Conclusions and future studies

Although the environmental, fluids and gravitational conditions on Mars and other planets are different from those on Earth, the principles governing mechanical and engineering aspects of rocks on other planets should be quite similar to those developed for rocks and rock discontinuities of Earth. Therefore, the next generations of our discipline will definitely see its extension to the rocks, rock discontinuities, rock masses and rock engineering aspects of other planets. Future studies are necessary to explain:

1 The stress state of planets by empirical, analytical and computational methods, as well as *in-situ* measurements.
2 Why mountains are higher in Mars and lower in Venus compared with those on Earth.
3 Why tectonism is less pronounced on Mars and Venus.
4 The properties of rocks, discontinuities and rock masses of the Moon and other planets.
5 The effect of weathering on each planet.

Acknowledgments

The author gratefully acknowledges NASA and the people involved in the development and operation of the Moon, Mars and other planets exploration programs and processing and releasing their images on related websites.

References

Anderson, D.L., Miller, W.F., Latham, G.V., Nakamura, Y., Toksöz, M.N., Dainty, A.M., Duennebier, F.K., Lazarewicz, A.R., Kovach, R.L. & Knight, T.C.D. (1977) Seismology on Mars. *Journal of Geophysical Research*, 82(28), 4524–4546.
Aydan, Ö. (1989) *The Stabilisation of Rock Engineering Structures by Rockbolts*. Doctorate Thesis, Nagoya University, Faculty of Engineering.
Aydan, Ö. (2016) Some thoughts about rock mechanics aspects of Mars. *UNSW*, 3rd Off Earth Mining Forum, 2017OEMF.
Aydan, Ö., Ichikawa, Y., Shimizu, Y. & Murata, K. (1991) An integrated system for the stability of rock slopes. *The 5th Int. Conf. on Computer Methods and Advances in Geomechanics, Cairns, 1.* pp. 469–465.
Brunetti, M. (2014) *Statistics of Terrestrial and Extraterrestrial Landslides*. Doctoral Thesis, Universita Degli Studi di Perugia. DOI:10.13140/2.1.4107.3444. 109p.
Christiansen, E.H. & Spilker, B. (2018) *Exploring the Planets*. Published by Prentice Hall in 1990, 1995. http://explanet.info.
Conway, S.J., Balme, M.R., Lamb, M.P., Towner, M.C. & Murray, J.B. (2011) Enhanced runout and erosion by overland flow under subfreezing and low pressure conditions: Experiments and application to Mars. *Icarus*, 211(1), 443–457.
Crosta, G.B., Utili, S., Blasio, F.V. & Riccardo Castellanza, R. (2014) Reassessing rock mass properties and slope instability triggering conditions in Valles Marineris, Mars. *Earth and Planetary Science Letters*, 388, 329–342.
Irvine, T. (2002) Moonquakes. March, Newsletter. Vibrationdata.com. pp. 4–13.
Johnston, D.H. & Toksöz, M.N. (1977) Internal structure and properties of Mars. *Icarus*, 32, 73–84.
Latham, G., Ewing, M., Press, F., Sutton, G., Dorman, J., Nakamura, Y., Lammlein, D., Duennebier, F. & Toksöz, N. (1971). Moonquakes, *Science*, 174, 687–692.

Lucchitta, B.K. (1979) Landslides in Vallis Marineris, Mars. *Journal of Geophysical Research*, 84, 8097–8113.

Nakamura, Y. (2003) New identification of deep moonquakes in the Apollo lunar seismic data. *Physics of Earth and Planetary Interiors*, 139, 197–205.

NASA (2008) *Exploration: NASA's Plans to Explore the Moon, Mars and Beyond.* www.nasa.gov/.

NASA: *Images.* https://mars.nasa.gov/mer/gallery/images.html (Accessed October 2019).

Neuffer, P.D. & Schultz, R.A. (2006) Mechanisms of slope failure in Valles Marineris, Mars. *Quarterly Journal of Engineering Geology and Hydrogeology*, 39(3), 227–240.

Schultz, R.A. (2002) Stability of rock slopes in Valles Marineris, Mars. *Geophysical Research Letters*, 29, 1932.

Scott, David H., and Carr, Michael H. (1978). Geologic Map of Mars: U.S. Geological Survey Investigations Series I-1083, scale 1:25000000, http://pubs.er.usgs.gov/publication/i1083

Sohl, F. & Spohn, T. (1997) The interior structure of Mars: Implications from SNC meteorites. *Journal Geophysical Research and Planets*, 102(E1), 1613–1635.

Steinberger, B., Werner, S.C. & Torsvik, T.H. (2010) Deep versus shallow origin of gravity anomalies, topography and volcanism on Earth, Venus and Mars. *Icarus*, 207, 564–577.

Toksöz, M.N. & Hsui, A.T. (1978) Thermal history and evolution of Mars. *Icarus*, 34(3), 537–547.

Toksöz, M.N., Goins, N.R. & Cheng, C.H. (1977) Moonquakes: Mechanisms and relation to tidal stresses. *Science*, 196, 979–981.

Watters, T.R. (1995) *Planets: A Smithsonian Guide.* MacMillan, New York, USA. 256p.

Appendices

Definitions of scalars, vectors and tensors and associated operations

A1.1 Scalar

Scalar is a quantity having a magnitude, and it remains the same irrespective of direction. It is defined as a rank-0 tensor. Examples of scalar quantities are volume, density, mass, temperature, energy, pressure. It can be easily added, subtracted, multiplied and divided.

A1.2 Vector

Scalar is a quantity having both magnitude and direction, as illustrated in Figure A1.1. It is defined as a rank-1 tensor. Examples of vectors are force, velocity, displacement, acceleration, moment.

A1.3 Vector operations

A1.3.1 Addition and subtraction

The addition and subtraction of vectors obey the geometrical parallelogram rule, and they are illustrated as shown in Figure A1.2 and are mathematically expressed as follows:

$$\mathbf{c} = \mathbf{a} + \mathbf{b} \tag{A1.1}$$

$$\mathbf{d} = \mathbf{a} - \mathbf{b} \tag{A1.2}$$

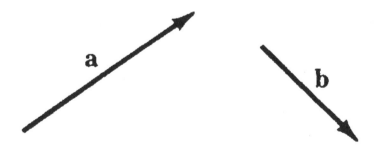

Figure A1.1 Illustration of vectors

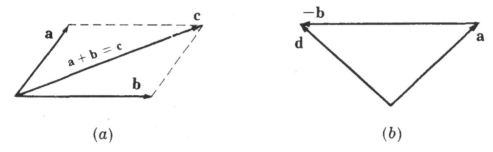

Figure A1.2 Geometrical illustration of addition and subtraction of vectors

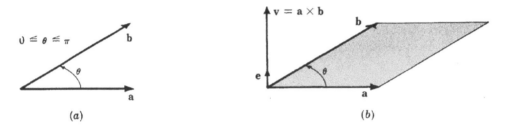

Figure A1.3 Illustration of cross product of two vectors

A1.3.2 Dot product

The dot product of two vectors is defined as follows:

$$\mathbf{a} \cdot \mathbf{b} = \|\mathbf{a}\|\|\mathbf{b}\| \cos \theta \tag{A1.3}$$

The preceding quantity is geometrically interpreted as the projection of vector ($\mathbf{a}$) in the direction of vector ($\mathbf{b}$).

A1.3.3 Cross product

The cross product of two vectors is defined as (Fig. A1.3).

$$\mathbf{a} \times \mathbf{b} = \mathbf{c} \tag{A1.4}$$

The magnitude of this quantity corresponds to the area bounded by vectors ($\mathbf{a}$) and ($\mathbf{b}$), and the direction of resulting vector ($\mathbf{c}$) is perpendicular to the plane constituted by vectors ($\mathbf{a}$) and ($\mathbf{b}$):

$$\|\mathbf{c}\| = \|\mathbf{a}\|\|\mathbf{b}\| \sin\theta \tag{A1.5}$$

A1.3.4 Unit vector

Unit vector is defined as a vector whose magnitude is 1. This vector is utilized to define the vectorial quantities.

A1.3.5 Coordinate systems and base vectors

Fundamentally there are different coordinate systems such as Cartesian, cylindrical, spherical coordinate systems. (See Figure A1.4.) Here, an orthogonal coordinate system (x_1, x_2, x_3) is considered. Three unit vectors (e_1, e_2, e_3) are assumed to be parallel to the coordinate axes, and they are called base-vectors. For the chosen coordinate system, the following relations hold for dot product operations:

$$\begin{aligned}
&\mathbf{e}_1 \cdot \mathbf{e}_1 = 1, \quad \mathbf{e}_2 \cdot \mathbf{e}_2 = 1, \quad \mathbf{e}_3 \cdot \mathbf{e}_3 = 1, \\
&\mathbf{e}_1 \cdot \mathbf{e}_2 = 0, \quad \mathbf{e}_2 \cdot \mathbf{e}_3 = 0, \quad \mathbf{e}_3 \cdot \mathbf{e}_1 = 0
\end{aligned} \tag{A1.6}$$

As for the cross-product, the relations hold among base vectors:

$$\begin{aligned}
&\mathbf{e}_1 \times \mathbf{e}_1 = \mathbf{0}, \quad \mathbf{e}_2 \times \mathbf{e}_2 = \mathbf{0}, \quad \mathbf{e}_3 \times \varepsilon_3 = \mathbf{0}, \\
&\mathbf{e}_1 \times \mathbf{e}_2 = \mathbf{e}_3, \quad \mathbf{e}_2 \times \mathbf{e}_3 = \mathbf{e}_1, \quad \mathbf{e}_3 \times \mathbf{e}_1 = \mathbf{e}_2, \\
&\mathbf{e}_2 \times \mathbf{e}_1 = -\mathbf{e}_3, \quad \mathbf{e}_3 \times \mathbf{e}_2 = -\mathbf{e}_1, \quad \mathbf{e}_1 \times \mathbf{e}_3 = -\mathbf{e}_2
\end{aligned} \tag{A1.7}$$

In the chosen coordinate system, vector a can be expressed as follow

$$\mathbf{a} = a_1 \mathbf{e}_1 + a_2 \mathbf{e}_2 + a_3 \mathbf{e}_3 = \sum_{i=1}^{3} a_i \mathbf{e}_i = a_i \mathbf{e}_i \quad (i = 1, 2, 3) \tag{A1.8}$$

The component of vector **a** can be give as follows:

$$a_1 = \mathbf{a} \cdot \mathbf{e}_1, \quad a_2 = \mathbf{a} \cdot \mathbf{e}_2, \quad a_3 = \mathbf{a} \cdot \mathbf{e}_3, \quad a_i = \mathbf{a} \cdot \mathbf{e}_i \tag{A1.9}$$

A1.4 Vector operations on a Cartesian coordinate system

A1.4.1 Addition and subtraction

Addition and subtraction vectors may be given in the following form using base vectors and their components:

$$\begin{aligned}
\mathbf{a} \pm \mathbf{b} &= (a_1 \pm b_1)\mathbf{e}_1 + (a_2 \pm b_2)\mathbf{e}_2 + (a_3 \pm b_3)\mathbf{e}_3 = \sum_{i=1}^{3}(a_i \pm b_i)\mathbf{e}_i \\
&= (a_i \pm b_i)\mathbf{e}_i \quad (i = 1, 2, 3)
\end{aligned} \tag{A1.10}$$

A1.4.2 Dot product

The dot product of two vectors may be given in the following form using base vectors and their components:

$$\mathbf{a} \cdot \mathbf{b} = a_1 b_1 + a_2 b_2 + a_3 b_3 = \sum_{i=1}^{3} a_i b_i = a_i b_i \quad (i = 1, 2, 3) \tag{A1.11}$$

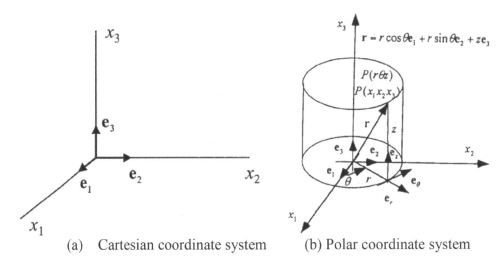

(a)   Cartesian coordinate system   (b) Polar coordinate system

Figure A1.4 Illustration of some of coordinate systems

A1.4.3   Cross product

The cross product of two vectors may be given in the following form using base vectors and their components:

$$\mathbf{a} \times \mathbf{b} = (a_2 b_3 - a_3 b_2)\,\mathbf{e}_1 + (a_3 b_1 - a_1 b_3)\,\mathbf{e}_2 + (a_1 b_2 - a_2 b_1)\,\mathbf{e}_3 \qquad (A1.12)$$

A1.5   Tensors of rank *n*

Generally, tensors of rank *n* have magnitude and directional components. Although it is difficult to visualize geometrically, examples are stress, strain, elasticity, viscosity relations.

A1.5.1   Definition of tensors of rank n

The rank or order of a tensor is defined in terms of the number of independent (nonrepeated) base vectors and magnitude. Examples are as follows:

1   Tensor of second order

$$\mathbf{D} = D_{11}\mathbf{e}_1\mathbf{e}_1 + D_{12}\mathbf{e}_1\mathbf{e}_2 + D_{13}\mathbf{e}_1\mathbf{e}_3 + D_{21}\mathbf{e}_2\mathbf{e}_1 + D_{22}\mathbf{e}_2\mathbf{e}_2 + D_{23}\mathbf{e}_2\mathbf{e}_3 + D_{31}\mathbf{e}_3\mathbf{e}_1$$
$$+ D_{32}\mathbf{e}_3\mathbf{e}_2 + D_{33}\mathbf{e}_3\mathbf{e}_3 \qquad\qquad (A1.13a)$$

or

$$\mathbf{D} = D_{11}\mathbf{e}_1\mathbf{e}_1 + D_{12}\mathbf{e}_1 \otimes \mathbf{e}_2 + D_{13}\mathbf{e}_1 \otimes \mathbf{e}_3 + D_{21}\mathbf{e}_2 \otimes \mathbf{e}_1 + D_{22}\mathbf{e}_2 \otimes \mathbf{e}_2 + D_{23}\mathbf{e}_2 \otimes \mathbf{e}_3$$
$$+ D_{31}\mathbf{e}_3 \otimes \mathbf{e}_1 + D_{32}\mathbf{e}_3 \otimes \mathbf{e}_2 + D_{33}\mathbf{e}_3 \otimes \mathbf{e}_3 \qquad\qquad (A1.13b)$$

or

$$\mathbf{D} = \sum_{i=1}^{3}\sum_{j=1}^{3}D_{ij}\mathbf{e}_i\mathbf{e}_j = \sum_{i=1}^{3}\sum_{j=1}^{3}D_{ij}\mathbf{e}_i \otimes \mathbf{e}_j = D_{ij}\mathbf{e}_i\mathbf{e}_j = D_{ij}\mathbf{e}_i \otimes \mathbf{e}_j \qquad \text{(A1.13c)}$$

$$(i = 1, 2, 3), (j = 1, 2, 3)$$

where $\otimes$ is called the tensor product, and $\mathbf{e}_i\mathbf{e}_j = \mathbf{e}_i \otimes \mathbf{e}_j$ are fundamentally the same and are called dyad.

2   Kronecker Delta Tensor

The Kronecker delta tensor is known as the identity tensor, whose normal components have the value of 1 and are given as follows:

$$\mathbf{I} = \delta_{ij}\,\mathbf{e}_i\mathbf{e}_j \quad \delta_{ij} = 1\, i = j \quad \delta_{ij} = 0\, i \neq j \qquad \text{(A1.14)}$$

3   Tensor of rank 3

$$\mathbf{C} = \sum_{i=1}^{3}\sum_{j=1}^{3}\sum_{k=1}^{3}C_{ijk}\mathbf{e}_i\mathbf{e}_j\mathbf{e}_k = \sum_{i=1}^{3}\sum_{j=1}^{3}\sum_{k=1}^{3}C_{ijk}\,\mathbf{e}_i \otimes \mathbf{e}_j \otimes \mathbf{e}_k \qquad \text{(A1.15)}$$

$$\mathbf{C} = C_{ijk}\mathbf{e}_i\mathbf{e}_j\mathbf{e}_k = C_{ijk}\mathbf{e}_i \otimes \mathbf{e}_j \otimes \mathbf{e}_k \quad (i = 1, 2, 3), (j = 1, 2, 3), (k = 1, 2, 3) \quad \text{(A1.16)}$$

4   Tensor of rank 4

$$\mathbf{E} = \sum_{i=1}^{3}\sum_{j=1}^{3}\sum_{k=1}^{3}\sum_{l=1}^{3}E_{ijkl}\mathbf{e}_i\mathbf{e}_j\mathbf{e}_k\mathbf{e}_l = \sum_{i=1}^{3}\sum_{j=1}^{3}\sum_{k=1}^{3}\sum_{l=1}^{3}E_{ijkl}\mathbf{e}_i \otimes \mathbf{e}_j \otimes \mathbf{e}_k \otimes \mathbf{e}_l \quad \text{(A1.17)}$$

$$\mathbf{E} = E_{ijkl}\mathbf{e}_i\mathbf{e}_j\mathbf{e}_k\mathbf{e}_l = E_{ijkl}\mathbf{e}_i \otimes \mathbf{e}_j \otimes \mathbf{e}_k \otimes \mathbf{e}_l \quad (i = 1, 2, 3), (j = 1, 2, 3), \quad \text{(A1.18)}$$

$$(k = 1, 2, 3), (l = 1, 2, 3)$$

A1.5.2   Tensor operations

A1.5.2.1   Multiplication of a tensor with a scalar

The multiplication of a tensor with scalar results in a tensor with the same order, whose components are magnified by the value of scalar value as given here:

$$\mathbf{E} = \lambda\mathbf{D} \qquad \text{(A1.19)}$$

A1.5.2.2   Operation of tensor with vectors

1   The dot product of vector and tensor of rank 2 can be given in the following form:

$$\mathbf{c} = \mathbf{a} \cdot \mathbf{D} = (a_i\mathbf{e}_i) \cdot (D_{ij}\mathbf{e}_i\mathbf{e}_j) = a_iD_{ij}(\mathbf{e}_i \cdot \mathbf{e}_i)\mathbf{e}_j = a_iD_{ij}\mathbf{e}_j \qquad \text{(A1.20)}$$

$$\mathbf{d} = \mathbf{D} \cdot \mathbf{a} = (D_{ij}\mathbf{e}_i\mathbf{e}_j) \cdot (a_j\mathbf{e}_j) = D_{ij}a_j(\mathbf{e}_i)\mathbf{e}_j \cdot \mathbf{e}_j = D_{ij}a_j\mathbf{e}_i \qquad \text{(A1.21)}$$

Fundamentally, the resulting products are vectors, having different quantities unless the tensor of rank 2 is symmetric.

2   The dot product of two tensors of rank 2 is written in the following form:

$$\mathbf{C} = \mathbf{D} \cdot \mathbf{E} = (D_{ij}\mathbf{e}_i\mathbf{e}_j) \cdot (E_{jk}\mathbf{e}_j\mathbf{e}_k) = D_{ij}E_{jk}\mathbf{e}_i(\mathbf{e}_j \cdot \mathbf{e}_j)\mathbf{e}_k = D_{ij}E_{jk}\mathbf{e}_i\mathbf{e}_k \qquad (A1.22)$$

The rank of the resulting tensorial quantity is 2.

3   The double dot product of two tensors of rank 2 is written in the following form:

$$W = \mathbf{D} : \mathbf{E} = (D_{ij}\mathbf{e}_i\mathbf{e}_j) : (E_{ij}\mathbf{e}_i\mathbf{e}_j) = D_{ij}E_{ij}(\mathbf{e}_i \cdot \mathbf{e}_i)(\mathbf{e}_j \cdot \mathbf{e}_j) = D_{ij}E_{ij} \qquad (A1.23)$$

The resulting quantity is a scalar.

4   Tensor product of two tensors of rank 2

The resulting tensorial quantity of the tensor product of two tensors of rank 2 are written in the following form, and its rank is 4.

$$\mathbf{F} = \mathbf{D} \otimes \mathbf{E} = (D_{ij}\mathbf{e}_i\mathbf{e}_j) \otimes (E_{kl}\mathbf{e}_k\mathbf{e}_l) = D_{ij}E_{kl}\mathbf{e}_i\mathbf{e}_j\mathbf{e}_k\mathbf{e}_l = F_{ijkl}\mathbf{e}_i\mathbf{e}_j\mathbf{e}_k\mathbf{e}_l \qquad (A1.24)$$

A1.6   Matrix representation of tensors

Base vectors may not be utilized when vectors or tensors are expressed. Herein, matrix operations are introduced to represent some vectorial or tensorial operations.

A1.6.1   Matrix representation of vectors

A vector can be represented using either a horizontal (1×3) or a vertical matrix (3×1) as given here:

$$a = [a_1, a_2, a_3], \quad a = \begin{bmatrix} a_1 \\ a_2 \\ a_3 \end{bmatrix} \qquad (A1.25)$$

The dot product of two vectors may be represented using the matrix operations as given here:

$$\mathbf{a} \cdot \mathbf{b} = [a_1, a_2, a_3] \begin{bmatrix} b_1 \\ b_2 \\ b_3 \end{bmatrix} \qquad (A1.26)$$

A1.6.2   Matrix representation of tensors

The tensor of rank 2 may be given in a matrix form as:

$$\mathbf{A} = \begin{bmatrix} A_{11} & A_{12} & A_{13} \\ A_{21} & A_{22} & A_{23} \\ A_{31} & A_{32} & A_{33} \end{bmatrix} \qquad (A1.27)$$

Similarly, the Kronecker delta tensor can be written as:

$$\mathbf{I} = \begin{bmatrix} 1 & 0 & 0 \\ 0 & 1 & 0 \\ 0 & 0 & 1 \end{bmatrix} \tag{A1.28}$$

The dot product of the vector and tensor of rank 2 is expressed using the matrix operations as follows:

$$\mathbf{a} \cdot \mathbf{A} = [a_1, a_2, a_3] \begin{bmatrix} A_{11} & A_{12} & A_{13} \\ A_{21} & A_{22} & A_{23} \\ A_{31} & A_{32} & A_{33} \end{bmatrix} \tag{A1.29}$$

$$\mathbf{A} \cdot \mathbf{a} = \begin{bmatrix} A_{11} & A_{12} & A_{13} \\ A_{21} & A_{22} & A_{23} \\ A_{31} & A_{32} & A_{33} \end{bmatrix} \begin{bmatrix} a_1 \\ a_2 \\ a_3 \end{bmatrix} \tag{A1.30}$$

A1.7 Coordinate transformation

Let us consider the position of point P in two coordinates systems (oxy, $ox'y'$). (See Figure A1.5.) From the geometry, the following relations can be obtained as follows:

$$\begin{bmatrix} x'_1 \\ x'_2 \end{bmatrix} = \begin{bmatrix} \cos\theta & \sin\theta \\ -\sin\theta & \cos\theta \end{bmatrix} \begin{bmatrix} x_1 \\ x_2 \end{bmatrix} \tag{A1.31}$$

The inverse of the preceding relation can be shown to be:

$$\begin{bmatrix} x_1 \\ x_2 \end{bmatrix} = \begin{bmatrix} \cos\theta & -\sin\theta \\ \sin\theta & \cos\theta \end{bmatrix} \begin{bmatrix} x'_1 \\ x'_2 \end{bmatrix} \tag{A1.32}$$

Let us replace the components of the matrix in the following manner:

$$\beta_{11} = \cos\theta, \quad \beta_{12} = \sin\theta, \quad \beta_{21} = -\sin\theta, \quad \beta_{22} = \cos\theta \tag{A1.33}$$

Equation (A1.31) and Equation (A1.32) may be rewritten as follows:

$$x'_i = \beta_{ij} x_j \tag{A1.34}$$

$$x_j = \beta_{jk} x'_k \tag{A1.35}$$

If Equation (A1.35) is inserted into Equation (A1.34), the following relation is obtained:

$$x'_i = \beta_{ij} \beta_{jk} x_{k'} = \delta_{ik} x'_k \tag{A1.36}$$

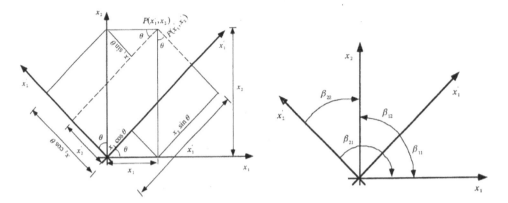

Figure A1.5 Coordinate systems

The matrix operation $\beta_{ij}\beta_{jk} = \delta_{ik}$ fundamentally corresponds to identity tensor and specifically is given as:

$$
\begin{bmatrix} \cos\theta & \sin\theta \\ -\sin\theta & \cos\theta \end{bmatrix} \begin{bmatrix} \cos\theta & -\sin\theta \\ \sin\theta & \cos\theta \end{bmatrix} = \begin{bmatrix} 1 & 0 \\ 0 & 1 \end{bmatrix} \tag{A1.37}
$$

A1.8   Derivation

A1.8.1   Derivative of a scalar function

The derivative of a scalar function is given here and is called directional derivation:

$$
df = \frac{\partial f}{\partial x_1} dx_1 + \frac{\partial f}{\partial x_2} dx_2 + \frac{\partial f}{\partial x_3} dx_3 \tag{A1.38}
$$

This expression can also be written as:

$$
df = \left(\frac{\partial f}{\partial x_1}\,\mathbf{e}_1 + \frac{\partial f}{\partial x_2}\,\mathbf{e}_2 + \frac{\partial f}{\partial x_3}\,\mathbf{e}_3 \right) \cdot (dx_1\mathbf{e}_1 + dx_2\mathbf{e}_1 + dx_3\mathbf{e}_3) \tag{A1.39}
$$

with the following definitions:

$$
\nabla = \frac{\partial}{\partial x_1}\,\mathbf{e}_1 + \frac{\partial}{\partial x_2}\,\mathbf{e}_2 + \frac{\partial}{\partial x_3}\,\mathbf{e}_3 \tag{A1.40}
$$

$$
d\mathbf{x} = dx_1\mathbf{e}_1 + dx_2\mathbf{e}_1 + dx_3\mathbf{e}_3 \tag{A1.41}
$$

Equation (A1.39) may also be written in the following form:

$$df = (\nabla f) \cdot d\mathbf{x} \tag{A1.42}$$

AI.8.2 Divergence

The dot product between the directional derivation operator (∇) and a vector ($\mathbf{a}$) results in the following form and is interpreted as the divergence of vector ($\mathbf{a}$):

$$div\,\mathbf{a} = (\nabla \cdot \mathbf{a} = \frac{\partial}{\partial x_1}\mathbf{e}_1 + \frac{\partial}{\partial x_2}\mathbf{e}_2 + \frac{\partial}{\partial x_3}\mathbf{e}_3) \cdot (a_1\mathbf{e}_1 + a_2\mathbf{e}_1 + a_3\mathbf{e}_3) = \frac{\partial a_1}{\partial x_1} + \frac{\partial a_2}{\partial x_2} + \frac{\partial a_3}{\partial x_3}$$

$$\tag{A1.42}$$

AI.8.3 Rotation

The cross product between directional the derivation operator (∇) and a vector ($\mathbf{a}$) results in the following form and is interpreted as the rotation of vector ($\mathbf{a}$):

$$curl\,\mathbf{a} = \nabla \times \mathbf{a} = \left[\frac{\partial a_3}{\partial x_2} - \frac{\partial a_2}{\partial x_3}\right]\mathbf{e}_1 + \left[\frac{\partial a_1}{\partial x_3} - \frac{\partial a_3}{\partial x_1}\right]\mathbf{e}_2 + \left[\frac{\partial a_2}{\partial x_1} - \frac{\partial a_1}{\partial x_2}\right]\mathbf{e}_3 \tag{A1.43}$$

AI.8.4 Gradient of a vector: second-order tensor

The tensor product between the directional derivation operator (∇) and a vector ($\mathbf{a}$) can be carried out in two ways and results in the following form. It is interpreted as the gradient of vector ($\mathbf{a}$):

$$\mathbf{D} = \nabla \mathbf{a} = \frac{\partial}{\partial x_1}\mathbf{e}_1 + \frac{\partial}{\partial x_2}\mathbf{e}_2 + \frac{\partial}{\partial x_3}\mathbf{e}_3\,)(a_1\mathbf{e}_1 + a_2\mathbf{e}_1 + a_3\mathbf{e}_3)$$

$$= \frac{\partial a_1}{\partial x_1}\mathbf{e}_1\mathbf{e}_1 + \frac{\partial a_2}{\partial x_1}\mathbf{e}_1\mathbf{e}_2 + \frac{\partial a_3}{\partial x_1}\mathbf{e}_1\mathbf{e}_3 + \frac{\partial a_1}{\partial x_2}\mathbf{e}_2\mathbf{e}_1 + \frac{\partial a_2}{\partial x_2}\mathbf{e}_2\mathbf{e}_2 + \frac{\partial a_3}{\partial x_2}\mathbf{e}_2\mathbf{e}_3 + \frac{\partial a_1}{\partial x_3}\mathbf{e}_3\mathbf{e}_1 + \frac{\partial a_2}{\partial x_3}\mathbf{e}_3\mathbf{e}_2 + \frac{\partial a_3}{\partial x_3}\mathbf{e}_3\mathbf{e}_3$$

$$= \frac{\partial a_j}{\partial x_i}\mathbf{e}_i\mathbf{e}_j = a_{j,i}\mathbf{e}_i\mathbf{e}_j \tag{A1.44a}$$

$$\mathbf{E} = \mathbf{a}\nabla = (a_1\mathbf{e}_1 + a_2\mathbf{e}_1 + a_3\mathbf{e}_3)(\frac{\partial}{\partial x_1}\mathbf{e}_1 + \frac{\partial}{\partial x_2}\mathbf{e}_2 + \frac{\partial}{\partial x_3}\mathbf{e}_3)$$

$$\frac{\partial a_1}{\partial x_1}\mathbf{e}_1\mathbf{e}_1 + \frac{\partial a_1}{\partial x_2}\mathbf{e}_1\mathbf{e}_2 + \frac{\partial a_1}{\partial x_3}\mathbf{e}_1\mathbf{e}_3 + \frac{\partial a_2}{\partial x_1}\mathbf{e}_2\mathbf{e}_1 + \frac{\partial a_2}{\partial x_2}\mathbf{e}_2\mathbf{e}_2 + \frac{\partial a_2}{\partial x_3}\mathbf{e}_2\mathbf{e}_3 + \frac{\partial a_3}{\partial x_1}\mathbf{e}_3\mathbf{e}_1 + \frac{\partial a_3}{\partial x_2}\mathbf{e}_3\mathbf{e}_2 + \frac{\partial a_3}{\partial x_3}\mathbf{e}_3\mathbf{e}_3$$

$$= \frac{\partial a_i}{\partial x_j}\mathbf{e}_i\mathbf{e}_j = a_{i,j}\mathbf{e}_i\mathbf{e}_j \tag{A1.44b}$$

A1.8.5 Divergence of a tensor (second-order tensor)

The dot product between the directional derivation operator (∇) and a tensor $(\mathbf{D})$ of rank 2 results in the following form and is interpreted as the divergence of a tensor $(\mathbf{D})$ of rank 2:

$$\mathbf{f} = (\nabla \cdot \mathbf{D} = \frac{\partial}{\partial x_1}\mathbf{e}_1 + \frac{\partial}{\partial x_2}\mathbf{e}_2 + \frac{\partial}{\partial x_3}\mathbf{e}_3) \cdot (D_{11}\mathbf{e}_1\mathbf{e}_1 + D_{12}\mathbf{e}_1\mathbf{e}_2 + D_{13}\mathbf{e}_1\mathbf{e}_3 +$$

$$D_{21}\mathbf{e}_2\mathbf{e}_1 + D_{22}\mathbf{e}_2\mathbf{e}_2 + D_{23}\mathbf{e}_2\mathbf{e}_3 + D_{31}\,\mathbf{e}_3\mathbf{e}_1 + D_{32}\mathbf{e}_3\mathbf{e}_2 + D_{33}\mathbf{e}_3\mathbf{e}_3) \tag{A1.45a}$$

or

$$(\frac{\partial}{\partial x_j}\,\mathbf{e}_j) \cdot (D_{ji}\mathbf{e_j}\mathbf{e_i}) = \frac{\partial D_{ji}}{\partial x_j}\mathbf{e}_i = D_{ji,j}\mathbf{e}_i \tag{A1.45b}$$

Stress analysis

A2.1 Definition of stress vector

Stress vector is defined as the limit of an infinitesimal force acting over an infinitesimal area with a unit normal vector as given here (Fig. A2.1(a)):

$$\mathbf{t}^{(\mathbf{n})} = \lim_{\Delta S \to 0} \frac{\Delta \mathbf{f}}{\Delta S} = \frac{d\mathbf{f}}{dS} \tag{A2.1}$$

Newton's action-and-reaction law requires at a given surface within a body of equilibrium (Fig. A2.1(b)):

$$\mathbf{t}^{(\mathbf{n})} - \mathbf{t}^{(-\mathbf{n})} = \mathbf{0} \tag{A2.2}$$

The stress vector is also known as the traction vector.

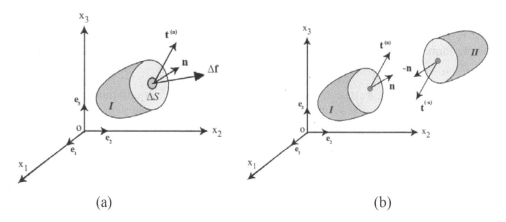

(a)   (b)

Figure A2.1 Illustration of (a) traction vector, (b) unit normal vectors

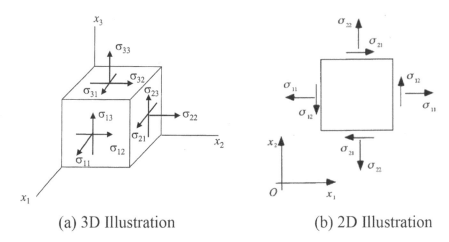

(a) 3D Illustration   (b) 2D Illustration

Figure A2.2 Illustration of stress tensor components

A2.2   Stress tensor

A stress vector acting on a cubic body results in nine components of the second-order stress tensor (Fig. A2.2). In two-dimensional space, the stress tensor has four components. It is given as using tensorial notation:

$$\boldsymbol{\sigma} = \sigma_{ij}\mathbf{e}_i\mathbf{e}_j \tag{A2.3}$$

The first and second subscripts of a component of stress tensor correspond to the surface and the axis, respectively.

A2.3   Relationship between stress vector and stress tensor: Cauchy relation

Cauchy states that the stress vector and stress tensor can be related to each other in the following form:

$$\mathbf{t} = \mathbf{n} \cdot \boldsymbol{\sigma} = \sigma_{ji}n_j\mathbf{e}_i \tag{A2.4}$$

To derive this relation, let us consider a two-dimensional body in equilibrium. The effect of the top part is taken into account as the traction acting on the plane with surface area Δs (see Figure A2.2.).

The unit normal and traction vectors and stress tensor can be given as:

$$\mathbf{n} = n_1\mathbf{e}_1 + n_2\mathbf{e}_2 \tag{A2.5}$$

$$\mathbf{t} = t_1\mathbf{e}_1 + t_2\mathbf{e}_2 \tag{A2.6}$$

$$\boldsymbol{\sigma} = \sigma_{11}\mathbf{e}_1\mathbf{e}_1 + \sigma_{12}\mathbf{e}_1\mathbf{e}_2 + \sigma_{21}\mathbf{e}_2\mathbf{e}_1 + \sigma_{22}\mathbf{e}_2\mathbf{e}_2 \tag{A2.7}$$

where

$$n_1 = \cos\theta, n_2 = \cos(90 - \theta) = \sin\theta$$

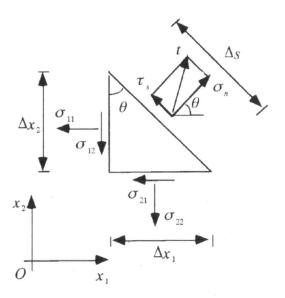

Figure A2.3

The force equilibrium in directions x_1 and x_2 can be written as:

x_1-direction

$$\sum F_{x_1} = t_1 \Delta s - \sigma_{11} n_1 \Delta s + \sigma_{21} n_2 \Delta s = 0, \quad t_1 = \sigma_{11} n_1 + \sigma_{21} n_2 \qquad \text{(A2.8a)}$$

x_2-direction

$$\sum F_{x_2} = t_2 \Delta s - \sigma_{12} n_1 \Delta s + \sigma_{22} n_2 \Delta s = 0, \quad t_2 = \sigma_{12} n_1 + \sigma_{22} n_2 \qquad \text{(A2.8b)}$$

Utilizing Equations (A2.5) to (A2.7) together with Equation (A2.8), the following relation can be written:

$$\mathbf{t} = (n_1 \mathbf{e}_1 + n_2 \mathbf{e}_2) \cdot (\sigma_{11} \mathbf{e}_1 \mathbf{e}_1 + \sigma_{12} \mathbf{e}_1 \mathbf{e}_2 + \sigma_{21} \mathbf{e}_2 \mathbf{e}_1 + \sigma_{22} \mathbf{e}_2 \mathbf{e}_2) \qquad \text{(A2.9a)}$$

or

$$\mathbf{t} = \mathbf{n} \cdot \boldsymbol{\sigma} = \sigma_{ji} n_j \mathbf{e}_i \quad (j = 1, 2, i = 1, 2) \qquad \text{(A2.9b)}$$

Thus Equation (A2.9) corresponds to Cauchy's relation.

Normal and shear components of traction vector on the plane can be given by the following relations:

$$\sigma_N = \mathbf{t} \cdot \mathbf{n} \qquad \text{(A2.11)}$$

$$\sigma_S = \mathbf{t} \cdot \mathbf{s} \qquad \text{(A2.12)}$$

Where

$$\mathbf{s} = s_1 \mathbf{e}_1 + s_2 \mathbf{e}_2, \ s_1 = -\sin\theta, \ s_2 = \cos\theta \qquad \text{(A2.13)}$$

A2.4   Stress transformation

Stress transformation becomes necessary when stress tensors are to be related to each other in different coordinate systems. This transformation law may be derived by requiring the stress vector to be the same independent of the coordinate system. Let us introduce coordinate systems $ox_1x_2x_3$ and $ox_1'x_2'x_3'$ together with stress tensors (σ_{ij} and σ_{km}'), as shown in Figure A2.4.

Thus, the following relation may be written:

$$\mathbf{t} = \sigma_{ji}n_j\mathbf{e}_i = \sigma_{km}'n_k'\mathbf{e}_m' \tag{A2.14}$$

Taking the dot products of the both sides of Equation (A2.14) with $\mathbf{e}_m'$ yields:

$$\sigma_{ji}n_j(\mathbf{e}_m' \cdot \mathbf{e}_i) = \sigma_{km}'n_k' \quad \text{or} \quad \sigma_{ji}n_j\beta_{mi} = \sigma_{km}'n_k' \tag{A2.15}$$

The unit normal vector on an undashed coordinate system can be related to a dashed coordinate system through the transformation law:

$$n_j = \beta_{kj}n_k' \tag{A2.16}$$

Inserting Equation (16) into Equation (15) yields the following relation:

$$(\sigma_{ji}\beta_{kj}\beta_{mi} - \sigma_{km}')n_k' = 0 \tag{A2.17}$$

As n_k' is arbitrary chosen, Equation (17) requires the following identity:

$$\sigma_{km}' = \sigma_{ji}\beta_{kj}\beta_{mi} \tag{A2.18}$$

Equation (18) may also be represented in the matrix form as follows:

$$[\sigma'] = [\beta][\sigma][\beta]^T \tag{A2.19}$$

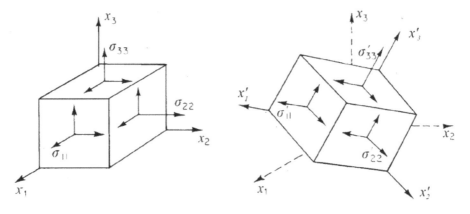

Figure A2.4 Stress components in two coordinate systems

Using a similar procedure, the stress tensor in undashed system can be related to that in dashed system as follows:

$$\sigma_{ij} = \sigma'_{km}\beta_{jk}\beta_{im} \tag{A2.20}$$

In matrix form, it is written as:

$$[\sigma] = [\beta]^T[\sigma'][\beta] \tag{A2.21}$$

A2.5   Principal stresses, stress invariants

The stress tensor can be represented by three orthogonal components as shear stress components disappear at the planes on which principal stresses act. Thus, principal stresses can be related to the stress tensor on a given coordinate system by requiring the stress vectors for both situations to be equivalent:

$$\mathbf{t} = \sigma_{ji}n_j\mathbf{e}_i = \sigma n_i\mathbf{e}_i \quad \text{or} \quad (\sigma_{ji}n_j - \sigma n_i)\mathbf{e}_i = \mathbf{0} \tag{A2.22}$$

with the following relation:

$$n_i = \delta_{ij}n_j \tag{A2.23}$$

Equation (A2.22) can be rewritten as:

$$(\sigma_{ji} - \sigma\delta_{ij})n_j\mathbf{e}_i = \mathbf{0} \tag{A2.24}$$

Equation (A2.24) requires the following condition to be satisfied:

$$|\sigma_{ji} - \sigma\delta_{ij}| = 0 \tag{A2.25}$$

Taking the determinant given in Equation (A2.25) yields the following relation, which yields three roots that correspond to the values of principal stresses:

$$\sigma^3 - I_1\sigma^2 + I_2\sigma - I_3 = 0 \tag{A2.26}$$

where

$$I_1 = \sigma_{ii} = tr(\boldsymbol{\sigma}) = \sigma_{11} + \sigma_{22} + \sigma_{33}, \quad I_2 = \tfrac{1}{2}(\sigma_{ii}\sigma_{jj} - \sigma_{ij}\sigma_{ij}), \quad I_3 = |\sigma_{ij} = \det(\boldsymbol{\sigma})|$$
$$(i = 1,2,3; j = 1,2,3)$$

A2.6   Representation of stress tensor on Mohr Circle for 2-D condition

Mohr (1882) devised a method to represent graphically the stress components using the Mohr-Circle method (Fig. A2.5):

$$\begin{bmatrix} \sigma'_{11} & \sigma'_{12} \\ \sigma'_{21} & \sigma'_{22} \end{bmatrix} = \begin{bmatrix} \beta_{11} & \beta_{12} \\ \beta_{21} & \beta_{22} \end{bmatrix}\begin{bmatrix} \sigma'_{11} & \sigma'_{12} \\ \sigma'_{21} & \sigma'_{22} \end{bmatrix}\begin{bmatrix} \beta_{11} & \beta_{21} \\ \beta_{12} & \beta_{22} \end{bmatrix} \tag{A2.27}$$

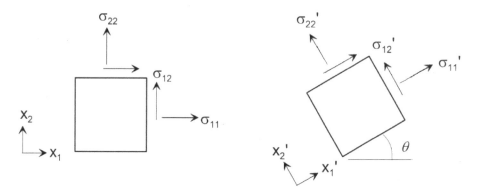

Figure A2.5 Stress components in two-coordinate systems

where

$$\begin{bmatrix} \beta_{11} & \beta_{12} \\ \beta_{21} & \beta_{22} \end{bmatrix} = \begin{bmatrix} c & s \\ -s & c \end{bmatrix}, \quad c = \cos\theta, \quad s = \sin\theta, \quad \cos^2\theta = \frac{1}{2}(1 + \cos 2\theta) \quad \text{(A2.28a)}$$

$$\sin^2\theta = \frac{1}{2}(1 - \cos 2\theta), \sin 2\theta = 2\cos\theta \sin\theta \quad \text{(A2.28b)}$$

Carrying out the matrix operation given in Equation (A2.27) together with Equation (A2.28) and the symmetry property of the stress tensor, one easily gets:

$$\sigma'_{11} = \sigma_{11}c^2 + \sigma_{22}s^2 + 2\sigma_{12}cs \quad \text{or} \quad \sigma'_{11} = \frac{\sigma_{11} + \sigma_{22}}{2} + \frac{\sigma_{11} - \sigma_{22}}{2}\cos 2\theta + \sigma_{12}\sin 2\theta$$

$$\text{(A2.29a)}$$

$$\sigma'_{22} = \sigma_{11}s^2 + \sigma_{22}c^2 - 2\sigma_{12}cs \quad \text{or} \quad \sigma'_{22} = \frac{\sigma_{11} + \sigma_{22}}{2} - \frac{\sigma_{11} - \sigma_{22}}{2}\cos 2\theta - \sigma_{12}\sin 2\theta$$

$$\text{(A2.29b)}$$

$$\sigma'_{12} = (\sigma_{22} - \sigma_{11})cs + \sigma_{12}(c^2 - s^2) \quad \text{or} \quad \sigma'_{12} = -\frac{\sigma_{11} - \sigma_{22}}{2}\sin 2\theta + \sigma_{12}\cos 2\theta$$

$$\text{(A2.29c)}$$

As the shear stress should disappear in order to obtain principal stresses, the angle of rotation should take the following form from Equation (A2.29c):

$$\theta = \frac{1}{2}\tan^{-1}\left(\frac{2\sigma_{12}}{\sigma_{11} - \sigma_{22}}\right) \quad \text{(A2.30)}$$

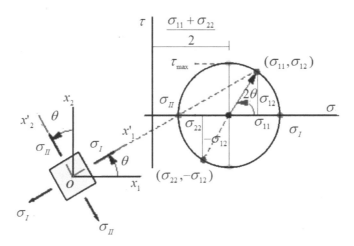

Figure A2.6 Graphical representation of stress tensor components on Mohr's circle

Furthermore, one can also derive the following relations for trigonometric relations in terms of stress tensor components:

$$\cos 2\theta = \frac{1}{\sqrt{\left(\frac{\sigma_{11}-\sigma_{22}}{2}\right)^2 + \sigma_{12}^2}} \frac{\sigma_{11} - \sigma_{22}}{2}, \sin 2\theta = \frac{\sigma_{12}}{\sqrt{\left(\frac{\sigma_{11}-\sigma_{22}}{2}\right)^2 + \sigma_{12}^2}} \quad (A2.31)$$

Using Equation (A2.31) in Equations (A2.29a) and (A2.29b), one can easily obtain the following relations for principal stresses:

$$\sigma_{11}' = \sigma_I = \frac{\sigma_{11} + \sigma_{22}}{2} + \sqrt{\left(\frac{\sigma_{11}-\sigma_{22}}{2}\right)^2 + \sigma_{12}^2}, \ \sigma_{22}' = \sigma_{II}$$

$$= \frac{\sigma_{11} + \sigma_{22}}{2} - \sqrt{\left(\frac{\sigma_{11}-\sigma_{22}}{2}\right)^2 + \sigma_{12}^2} \quad (A2.32)$$

Accordingly, the maximum shear stress is obtained using Equation (A2.32) as:

$$\tau_{max} = \frac{\sigma_I - \sigma_{II}}{2} = \sqrt{\left(\frac{\sigma_{11}-\sigma_{22}}{2}\right)^2 + \sigma_{12}^2} \quad (A2.33)$$

Figure A2.6 shows the graphical presentation of the components of stress tensor and principal stresses on Mohr's circle for the 2-D condition.

Appendix 3

Deformation and strain

A3.1 Preliminaries

Let us consider a body in the space of two coordinate systems represented by OX_1X_2 and ox_1x_2, as shown in Figure A3.1. Coordinate system OX_1X_2 is introduced at time step ($t = A$) before the deformation of the body. On the other hand, the coordinate system ox_1x_2 is introduced at time step ($t = B$) after the deformation of the body. Let us also introduce base vectors $\mathbf{E}_i$, $\mathbf{e}_i$ associated with each coordinate system.

If we describe the deformation of the body using the coordinate system OX_1X_2, it is called a Lagrangian description. On the other hand, if we describe the deformation of

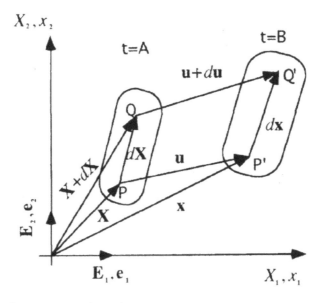

Figure A3.1 Coordinate system and notations

the body using the coordinate system, ox_1x_2, it is called a Eulerian description. Herein, the derivation of strain tensors is derived in a two-coordinate system.

Let us introduce the following preliminary relations:

Lagrangian description

Position vectors

$$\mathbf{X} = X_1\mathbf{E}_1 + X_2\mathbf{E}_2 \tag{A3.1}$$

$$\mathbf{x} = x_1\mathbf{E}_1 + x_2\mathbf{E}_2 \tag{A3.2}$$

Displacement vector

$$\mathbf{u} = u_1\mathbf{E}_1 + u_2\mathbf{E}_2 \tag{A3.3}$$

Eulerian description

Position vectors

$$\mathbf{X} = X_1\mathbf{e}_1 + X_2\mathbf{e}_2 \tag{A3.4}$$

$$\mathbf{x} = x_1\mathbf{e}_1 + x_2\mathbf{e}_2 \tag{A3.5}$$

Displacement vector

$$\mathbf{u} = u_1\mathbf{e}_1 + u_2\mathbf{e}_2 \tag{A3.6}$$

The relation position vectors before and after deformation may be given as:

$$\mathbf{x} = \mathbf{X} + \mathbf{u} \tag{A3.7}$$

$$d\mathbf{x} = d\mathbf{X} + d\mathbf{u} \tag{A3.8}$$

Tensorial operation

$$(\mathbf{a}\,\mathbf{b}) \cdot \mathbf{c} = \mathbf{a}(\mathbf{b} \cdot \mathbf{c}) \tag{A3.9}$$

A3.2 Derivation of strain tensor using Lagrangian description

In this subsection, strain tensor is derived using the Lagrangian description. The position vector of an arbitrary point in the body after deformation using the coordinate system OX_1X_2 may be given as follows:

$$x_1 = x_1(X_1, X_2), \quad x_2 = x_2(X_1, X_2) \tag{A3.10}$$

Points P and Q in the body before deformation move to a new position denoted by P' and Q'. The power of length before and after deformation can be written as:

$$dS^2 = d\mathbf{X} \cdot d\mathbf{X} = dX_1 dX_1 + dX_2 dX_2 = dX_i dX_i, \quad i = 1, 2 \tag{A3.11}$$

$$ds^2 = d\mathbf{x} \cdot d\mathbf{x} = dx_1 dx_1 + dx_2 dx_2 = dx_i dx_i, \quad i = 1, 2 \tag{A3.12}$$

One may write the following relation for length change:

$$ds^2 - dS^2 = d\mathbf{x} \cdot d\mathbf{x} - d\mathbf{X} \cdot d\mathbf{X} \tag{A3.13a}$$

Or

$$ds^2 - dS^2 = (dx_1 dx_1 + dx_2 dx_2) - (dX_1 dX_1 + dX_2 dX_2) = dx_i dx_i - dX_i dX_i, \quad i = 1, 2 \tag{A3.13b}$$

The length vector after deformation can be given in Lagrangian description as follows:

$$d\mathbf{x} = \nabla_X \mathbf{x} \cdot d\mathbf{X} = \left(\frac{\partial x_1}{\partial X_1} \mathbf{E}_1 \mathbf{E}_1 + \frac{\partial x_1}{\partial X_2} \mathbf{E}_1 \mathbf{E}_2 + \frac{\partial x_2}{\partial X_1} \mathbf{E}_2 \mathbf{E}_1 + \frac{\partial x_2}{\partial X_2} \mathbf{E}_2 \mathbf{E}_2 \right) \cdot (dX_1 \mathbf{E}_1 + dX_2 \mathbf{E}_2) \tag{A3.14}$$

or

$$d\mathbf{x} = \mathbf{F} \cdot d\mathbf{X} = \left(\frac{\partial x_1}{\partial X_1} dX_1 + \frac{\partial x_1}{\partial X_2} dX_2 \right) \mathbf{E}_1 + \left(\frac{\partial x_2}{\partial X_1} dX_1 + \frac{\partial x_2}{\partial X_2} dX_2 \right) \mathbf{E}_2 \tag{A3.15}$$

where

$$\mathbf{F} = \frac{\partial x_1}{\partial X_1} \mathbf{E}_1 \mathbf{E}_1 + \frac{\partial x_1}{\partial X_2} \mathbf{E}_1 \mathbf{E}_2 + \frac{\partial x_2}{\partial X_1} \mathbf{E}_2 \mathbf{E}_1 + \frac{\partial x_2}{\partial X_2} \mathbf{E}_2 \mathbf{E}_2 \tag{A3.16}$$

F is called a deformation gradient. Variation of the displacement vector is similarly given as:

$$d\mathbf{u} = \nabla_X \mathbf{u} \cdot d\mathbf{X}$$
$$= \left(\frac{\partial u_1}{\partial X_1} \mathbf{E}_1 \mathbf{E}_1 + \frac{\partial u_1}{\partial X_2} \mathbf{E}_1 \mathbf{E}_2 + \frac{\partial u_2}{\partial X_1} \mathbf{E}_2 \mathbf{E}_1 + \frac{\partial u_2}{\partial X_2} \mathbf{E}_2 \mathbf{E}_2 \right) \cdot (dX_1 \mathbf{E}_1 + dX_2 \mathbf{E}_2) \tag{A3.17}$$

or

$$d\mathbf{u} = \mathbf{H} \cdot d\mathbf{X} = \left(\frac{\partial u_1}{\partial X_1} dX_1 + \frac{\partial u_1}{\partial X_2} dX_2 \right) \mathbf{E}_1 + \left(\frac{\partial u_2}{\partial X_1} dX_1 + \frac{\partial u_2}{\partial X_2} dX_2 \right) \mathbf{E}_2 \tag{A3.18}$$

where

$$\mathbf{H} = \frac{\partial u_1}{\partial X_1} \mathbf{E}_1 \mathbf{E}_1 + \frac{\partial u_1}{\partial X_2} \mathbf{E}_1 \mathbf{E}_2 + \frac{\partial u_2}{\partial X_1} \mathbf{E}_2 \mathbf{E}_1 + \frac{\partial u_2}{\partial X_2} \mathbf{E}_2 \mathbf{E}_2 \tag{A3.19}$$

H is called the displacement gradient tensor. If the preceding relation is inserted into Equation (A3.13), the following relation may be written:

$$ds^2 - dS^2 = (\mathbf{F} \cdot d\mathbf{X}) \cdot (\mathbf{F} \cdot d\mathbf{X}) - d\mathbf{X} \cdot d\mathbf{X} = (d\mathbf{X} \cdot \mathbf{F}_c) \cdot (\mathbf{F} \cdot d\mathbf{X}) - d\mathbf{X} \cdot d\mathbf{X} \tag{A3.20}$$

provided that:

$$\mathbf{F}_c = \mathbf{F}^T, \quad \mathbf{I}_c = \mathbf{I}, \quad d\mathbf{X} = \mathbf{I} \cdot d\mathbf{X} \tag{A3.21}$$

Equation (A3.20) may be rewritten as:

$$ds^2 - dS^2 = d\mathbf{X} \cdot (\mathbf{F}_c \cdot \mathbf{F}) \cdot d\mathbf{X} - d\mathbf{X} \cdot (\mathbf{I}_c \cdot \mathbf{I}) \cdot d\mathbf{X} = d\mathbf{X} \cdot (\mathbf{F}_c \cdot \mathbf{F} - \mathbf{I}_c \cdot \mathbf{I}) \cdot d\mathbf{X} \tag{A3.22}$$

If we use the following relations:

$$d\mathbf{x} = d\mathbf{X} + d\mathbf{u} = \mathbf{I} \cdot d\mathbf{X} + \mathbf{H} \cdot d\mathbf{X} = (\mathbf{H} + \mathbf{I}) \cdot d\mathbf{X} \tag{A3.23}$$

$$\mathbf{F}_c \cdot \mathbf{F} = \mathbf{F}^T \cdot \mathbf{F} = (\mathbf{H} + \mathbf{I})^T \cdot (\mathbf{H} + \mathbf{I}) \tag{A3.24}$$

$$(\mathbf{H} + \mathbf{I})^T = \mathbf{H}^T + \mathbf{I}^T = \mathbf{H}^T + \mathbf{I} \tag{A3.25}$$

$$\mathbf{F}^T \cdot \mathbf{F} = \mathbf{H}^T \cdot \mathbf{H} + \mathbf{H}^T \cdot \mathbf{I} + \mathbf{I} \cdot \mathbf{H} + \mathbf{I} \cdot \mathbf{I} = \mathbf{H}^T \cdot \mathbf{H} + \mathbf{H}^T + \mathbf{H} + \mathbf{I} \tag{A3.26}$$

Lagrangian Strain tensor is defined as:

$$\mathbf{L} = \frac{1}{2}[\mathbf{F}^T \cdot \mathbf{F} - \mathbf{I}] = \frac{1}{2}[\mathbf{H} + \mathbf{H}^T + \mathbf{H}^T \cdot \mathbf{H}] \tag{A3.27a}$$

or

$$\mathbf{L} = \frac{1}{2}[(\nabla_X \mathbf{u}) + (\nabla_X \mathbf{u})^T + (\nabla_X \mathbf{u})^T \cdot (\nabla_X \mathbf{u})] \tag{A3.27b}$$

The Lagrangian strain tensor is also known the Green strain tensor. It is interpreted as a finite strain tensor. In index notation, it is expressed as follows:

$$L_{jk} = \frac{1}{2}\left[\frac{\partial u_j}{\partial X_k} + \frac{\partial u_k}{\partial X_j} + \frac{\partial u_i}{\partial X_j}\frac{\partial u_i}{\partial X_k}\right] = \frac{1}{2}[u_{j,k} + u_{k,j} + u_{i,j}u_{i,k}] \tag{A3.28}$$

A3.3   Derivation of strain tensor using Eulerian description

In this subsection, strain tensor is derived using Eulerian description. The position vector of an arbitrary point in the body after deformation using the coordinate system ox_1x_2 may be given as follows:

$$X_1 = X_1(x_1, x_2), \quad X_2 = X_2(x_1, x_2) \tag{A3.29}$$

The length vector before deformation can be given in Eulerian description as follows:

$$d\mathbf{X} = \nabla_x \mathbf{X} \cdot d\mathbf{x} = \left(\frac{\partial X_1}{\partial x_1}\mathbf{e}_1\mathbf{e}_1 + \frac{\partial X_1}{\partial x_2}\mathbf{e}_1\mathbf{e}_2 + \frac{\partial X_2}{\partial x_1}\mathbf{e}_2\mathbf{e}_1 + \frac{\partial X_2}{\partial x_2}\mathbf{e}_2\mathbf{e}_2\right) \cdot (dx_1\mathbf{e}_1 + dx_2\mathbf{e}_2) \tag{A3.30}$$

or

$$dX = J \cdot dx = \left(\frac{\partial X_1}{\partial x_1} dx_1 + \frac{\partial X_1}{\partial x_2} dx_2 \right) e_1 + \left(\frac{\partial X_2}{\partial x_1} dx_1 + \frac{\partial X_2}{\partial x_2} dx_2 \right) e_2 \qquad (A3.31)$$

where

$$J = \frac{\partial X_1}{\partial x_1} e_1 e_1 + \frac{\partial X_1}{\partial x_2} e_1 e_2 + \frac{\partial X_2}{\partial x_1} e_2 e_1 + \frac{\partial X_2}{\partial x_2} e_2 e_2 \qquad (A3.32)$$

J is denoted as the deformation gradient tensor. The variation of displacement is given using Eulerian description as follows:

$$du = \nabla_x u \cdot dx = \left(\frac{\partial u_1}{\partial x_1} e_1 e_1 + \frac{\partial u_1}{\partial x_2} e_1 e_2 + \frac{\partial u_2}{\partial x_1} e_2 e_1 + \frac{\partial u_2}{\partial x_2} e_2 e_2 \right) \cdot (dx_1 e_1 + dx_2 e_2)$$

$$(A3.33)$$

or

$$du = K \cdot dx = \left(\frac{\partial u_1}{\partial x_1} dx_1 + \frac{\partial u_1}{\partial x_2} dx_2 \right) e_1 + \left(\frac{\partial u_2}{\partial x_1} dx_1 + \frac{\partial u_2}{\partial x_2} dx_2 \right) e_2 \qquad (A3.34)$$

where

$$K = \frac{\partial u_1}{\partial x_1} e_1 e_1 + \frac{\partial u_1}{\partial x_2} e_1 e_2 + \frac{\partial u_2}{\partial x_1} e_2 e_1 + \frac{\partial u_2}{\partial x_2} e_2 e_2 \qquad (A3.35)$$

K is the denoted displacement gradient in Eulerian description. If it is inserted into Equation (A3.13), the following relation may be written as:

$$ds^2 - dS^2 = dx \cdot dx - (J \cdot dx) \cdot (J \cdot dx) = dx \cdot dx - (dX \cdot J_c) \cdot (J \cdot dX)$$

$$(A3.36)$$

provided that

$$J_c = J^T, \quad I_c = I, \quad dx = I \cdot dx \qquad (A3.37)$$

Equation (A3.36) can be rewritten as:

$$ds^2 - dS^2 = dx \cdot (I_c \cdot I) \cdot dx - dx \cdot (J_c \cdot J) \cdot dx = dx \cdot (I_c \cdot I - J_c \cdot J) \cdot dx$$

$$(A3.38)$$

Introducing the following relations:

$$dX = dx - du = I \cdot dx - K \cdot dx = (I - K) \cdot dx \qquad (A3.39)$$

$$J_c \cdot J = J^T \cdot J = (I - K)^T \cdot (I - K) \qquad (A3.40)$$

$$(I - K)^T = I^T - K^T = I - K^T \qquad (A3.41)$$

$$J^T \cdot J = K^T \cdot K - K^T \cdot I - I \cdot K + I \cdot I = K^T \cdot K - K^T - K + I \qquad (A3.42)$$

the Eulerian strain tensor is defined as:

$$E = \frac{1}{2}[I - J^T \cdot J] = \frac{1}{2}[K + K^T - K^T \cdot K] \qquad (A3.43a)$$

or

$$E = \frac{1}{2}[(\nabla_x u) + (\nabla_x u)^T - (\nabla_x u)^T \cdot (\nabla_x u)] \qquad (A3.43b)$$

The Eulerian strain tensor is also known the Almani strain tensor, and it is a finite strain tensor used for the large deformation of materials. In index notation, it is rewritten as:

$$E_{jk} = \frac{1}{2}\left[\frac{\partial u_j}{\partial x_k} + \frac{\partial u_k}{\partial x_j} - \frac{\partial u_i}{\partial x_j}\frac{\partial u_i}{\partial x_k}\right] = \frac{1}{2}[u_{j,k} + u_{k,j} - u_{i,j}u_{i,k}] \qquad (A3.44)$$

A3.4 Relation between small strain theory and finite strain theory

As noted from the strain definitions given by (A3.28) and (A3.44), the term corresponding to the power of strain components is noted. Therefore, the finite strain tensors are geometrically nonlinear, and their use in practice becomes troublesome. When the strain is small, say, less than 10%, the nonlinear components may be omitted. Furthermore, if coordinate systems are assumed to be the same, say, $x = X$, strain tensors given by (A3.28) and (A3.44) reduced to the following form:

$$E_{jk} = L_{jk} = \frac{1}{2}\left[\frac{\partial u_j}{\partial x_k} + \frac{\partial u_k}{\partial x_j}\right] = \frac{1}{2}[u_{j,k} + u_{k,j}] \qquad (A3.45)$$

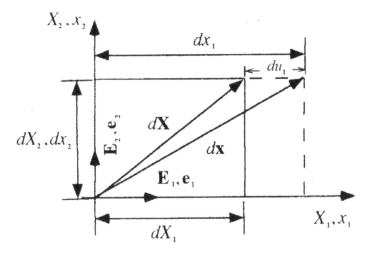

Figure A3.2 One-dimensional normal deformation

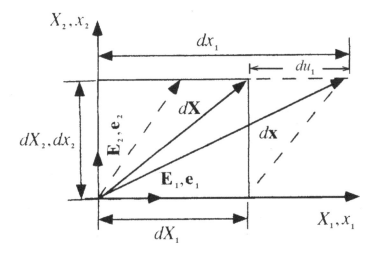

Figure A3.3 Simple shear deformation

A3.5   Geometrical interpretations of strain tensor

A3.5.1   Uniaxial deformation

Let us consider a body deformed uniaxially as shown in Figure A3.2. Length change for this example may be written as:

$$ds^2 - dS^2 = 2L_{11}dX_1^2 \tag{A3.46}$$

Provided that:

$$ds^2 = (dX_1 + du_1)^2 + dX_2^2, \quad dS^2 = dX_1^2 + dX_2^2$$

$$dX_1du_1 + du_1^2 = 2L_{11}dX_1^2 \tag{A3.47}$$

and $du_1^2 \approx 0$, one can obtain the following:

$$\frac{du_1}{dX_1} = L_{11} \tag{A3.48}$$

A3.5.2   Simple shear deformation

Let us consider a body deformed in simple shear as shown in Figure A3.3. Length change for this example may be written as:

$$ds^2 - dS^2 = 2L_{12}dX_1dX_2 \tag{A3.49}$$

Provided that:

$$ds^2 = (dX_1 + du_1)^2 + dX_2^2, \quad dS^2 = dX_1^2 + dX_2^2$$

$$dX_1du_1 + du_1^2 = 2L_{12}dX_1dX_2 \tag{A3.50}$$

and $du_1^2 \approx 0$, the following relation may be written:

$$\frac{du_1}{dX_2} = L_{12} \tag{A3.51}$$

This relation can be easily interpreted as angle variation

Appendix 4

Gauss divergence theorem

A4.1 One-dimensional (1-D) Gauss theorem

Gauss theorem is written as for 1-D case:

$$\int_\Omega \frac{\partial f}{\partial x} d\Omega = \int_\Gamma f_n d\Gamma \tag{A4.1}$$

The preceding expression is explicitly written as:

$$\int_\Omega \frac{\partial f(x)}{\partial x} \Delta x \Delta y \Delta z = \int_\Gamma (f \cdot n)|_{x=a}^{x=b} \Delta y \Delta z \tag{A2.2}$$

As $n_a = \cos 180° = -1$, $n_b = \cos 0° = 1$ and $f(x=a) = f_a$, $f(x=b) = f_b$, the preceding expression is rewritten as:

$$\int_\Omega \frac{\partial f(x)}{\partial x} \Delta x \Delta y \Delta z = \int_\Gamma f_b \Delta y \Delta z - \int_\Gamma f_a \Delta y \Delta z \tag{A3.3}$$

Figure A4.1 illustrates the geometrical interpretation of the Gauss divergence theorem.

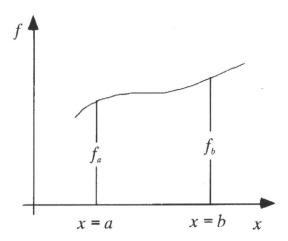

Figure A4.1 Geometrical illustration of Gauss divergence theorem

A4.2   Three-dimensional (3-D) Gauss theorem

To get 3-D version of Equation (A4.1), let us introduce the replacements:

$$\frac{\partial}{\partial x} \rightarrow \nabla = \frac{\partial}{\partial x}\mathbf{e}_x + \frac{\partial}{\partial y}\mathbf{e}_y + \frac{\partial}{\partial z}\mathbf{e}_z \tag{A4.4}$$

$$n \rightarrow \mathbf{n} = n_x\mathbf{e}_x + n_y\mathbf{e}_y + n_z\mathbf{e}_z \tag{A4.5}$$

where $n_x = \cos\alpha$, $n_y = \cos\beta$ and $n_z = \cos\gamma$.

   If the integrand is a scalar function (f), Equation (A4.1) takes the following form in the 3-D case:

$$\int_\Omega \nabla f d\Omega = \int_\Gamma f\mathbf{n}d\Gamma \tag{A4.6}$$

If the integrand is a vector (v), Equation (A4.1) takes the following form in the 3-D case:

$$\int_\Omega \nabla \cdot v d\Omega = \int_\Gamma \mathbf{v} \cdot \mathbf{n}d\Gamma \tag{A4.7}$$

If the integrand is a tensor (σ), Equation (A4.1) takes the following form in the 3-D case:

$$\int_\Omega \nabla \cdot \sigma d\Omega = \int_\Gamma \boldsymbol{\sigma} \cdot \mathbf{n}d\Gamma \tag{A4.8}$$

Geometrical interpretation of Taylor expansion

A scalar function ϕ at a given coordinate $x + \Delta x$ can be expressed using the Taylor expansion as:

$$\varphi_{x+\Delta x} = \varphi_x + \frac{\partial \varphi}{\partial x}\Delta x + \frac{\partial^2 \varphi}{\partial^2 x}\Delta x^2 + \cdots \frac{\partial^{(n)}\varphi}{\partial^{(n)}x}\Delta x^n + \cdots \tag{A5.1}$$

The first term on the right-hand side is the value of function ϕ at position (x). The second term involves the gradient of function ϕ at position (x) multiplied by the position increment Δx, which corresponds to $\Delta\phi$, which is the increment of function ϕ. As noted from the figure, there is deviation between the exact value at $x + \Delta x$. If the higher terms of function ϕ are possible, the use of higher-order derivatives is expected to yield better estimations. However, the linear term is often utilized in the derivation of governing equations in many applications of mechanics. Therefore, the mechanics are called linear mechanics.

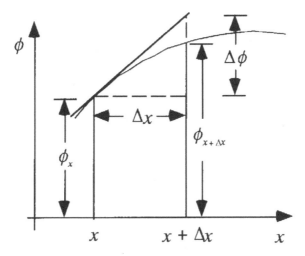

Figure A5.1

Appendix 6

Reynolds transport theorem

Inserting the time derivation operator into the integral operator:

$$\frac{d}{dt}\int_{\Omega}()d\Omega = \int_{\Omega}\frac{d()}{dt}d\Omega + \int_{\Omega}()\frac{d(d\Omega)}{dt} = \int_{\Omega}\left[\frac{d()}{dt} + ()\nabla \cdot \mathbf{v}\right]d\Omega \qquad (A6.1)$$

Equation (A6.1) is also known the Reynolds transport theorem. The time derivative of infinitely small volume $d\Omega$ takes the following form in 3-D and 1-D:

In 3-D

$$d\Omega = Jd\Omega_o = (\nabla_{x_o} \cdot \mathbf{x})d\Omega_o \qquad (A6.2)$$

where J is Jacobian.

In 1-D

$$dx = \frac{\partial x}{\partial x_o}dx_o = Jdx_o \qquad (A6.3)$$

Detailed supplementary explanation for Equation (A6.3)

$$\frac{d(dx)}{dt} = \frac{d}{dt}\left(\frac{\partial x}{\partial x_o}dx_o\right) = \frac{d}{dt}\left(\frac{\partial x}{\partial x_o}\right)dx_o + \frac{\partial x}{\partial x_o}\frac{d(dx_o)}{dt} \qquad (A6.4)$$

The time derivative of the second term is nil as the initial control element length (dx_o) is constant. Thus we have:

$$\frac{d(dx)}{dt} = \frac{d}{dt}\left(\frac{\partial x}{\partial x_o}\right)dx_o \Rightarrow \frac{\partial}{\partial x_o}\left(\frac{dx}{dt}\right)dx_o \Rightarrow \frac{\partial v}{\partial x_o}dx_o \Rightarrow \frac{\partial v}{\partial x}\frac{\partial x}{\partial x_o}dx_o \Rightarrow \frac{\partial v}{\partial x}dx \qquad (A6.5)$$

Therefore, we may write the following relations given in 1-D and generalized to 3-D versions as follow:

1-D to 3-D

$$\frac{\partial}{\partial x} \Rightarrow \nabla \tag{A6.6a}$$

$$v \Rightarrow \mathbf{v} \tag{A6.6b}$$

$$dx \Rightarrow d\Omega \tag{A6.6c}$$

$$\frac{\partial v}{\partial x} dx \Rightarrow (\nabla \cdot \mathbf{v}) d\Omega \tag{A6.6d}$$

Index

action 4, 14, 16, 102, 300; chemical 4; fluid 16; infiltrating water 14; physical 4; pressure 16; relative slip 102; springs 300; temperature 16

airflow 54

analytical solution 125, 132, 188, 198, 200, 203, 215, 221, 224, 226, 228, 241, 245; circular hole 224; diffusion problems 245; equation 198; fluid flow 228; heat flow 241; solids 188, 198

atmospheric 138, 143, 146, 239, 245, 329; pressure 138, 143, 146, 239, 329; temperature 245

axisymmetric 53, 101; radial flow 53; rock sample 101

bar 125, 146, 198, 211, 299; divided 125, 146; one-dimensional 198; radius 211; steel 299

Barton, N. 29, 41, 111, 113, 145, 301

behavior 3, 4, 38, 47, 61, 68, 96, 104, 107, 120, 146, 153, 157, 162, 180, 191, 204, 213, 215, 220, 221, 242, 298, 299, 301, 333; brittle 107, 153; creep 120, 146; ductile 107, 153; elastic 153, 220, 333; elasto-plastic 68, 155, 157, 162, 213, 220, 299, 301, 333; hydro-mechanical 47; mechanical 3, 4, 61, 153, 157, 162; post-failure 104; time dependent 215; visco-elastic 214

Bieniawski, Z. 121, 145, 162, 183, 214, 255

blasting 167, 223

body force 44, 45, 51, 58, 213, 266, 273, 280

borehole 36, 37, 38, 40, 60

breakout 147, 162, 163, 167, 179, 181, 185, 223, 225

Brekke, T.L. 301

Brown, E.T. 3, 5, 83–87, 96, 99, 145, 147, 166–168, 172, 183, 204, 213, 221–224, 255

cavern 2, 93, 176, 182, 301

cavity 158, 166, 200, 203, 219–221; circular 166, 200, 203, 221; spherical 40, 203, 219, 220

chemical reaction 332

condition 2, 4, 9, 16, 19, 35, 41, 49, 55, 57, 64, 69, 73, 76, 77, 80, 86, 99, 102–106, 111, 113, 118, 119, 128, 129, 132, 133, 135–138, 141–143, 146, 147, 150, 151, 154, 162, 165, 168, 172, 177, 187, 188, 191, 196–198, 201–204, 206–207, 212, 215, 216, 221, 224, 225, 229, 232, 235, 237- 239, 243–244, 247, 250–251, 257–258, 262, 268, 271–273, 278–280, 288, 295, 297–298, 301, 306, 308, 315, 327, 333–335, 339, 341, 359, 361; alkaline 9; boundary 102, 103, 136, 141, 146, 187, 196–198, 201–203, 206–207, 212, 215, 221, 229, 232, 237, 239, 251, 257–258, 262, 268, 273, 279, 288, 295, 297; climatic 4, 308, 327; compatibility 204; consistency 64, 73, 77, 80; continuity 151, 207; creeping 243, 244; drained 9; dynamic 2, 50, 106; environmental 19, 99, 113, 247, 315, 333–334; flow 132; gravitational 341; ill-conditioning 301; initial 133, 135, 137, 138, 142, 143, 191, 198, 221, 235, 238, 247, 262, 268, 273, 279–280, 288; laboratory 118, 162, 172; loading 105, 106, 215; pure-shear 165; saturated 119; slip 306; spherical-symmetric 333; steady-state 76; stress 188, 215, 225, 333; undrained 49; unstrained 129; three-point bending 104; yield 86

Cook, N.G.W. 3, 5, 77, 86, 96, 102, 145, 151, 163, 165–167, 184, 187–188, 219, 221, 223, 255, 308

creep 65–68, 71, 79–80, 91, 96–97, 105, 117–124, 145–146, 214, 251, 256, 266; Brazilian 119; compression 119; device 105, 119, 120; experiment 80; failure 214; impression 119–120; in-situ 121–122; plate-bearing 122; primary 124; secondary 124, 214; shear 105, 119–120, 124; steady-state 65, 67; strength 214; tertiary 214; test 71, 117–119, 123, 214; transient 65, 67, 214